Oxford Cambridge and RSA Examinations

RECOGNISING ACHIEVEMENT

GCSE Mathematics

INTERMEDIATE COURSE

D1494370

SECOND EDITION

SERIES EDITOR BRIAN SEAGER

HOWARD BAXTER, MIKE HANDBURY, JOHN JESKINS, JEAN MATTHEWS, MARK PATMORE

Hodder & Stoughton
A MEMBER OF THE HODDER HEADLINE GROUP

Acknowledgements

The Publishers would like to thank the following individuals and companies for permission to reproduce photographs in this book:

Bubbles Photo Library: Frans Rombout page 252.
Paul Hart: page 224 (top).
Life File Photo Library: Mike Evans page 313, Ron Gregory page 224 (bottom), David Kampfner page 253 (bottom) and Lionel Moss page 251 (bottom).
Robert Harding Picture Library: pages 140, 155, 223, and Phil Robinson page 38.
The Photographers Library: pages 6, 31, 138, 144, 145, 148, 228 (both), 247, 251 (top), 253 (top) and 327.

Every effort has been made to trace ownership of copyright. The Publishers would be happy to make arrangements with any copyright holder whom it has not been possible to trace.

Orders: please contact Bookpoint Ltd, 78 Milton Park, Abingdon, Oxon OX14 4TD. Telephone: (44) 01235 827720, Fax: (44) 01235 400454. Lines are open from 9.00 - 6.00, Monday to Saturday, with a 24 hour message answering service. Email address: orders@bookpoint.co.uk

British Library Cataloguing in Publication Data
A catalogue record for this title is available from The British Library

ISBN 0 340 801 24 7

First published 2000
Second edition 2001

Impression number	10	9	8	7	6	5	4	3	2	1
Year	2006	2005	2004	2003	2002	2001				

Designed and typeset by Cambridge Publishing Management, a division of G&E 2000 Ltd.

Printed in Italy for Hodder & Stoughton Educational, a division of Hodder Headline Plc, 338 Euston Road, London NW1 3BH.

Contents

Introduction   1

Chapter one

Fractions, percentages and ratio   6
Equivalence of fractions, decimals and percentages   6
Expressing one quantity as a percentage of another   8
Percentage increase and decrease, profit and loss   9
Ratio   12
Key points   16

Chapter two

Drawing   17
Isometric drawings   17
Drawing and constructing   19
Drawing triangles   20
Plan and elevations   24
Key points   28

Chapter three

Sequences, symbols and equations   31
Letters for unknowns   31
Making it simple   33
Sequences   34
Brackets   36
Forming equations   37
Inequalities   42
Key points   43

Chapter four

Calculating and illustrating data   44
Grouping data   44
Pie charts   47
Scatter diagrams   51
Mean, median, mode and range of data   54
Comparing data   56
Stem and leaf tables   58
Key points   59

Chapter five

Solving problems and checking results I   61
Solving ratio problems   61
Checking your work   64
Solving money problems   66
Selecting suitable methods   69
Key points   72

Chapter six

Shapes and angles   73
Regular and irregular polygons   73
Some basic angle facts   77
Angles with parallel lines   81
Angles in polygons   84
Key points   87

Chapter seven

Coordinates, graphs and bearings   89
Points in all four quadrants   89
Harder straight-line equations   92
Quadratic graphs   93

Harder graphs   99
Bearings   103
Key points   107

Chapter eight

Transformations and symmetry   109
Transformations   109
Enlargements   116
Symmetry in shapes   117
Tessellations   118
Key points   123

Chapter nine

Whole numbers – multiples, factors, primes and roots   125
Key points   132

Chapter ten

Length, area, volume and units   135
Area of a triangle   135
Circumference of a circle   138
Area of a circle   141
Volume of a cuboid   143
Key points   146

Chapter eleven

Probability I   148
Probability of an outcome not happening   148
Relative frequency and probability   152
Key points   155

Chapter twelve

Numbers   157
Approximating numbers   157
Ordering numbers   162
Ratio   163
Negative numbers   164
Indices   166
Standard form   168
Powers and roots   169
Reciprocal   170
Key points   171

Chapter thirteen

Angle properties   172
Exterior angle of a triangle   172
Angle in a semicircle   175
Angles in polygons   177
Angles in circles   179
Key points   180

Chapter fourteen

Calculating and representing grouped data   181
Grouped discrete data   181
Representing continuous data   184
Calculating with grouped continuous data   186
Mean, median or mode?   190
Key points   191

Contents

Chapter fifteen

Transformations 193

Drawing reflections 193
Rotations 195
Translations 199
Enlargements 202
Combining transformations 206
Key points 209

Chapter sixteen

Calculations 211

Fractions 211
Mixed numbers 214
Multiplying and dividing fractions 217
Adding and subtracting negative numbers 218
Using a calculator 220
Ratio and proportion 222
Repeated proportional changes 225
Finding the value before a percentage change 229
Key points 231

Chapter seventeen

Equations and manipulation I 233

Substituting numbers in a formula 233
Collecting like terms and simplifying expressions 236
Multiplying out two brackets 237
Simplifying expressions using indices 239
Finding the nth term of a sequence 241
Factorising algebraic expressions 243
Factorising expressions of the type $x^2 + ax + b$ 244
Rearranging formulae 246
Inequalities 248
Forming equations and inequalities 250
Key points 254

Chapter eighteen

Questionnaires and cumulative frequency 256

Questionnaires 256
Analysing data 259
Box-and-whisker plots 261
Key points 263

Chapter nineteen

Solving problems and checking results 2 265

Checking answers by rounding to one significant figure 265
Compound interest 266
Insurance 267
Compound measures 269
Working to a reasonable degree of accuracy 272
Key points 273

Chapter twenty

Pythagoras' theorem and trigonometry 274

Pythagoras' theorem 274
Trigonometry 279
Key points 286

Chapter twenty-one

Equations and manipulation 2 287

Solving harder linear equations 287
Solving inequalities 289
Forming equations and inequalities 290
Simultaneous equations 292
Solving quadratic equations 296

Graphical methods of solving equations 298
Solving cubic equations by trial and improvement 301
Problems that lead to simultaneous or quadratic equations 303
Showing regions on graphs 305
Rearranging formulae 308
Key points 309

Chapter twenty-two

Measurement and compound units 312

Estimating measurements 312
Discrete and continuous measures 315
Bounds of measurement 317
Compound units 318
Key points 320

Chapter twenty-three

Interpreting graphs 321

Story graphs 321
Gradient 325
Straight-line graphs 329
Equations of curved graphs 330
Key points 334

Chapter twenty-four

Probability 2 336

Covering all the possibilities 336
Probability of event A or event B happening 342
Probability of event A and event B happening 343
Using tree diagrams for unequal probabilities 346
Key points 348

Chapter twenty-five

Length, area and volume 2 350

Area of a parallelogram 350
Area of a trapezium 354
Volume of a prism 357
Dimensions 359
Key points 361

Chapter twenty-six

Properties of transformations and similar shapes 364

Properties of transformations 364
Enlargements 368
Key points 374

Chapter twenty-seven

Comparing 376

Comparing data 376
Correlation 381
Time series 386
Key points 391

Chapter twenty-eight

Locus 393

Identifying a locus 393
Problems involving intersection of loci 397
Key points 400

Answers 402

Index 458

Contents

Introduction

About this book

This book covers the complete specification for the Intermediate Tier of GCSE Mathematics. It is particularly aimed at OCR Specification A but is also a suitable preparation for all Intermediate Tier GCSE Mathematics examinations.

The book aims to make the best of your performance in the examinations:

- Each chapter is presented in a style intended to help you understand the mathematics, with straightforward explanations and worked examples.
- At the start of each chapter is a list of what you need to know before you begin.
- There are plenty of exercises for you to work through and practise the skills.
- At the end of each chapter there is a list of key points and a revision exercise.
- Some exercises are designed to be done without a calculator so that you can practise for the non-calculator paper.
- Some of the answers are given at the end of the book to help you check your progress.
- Some chapters are designed to help you develop the necessary skills to undertake coursework.
- At frequent intervals throughout the book there are Examiner's tips, where the experienced examiners who have written this book offer advice and tips to improve your examination performance.
- Revision tests are provided in the Teacher's Book, after each term's work.

Part of the examination is now a calculator-free zone. You will have to do the first of the papers without a calculator and the questions are designed appropriately.

The marks on the two papers (out of 200) for the Assessment Objectives are:

25 marks   AO1 Using and Applying Mathematics
100 marks   AO2 Number and Algebra
50 marks   AO3 Shape, Space and Measures
25 marks   AO4 Handling Data

The remaining marks to balance AO1 and AO4 are awarded on the internal assessment (coursework).

Other changes are not so obvious from a study of the specification. Most of the marks given for Algebra are for 'manipulative' algebra. This includes simplifying algebraic expressions, factorising, solving equations and changing formulae. Some questions are also being set which offer you little help to get started. These are called 'unstructured' or 'multi-step' questions. Instead of the question having several parts, each of which leads to the next, you have to work out the necessary steps to find the answer. There will be examples of this kind of question in the end-of-term tests and past examination papers.

Top ten tips

Here are some general tips from the examiners to help you do well in your examinations.

Practise:

1. all aspects of **manipulative algebra** in the specification
2. answering questions **without** a calculator
3. answering questions which require **explanations**
4. answering **unstructured** questions
5. **accurate** drawing and construction
6. answering questions which **need a calculator**, trying to use it efficiently
7. **checking answers**, especially for reasonable size and degree of accuracy
8. making your work **concise** and well laid out
9. using the **formula sheet** before the examination
10. **rounding** numbers, but only at the appropriate stage.

Coursework

The GCSE Mathematics examinations will assess your ability to use your mathematics on longer problems than those normally found on timed written examination papers. Assessment of this type of work will account for 20% of your final mark. It will involve two 3-hour tasks taken during your final year. One task will be an investigation, the other a statistics task.

Each type of task has its own mark scheme in which marks are awarded in three categories or 'strands'. The titles of these strands give you clues about the important aspects of this work.

For the investigation tasks the strands are:

- Making and monitoring decisions – what you are going to do and how you will do it
- Communicating mathematically – explaining and showing exactly what you have done
- Developing the skills of mathematical reasoning – using mathematics to analyse and prove your results.

The table below gives some idea of what you will have to do and show. Look at this table whenever you are doing some extended work and try to include what it suggests you do.

Mark	Making and monitoring decisions	Communicating mathematically	Developing the skills of mathematical reasoning
1	organising work, producing information and checking results	discussing work using symbols and diagrams	finding examples that match a general statement
2	beginning to plan work, choosing your methods	giving reasons for choice of presentation of results and information	searching for a pattern using at least three results
3	finding out necessary information and checking it	showing understanding of the task by using words, symbols, diagrams	explaining reasoning and making a statement about the results found
4	simplifying the task by breaking it down into smaller stages	explaining what the words, symbols and diagrams show	testing generalisations by checking further cases
5	introducing new questions leading to a fuller solution	justifying the means of presentation	justifying solutions explaining why the results occur
6	using a range of techniques and reflecting on lines of enquiry and methods used	using symbolisation consistently	explaining generalisations and making further progress with the task
7	analysing lines of approach and giving detailed reasons for choices	using symbols and language to produce a convincing and reasoned argument	report includes mathematical justifications and explanations of the solutions to the problem
8	exploring extensively an unfamiliar context or area of mathematics and applying a range of appropriate mathematical techniques to solve a complex task	using mathematical language and symbols efficiently in presenting a concise reasoned argument	providing a mathematically rigorous justification or proof of the solution considering the conditions under which it remains valid

For the statistics tasks the strands are:

- Specify the problem and plan – choosing or defining a problem and outlining the approach to be followed
- Collect, process and represent data – explaining and showing what you have done
- Interpret and discuss results – use mathematical and statistical knowledge and techniques to analyse, evaluate and interpret your results and findings.

The table below gives some idea of what you will have to do and show. Look at this table whenever you are doing some extended work and try to include what it suggests you do.

The marks obtained from each task are added together to give a total out of 48.

Mark	Specify the problem and plan	Collect, process and represent data	Interpret and discuss results
1–2	choosing a simple problem and outlining a plan	collecting some data; presenting information, calculations and results	making comments on the data and results
3–4	choosing a problem which allows you to use simple statistics and plan the collection of data	collecting data and then processing it using appropriate calculations involving appropriate techniques; explaining what the words, symbols and diagrams show	explaining and interpreting the graphs and calculations and any patterns in the data
5–6	considering a more complex problem and using a range of techniques and reflecting on the method used	collecting data in a form that ensures they can be used; explaining statistical meaning through the consistent use of accurate statistics and giving a reason for the choice of presentation, explaining features selected	commenting on, justifying and explaining results and calculations; commenting on the methods used
7–8	analysing the approach and giving reasons for the methods used; using a range of appropriate statistical techniques to solve the problem	using language and statistical concepts effectively in presenting a convincing reasoned argument; using an appropriate range of diagrams to summarise the data and show how variables are related	correctly summarising and interpreting graphs and calculations and making correct and detailed inferences from the data; appreciating the significance of results obtained and, where relevant, allowing for the nature and size of the sample and any possible bias; evaluating the effectiveness of the overall strategy and recognising limitations of the work done, making suggestions for improvement

Advice

Starting a task

Ask yourself:

- what does the task tell me?
- what does it ask me?
- what can I do to get started?
- what equipment and materials do I need?

Working on the task

- Make sure you explain your method and present your results as clearly as possible.
- Break the task down into stages. For example in 'How many squares on a chessboard', begin by looking at 1×1 squares then 2×2 squares, then 3×3 squares. In a task asking for the design of a container, start with cuboids then nets, surface area, prisms … or in statistics you might want to start with a pilot survey or questionnaire.
- Write down questions that occur to you, for example, *what happens if you change the size of a rectangle systematically?* They may help you find out more about the work. In a statistical task you might wish to include different age groups or widen the type of data …
- Explore as many aspects of the task as possible.
- Develop the task into new situations and explore these thoroughly.
  - What connections are possible?
  - Is there a result to help me?
  - Is there a pattern?
  - Can the problem be changed? If so, how?

Explain your work

- Use appropriate words and suitable tables, diagrams, graphs, calculations …
- Link as much of your work together as possible, explaining, for example, why you chose the tables and charts you used and rejected others, or why the median is more appropriate than the mean in a particular statistical analysis, or why a pie chart is not appropriate. Don't just include diagrams to show identical information in different ways.
- Use algebra or symbols to give clear and efficient explanations; in investigations, you must use algebra to progress beyond about 4 marks. You will get more credit for writing $T = 5N + 1$ than for writing 'the total is five times the pattern number, plus one'.
- Present results and conclusions clearly.

State your findings

- Show the patterns and test conclusions.
- State general results in words and explain what they mean.
- Write formulae and explain how they have been found from the situations explored.
- Prove the results using efficient mathematical methods.
- Develop new results from previous work and use clear reasoning to prove conclusions.
- Make sure your reasoning is accurate and draws upon the evidence you've presented.
- Show findings in clear, relevant diagrams.
- Check you've answered the question.

Review/conclusion/extension

- Is the solution acceptable?
- Can the task be extended?
- What can be learned from it?

Example task

On the next page there is a short investigative task for you to try, in both 'structured' and 'unstructured' form. The unstructured form represents the usual style of a coursework task. The structured form leads you to an algebraic conclusion. This mirrors the sort of questions you would be expected to think of (and answer) if you were trying it as coursework.

Although the task in both forms directs you to investigate trapezium numbers, you would be expected to extend the investigation into considering other forms of number, such as pentagon numbers, to achieve the higher marks.

Other tasks

In chapters 9, 13, 18 and 20, the text has been written using a 'task' approach. More practice for coursework can be obtained by using some or all of these chapters in this way.

structured form

Trapezium numbers

These diagrams represent the first three trapezium numbers.

Each diagram always starts with two dots in the top row.

1st	2nd	3rd
••	••	••
	•••	•••
		••••
2 dots	5 dots	9 dots

So the third trapezium number is 9 because 9 dots can be arranged as a trapezium. There are two dots in the top row, three dots in the next row and four dots in the bottom row.

1. Write down the next two trapezium numbers.

2. (a) Draw a table, graph or a chart of all the trapezium numbers, from the first to the tenth.
    (b) Work out the 11th trapezium number.

3. The 19th trapezium number is 209. Explain how you could work out the 20th trapezium number without drawing any diagrams.

4. Find an expression for the number of dots in the bottom row of the *n*th trapezium number. Test your expression for a suitable value of *n*.

5. Find, giving an explanation, an expression for the number of dots in the bottom row of the diagram for the $(n + 1)$th trapezium number.

6. The *n*th trapezium number is *x*. Write down an expression in terms of *n* and *x* for the $(n + 1)$th trapezium number. Test your expression for a suitable value of *n*.

unstructured form

Trapezium numbers

These diagrams represent the first three trapezium numbers.

Each diagram starts with two dots in the top row.

1st	2nd	3rd
••	••	••
	•••	•••
		••••
2 dots	5 dots	9 dots

So the third trapezium number is 9 because nine dots can be arranged as a trapezium.

Investigate trapezium numbers

NB Although the task in this form directs you to investigate trapezium numbers, you have the freedom to – and are expected to – extend the investigation to consider other forms of number such as pentagon numbers.

Commentary

This question allows you to show understanding of the task, systematically obtaining information which **could** enable you to find an expression for trapezium numbers.

This question provides a structure, using symbols, words and diagrams, from which you should be able to derive an expression from either a table or a graph. Part (b) could be done as a 'predict and test'.

In the unstructured form you would not normally answer a question like this.

From here you are **directed** in the structured task, and **expected** in the unstructured task, to use algebra, testing the expression – the **generalisation**.

In the unstructured form this would represent the sort of 'new' question you might ask, to lead to a further solution and to demonstrate symbolic presentation and the ability to relate the work to the physical structure, rather than doing all the analysis from a table of values.

Introduction About this book

Fractions, percentages and ratio

Equivalence of fractions, decimals and percentages

You know already that $\frac{1}{2}$, 0.5 and 50% all mean the same.

You can say: 'Half the children in the school are girls.'
or   '0.5 of the children in the school are girls.'
or   '50% of the children in the school are girls.'

You may also realise that $\frac{1}{4}$, 0.25 and 25% all mean the same. In this section you will look at other fractions and their decimal and percentage equivalents. You will also see how to convert from one form to another.

Fraction to decimal

Looking back to the case of $\frac{1}{2}$, think of the fraction line between the 1 and 2 as a division ($\div$) sign. This means that $\frac{1}{2}$ is the same as $1 \div 2$ which is equal to 0.5.

So the decimal equivalent of $\frac{1}{2}$ is 0.5.

Decimal to percentage

Look again at the two cases above.

0.5 is the same as 50% because 'per cent' means 'out of 100' and 50 out of $100 = \frac{1}{2} = 0.5$.

0.25 is the same as 25% because 25 out of $100 = \frac{1}{4} = 0.25$.

Now $0.5 \times 100 = 50$ and $0.25 \times 100 = 25$.

So to change a decimal to a percentage, simply multiply the decimal by 100.

Example ❶ Convert $\frac{3}{5}$ to (a) a decimal   (b) a percentage.

(a) $\frac{3}{5} = 3 \div 5 = 0.6$ (This can be done on a calculator, or as $5\overline{)3.0}$ with 0.6 above)

(b) $0.6 \times 100 = 60\%$

NB It is possible to go straight from the fraction to the percentage using multiplication of fractions.

$$\tfrac{3}{5} \times 100 = \tfrac{3}{5} \times \tfrac{100}{1} = \tfrac{300}{5} = 60\%$$

Exercise 1.1a

1. Copy the following table and complete it.

Fraction $\frac{a}{b}$	Decimal $a \div b$	Percentage = decimal × 100
$\frac{7}{10}$		
$\frac{2}{5}$		
$\frac{3}{4}$		
$\frac{1}{3}$		
$\frac{2}{3}$		

2. Change the following decimals to percentages.
   (a) 0.37 (b) 0.83 (c) 0.08 (d) 0.345 (e) 1.25
3. Change the following fractions to decimals.
   (a) $\frac{1}{100}$ (b) $\frac{17}{100}$ (c) $\frac{2}{50}$ (d) $\frac{8}{5}$
4. On Wednesday, 0.23 of the population watched Westenders. What percentage is this?
5. At a matinée, $\frac{4}{5}$ of the audience at a pantomime were children. What percentage is this?
6. In a survey, $\frac{3}{10}$ of children liked cheese and onion crisps. What percentage is this?

Exercise 1.1b

1. Change the following fractions to decimals.
   (a) $\frac{1}{8}$ (b) $\frac{3}{8}$ (c) $\frac{3}{20}$ (d) $\frac{17}{40}$ (e) $\frac{5}{16}$
2. Change the decimals you found in question 1 to percentages.
3. Change the following fractions into percentages. Give your answers correct to one decimal place.
   (a) $\frac{1}{6}$ (b) $\frac{5}{6}$ (c) $\frac{5}{12}$ (d) $\frac{1}{15}$ (e) $\frac{3}{70}$
4. A kilometre is $\frac{5}{8}$ of a mile. What percentage is this?
5. The winning candidate in an election gained $\frac{7}{12}$ of the votes. What percentage is this? Give your answer correct to the nearest 1%.
6. Nicola spends $\frac{3}{7}$ of her pocket money on sweets and drinks. What percentage is this? Give your answer correct to the nearest 1%.
7. Imran has a part-time job. He saves $\frac{2}{9}$ of his wages. What percentage is this? Give your answer correct to the nearest 1%.

Examiner's tip

For the non-calculator paper it is worth learning some basic equivalents.

$\frac{1}{2} = 0.5 = 50\%$	$\frac{1}{3} = 0.333... = 33.3...\%$	$\frac{1}{5} = 0.2 = 20\%$
$\frac{1}{4} = 0.25 = 25\%$	(33% to nearest 1%)	$\frac{2}{5} = 0.4 = 40\%$
$\frac{3}{4} = 0.75 = 75\%$	$\frac{2}{3} = 0.666... = 66.6...\%$	$\frac{3}{5} = 0.6 = 60\%$
	(67% to nearest 1%)	$\frac{4}{5} = 0.8 = 80\%$

Notice the rounding of $\frac{1}{3}$ and $\frac{2}{3}$. It is a common error to assume $\frac{1}{3} = 0.3 = 30\%$ and $\frac{2}{3} = 0.66 = 66\%$ or even $\frac{2}{3} = 0.6 = 60\%$.

Expressing one quantity as a percentage of another

To express one quantity as a percentage of another, start by writing the first quantity as a fraction of the second.

Then use the methods already described to change the fraction to a percentage.

Express 4 as a percentage of 5.

First write 4 as a fraction of 5: $\frac{4}{5}$

Then change $\frac{4}{5}$ to a percentage: $\frac{4}{5} = 0.8 = 80\%$

Express £5 as a percentage of £30.

$\frac{5}{30} = 0.167 = 16.7\%$ to one decimal place

Express 70 cm as a percentage of 2.3 m.

First change 2.3 m to centimetres:

$2.3\,\text{m} = 230\,\text{cm}$

$\frac{70}{230} = 0.304 = 30.4\%$ to one decimal place

or

Change 70 cm to metres. $70\,\text{cm} = 0.7\,\text{m}$

$\frac{0.7}{2.3} = 0.304 = 30.4\%$ to one decimal place

Exercise 1.2a

1. In each case, express the first quantity as a percentage of the second.

	First quantity	Second quantity
(a)	16	100
(b)	12	50
(c)	1 m	4 m
(d)	£3	£10
(e)	73p	£1
(f)	8p	£1
(g)	£1.80	£2
(h)	40 cm	2 m
(i)	50p	£10
(j)	£2.60	£2

2. An article in a shop costs £5, then the shopkeeper increases the price by £1. What percentage increase is this?

3. A sailor has a rope 50 m long. He cuts off 12 m. What percentage of the rope has he cut off?

4. In a sales promotion a company offers an extra 150 ml of cola free. The bottles usually hold 1 litre. What percentage is 150 ml of 1 litre?

5. After thermal insulation was installed, Saima's central heating bill was reduced by £140 per year. If the bill was previously £500, what is the percentage reduction?

Exercise 1.2b

1. In each case, express the first quantity as a percentage of the second. Where appropriate give your answer to the nearest 1%.

	First quantity	Second quantity
(a)	11	16
(b)	9	72
(c)	1 m	7 m
(d)	£3	£18
(e)	73p	£3
(f)	17p	£2.50
(g)	£2.60	£7
(h)	40 cm	2.6 m
(i)	£3.72p	£12.96
(j)	£2.85	£2.47

2. A shopkeeper reduces the cost of an item by £2. If it originally cost £11, what is the percentage reduction? Give your answer to one decimal place.

3. A spring is originally 60 cm long. It is stretched by 17 cm. By what percentage is it stretched? Give your answer to one decimal place.

4. A school has 857 pupils and 70 of them go on a school holiday. What percentage is this? Give your answer to one decimal place.

5. A mathematics test has 45 questions. Abigail got 42 right. What percentage did Abigail get right? Give your answer to the nearest 1%.

Percentage increase and decrease, profit and loss

In this section you will solve problems, using methods you have learnt previously.

A shopkeeper buys an article for £25 and sells it for £30. What percentage profit is this?

Profit = £30 – £25 = £5

Percentage profit = $\frac{5}{25} \times 100 = 20\%$

A car drops in value from £7995 to £7000 in a year. What percentage decrease is this?

Decrease in value = £7995 – £7000 = £995

Percentage decrease = $\frac{995}{7995} \times 100 = 12.4\%$
to one decimal place

The rate of inflation is 3%. An engineer's salary increases at the same rate. If she earned £24 000 before the increase, what is her new salary?

Increase = $24\,000 \times \frac{3}{100} = 24\,000 \times 0.03 = 720$

New salary = £24 000 + £720 = £24 720

Examiner's tip

Percentage increases and decreases are worked out as percentages of the original amount, not the new amount. Percentage profit or loss is worked out as a percentage of the cost price, not the selling price.

Example 8

In a sale all prices are reduced by 15%. Find the new price of an article previously priced at £17.60.

You need to find 15% of £17.60 and subtract the answer from £17.60.

You need 100% of £17.60 – 15% of £17.60.

This is the same as 85% of £17.60 (since 100% – 15% = 85%).

The calculation now becomes £17.60 × $\frac{85}{100}$ = £17.60 × 0.85 = £14.96.

New price = £14.96

> Using this method, you carry out the percentage calculation and the subtraction in one step.

Examiner's tip

Although the method shown in Example 7 is correct, there is a shorter method which makes work much easier when doing repeated calculations and, in later work, when working back from an answer. The method is shown in Examples 8, 9 and 10. It is particularly useful on the calculator paper.

Example 9

£24 000 is invested for one year at 3% simple interest. Find the total amount at the end of the year.

You need to calculate 3% of £24 000 and add it on to the original £24 000.

You need 3% of 24 000 + 100% of 24 000.

This is the same as 103% of £24 000.

The calculation now becomes
£24 000 × $\frac{103}{100}$ = £24 000 × 1.03 = £24 720

> Using this method, you carry out the percentage calculation and the addition in one step.

Example 10

The bill for a caravan repair is £775.46. VAT at $17\frac{1}{2}$% is added. What is the total?

You need to find 100% + $17\frac{1}{2}$% of £775.46.

This is the same as $117\frac{1}{2}$% of £775.46.

The calculation now becomes 775.46 × $\frac{117.5}{100}$ = 775.46 × 1.175 = 991.1655
= £911.17 to the nearest penny

Example 11

Due to inflation, prices increase by 5% per year. An item costs £12 now. What will it cost in 2 years' time?

In 1 year the price will be £12 × 1.05 = £12.60.

In 2 years the price will be £12.60 × 1.05 = £13.23.

Alternatively, this repeated calculation could be worked out as:
£12.60 × 1.05 × 1.05 = £13.23

Exercise 1.3a

1. A shopkeeper buys an article for £10 and sells it for £13. What percentage profit does she make?

2. David earned £3 per hour. When the National Minimum Wage was introduced his pay increased to £3.60 per hour. What was the percentage increase in David's pay?

3. A season ticket for Newtown Rovers normally costs £500. If it is bought before 1 June it costs £420. What percentage reduction is this?

4. Lee bought a CD for £12.50. A year later he sold it for £5. What percentage of the value did he lose?

5. Ghalib's gas bill is £240 before VAT is added on. What is the bill after VAT at 5% is added on?

6. Kate's diet has led to her losing 15% of her starting weight. If she weighed 80 kg before starting her diet, what does she weigh now?

7. The rate of inflation is 8%. Ticket prices are increased by this rate. What is the new price for a £3 ticket?

8. In a sale, prices are reduced by 20%. What is the sale price if the original price was £13?

Example 12

Each year a car loses value by 12% of its value at the beginning of the year. If its starting value was £9000 find its value after 3 years.

$$100\% - 12\% = 88\%$$

So after 1 year the value $= £9000 \times 0.88$
$$= £7920$$

after 2 years the value $= £7920 \times 0.88$
$$= £6969.60$$

after 3 years the value $= £6969.60 \times 0.88$
$$= £6133.25 \text{ (to the nearest 1p)}$$

Alternatively, this repeated calculation could be worked out as:
$$£9000 \times 0.88 \times 0.88 \times 0.88 = £6133.25$$

Exercise 1.3b

1. A shopkeeper buys an article for £22 and sells it for £27. What is his percentage profit? Give your answer to the nearest 1%.

2. A rail company reduces the time for a journey from 55 minutes to 48 minutes. What percentage reduction is this? Give your answer to the nearest 1%.

3. Claire receives a pay increase from £135 per week to £145 per week. What percentage rise is this? Give your answer to one decimal place.

4. In a sale the price of a dress is reduced from £60 to £39. What percentage reduction is this?

5. A1 Electrics buys washing machines for £255 and sells them for £310. Bob's Budget Bargains buys washing machines for £270 and sells them for £330. Which company makes the greater percentage profit? You must show all your working.

6. To test the strength of a piece of wire it is stretched by 12% of its original length. If it was originally 1.5 m long, what will its length be after stretching?

7. Rushna pays 6% of her pay into a pension fund. If she earns £185 per week, what will her pay be after taking off her pension payments?

8. A mini-tower stereo system costs £280 before VAT is added on. What will it cost after VAT at 17.5% is added on?

9. Craig puts £240 into a savings account. Each year the savings earn interest at 6% of the amount in the account at the start of the year. What will his savings be worth after three years? Give your answer to the nearest penny.

10. Each year a car loses value by 11% of its value at the start of the year. If it was worth £8000 when it was new, what will it be worth after two years?

Ratio

When you are mixing quantities it is often important to keep the amounts in the same proportion.

For example, if you are mixing black paint and white paint to make a certain shade of grey, it will be important to keep the proportions of black and white paint the same.

If the colour you want is obtained by mixing 2 litres of black paint with 1 litre of white paint, then you will need 4 litres of black paint if you use 2 litres of white paint.

The ratio of black paint to white paint is 2 parts to 1 part.

Example 13

To mix the same shade of grey paint as above, how much white paint will you need to mix with 8 litres of black paint?

To keep track of ratio questions, it is often helpful to form a table.

Black paint   **White paint**

The multiplier for black paint is 4, so you must use the same multiplier for white paint: $1 \times 4 = 4$.

So you need 4 litres of white paint.

Example 14

To make pink paint, red paint is added to white paint in the ratio 3 parts red to 2 parts white.

(a) How much white paint should be mixed with 6 litres of red paint?

(b) How much red paint should be mixed with 10 litres of white paint?

   Red paint   **White paint**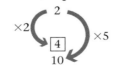

For (a) the multiplier is 2 so you need $2 \times 2 = 4$ litres of white paint.

For (b) the multiplier is 5 so you need $3 \times 5 = 15$ litres of red paint.

If the multiplier is not immediately obvious you may need to use division to find it.

> **15**
>
> To make light grey paint, Rosie mixes black and white paint in the ratio 1 part black to 5 parts white. How much black paint will she need to mix with 7 litres of white paint?
>
> **Black paint**   **White paint**
>
> 1) ×1.4   5) ×1.4
>
> 1.4   7
>
> Here the multiplier is 7 ÷ 5 = 1.4.
>
> So she needs 1 × 1.4 = 1.4 litres of black paint.

Exercise 1.4a

1. Sanjay is mixing light pink paint. He mixes red paint to white paint in the ratio 1 part red to 3 parts white.
   (a) How much white paint should he mix with 3 litres of red paint?
   (b) How much red paint should he mix with 12 litres of white paint?
2. Michelle is making mortar. To do this she mixes sand and cement in the ratio 5 parts sand to 1 part cement. She measures the quantities in bags.
   (a) How many bags of sand should she mix with 2 bags of cement?
   (b) How many bags of cement should she mix with 20 bags of sand?
3. Julia is making a cake. To do this she starts by mixing flour and fat in the ratio 8 parts flour to 3 parts fat.
   (a) How much fat should she mix with 800 grams of flour?
   (b) How much flour should she mix with 60 grams of fat?
4. Marco is making jam. To do this he mixes fruit and sugar in the ratio 2 parts fruit to 3 parts sugar.
   (a) How much sugar should he mix with 8 kg of fruit?
   (b) How much fruit should he mix with 15 kg of sugar?
5. A chemist is making a solution of a chemical. To do this he mixes 1 part of the chemical to 20 parts of water.
   (a) How much water should he mix with 10 ml of the chemical?
   (b) How much chemical should he mix with 2 litres of water?
       (1 litre = 1000 ml)
6. The same chemist is now making a weaker solution of the chemical. To do this he mixes 1 part of the chemical to 50 parts of water.
   (a) How much water should he mix with 10 ml of the chemical?
   (b) How much chemical should he mix with 2 litres of water?
       (1 litre = 1000 ml)

1. Graham is making pastry. To do this he starts by mixing flour and fat in the ratio 8 parts flour to 3 parts fat.
   (a) How much fat should he mix with 320 grams of flour?
   (b) How much flour should he mix with 150 grams of fat?

2. Kate is mixing dark pink paint. She mixes red and white paint in the ratio 4 parts red to 3 parts white.
   (a) How much white paint should she mix with 12 litres of red paint?
   (b) How much red paint should she mix with 15 litres of white paint?

3. Sarah and Heather share a flat. They agree to spend time doing the cleaning in the ratio 2 parts (Sarah) to 3 parts (Heather).
   (a) If Sarah spends 5 hours cleaning, how long should Heather spend?
   (b) If Heather spends 4 hours cleaning, how long should Sarah spend?

4. Regulations for school trips say that there must be at least two adults for every 30 pupils.
   (a) How many adults should there be for 75 pupils?
   (b) How many pupils can go on a trip if there are seven adults available?
   (c) How many adults should there be for 200 pupils?

5. Rashid fills his flower tubs with a mixture of 4 parts soil to 3 parts compost.
   (a) How much soil should he mix with 45 litres of compost?
   (b) How much compost should he mix with 68 litres of soil?

6. Orange squash needs to be mixed in the ratio 1 part concentrate to 6 parts water.
   (a) How much water should be mixed with 80 ml of concentrate?
   (b) How much concentrate should be mixed with 2 litres of water?
   (1 litre = 1000 ml)

Sharing in a given ratio

In the previous section you saw how to mix grey paint if you know how much black or white paint to use. In this section you will find out how much black and white paint to use, to make up the amount of grey paint you need.

In Example 13, black and white paint were mixed in the ratio 2 parts black to 1 part white. How much of each colour paint will you need, to mix 12 litres of grey paint?

2 litres of black paint and 1 litre of white paint will make 3 litres of grey paint.

You can solve this problem in a similar way, using a table with three columns.

Black paint	White paint	Mixture (grey)
2	1	3
$\times 4$	$\times 4$	$\times 4$
8	4	12

The multiplier is 4.

So you need: $2 \times 4 = 8$ litres of black paint
and $1 \times 4 = 4$ litres of white paint.

Now check that the answers add up to 12.

Example 16

Share £10 between Abigail and Becky in the ratio 3 parts to 2 parts.

Abigail	Becky	Total
3	2	5
×2 ↘	×2 ↘	×2 ↘
6	4	10

3 + 2 = 5

The multiplier is 2.

So Abigail gets 3 × 2 = £6 and Becky gets 2 × 2 = £4.

Check: £6 + £4 = £10

Example 17

A recipe for white sauce says, 'Mix butter, flour and milk in the ratio 1 part butter to 1 part flour to 10 parts milk.'

How much of each ingredient is needed to make 240 grams of white sauce?

Butter	Flour	Milk	Total (mixture)
1	1	10	12
×20 ↘	×20 ↘	×20 ↘	×20 ↘
20	20	200	240

The multiplier is 240 ÷ 12 = 20.

So you need 1 × 20 = 20 g each of butter and flour and 10 × 20 = 200 g of milk.

Check: 20 + 20 + 200 = 240

1. Share £20 in the ratio 2 parts to 3 parts.
2. A 20% solution is made by mixing 1 part of a chemical to 4 parts of water. How much chemical will there be in 100 ml of the solution?
3. Light pink paint is made by mixing red and white paint in the ratio 1 part red to 3 parts white. How much of each colour is needed to make 16 litres of light pink paint?
4. Dark grey paint is made by mixing black and white paint in the ratio 5 parts black to 2 parts white. How much of each colour is needed to make 35 litres of dark grey paint?
5. Asif is mixing mortar using sand and cement in the ratio 5 parts sand to 1 part cement. How much sand is needed to make 36 kg of the mixture?
6. Orange squash is mixed in the ratio 1 part concentrate to 6 parts water. How much concentrate is needed to make 3.5 litres of orange squash?
7. Share £18 among Eileen, Fiona and George in the ratio 4 parts to 2 parts to 3 parts.

Exercise 1.5b

1. Share £28 between Hamid and Ian in the ratio 5 parts to 3 parts.
2. In the audience for a pantomime, the ratio of adults to children is 1 part to 5 parts. If there are 630 in the audience, how many children are there?
3. Jane is cutting a rope into two pieces in the ratio 7 parts to 3 parts. If the length of the rope is 3 m, how long will each of the two pieces be?
4. Light grey paint is made by mixing black paint and white paint in the ratio 4 parts black to 11 parts white. How much of each colour will be needed to make 180 litres of light grey paint?
5. The number of members in a parliament is shared among three parties in the ratio 4 parts to 3 parts to 2 parts. If the total number of members is 657, how many of each party are there?
6. To get home Michael runs and walks. The distances he runs and walks are in the ratio 2 parts to 3 parts. If he lives 2 km from home, how far does he run?
7. Three children share £50 in the ratio of their ages. Sachin is 4 years old, Rehan is 7 years old and Samrina is 9 years old. How much do they each receive?

Key points

- To change a fraction into a decimal, divide the top of the fraction by the bottom.
$$\frac{a}{b} = a \div b$$
- To change a decimal into a percentage, multiply by 100.
- To find one quantity as a percentage of another, write the first quantity as a fraction of the second. Then change to a percentage as above. (Both quantities must be in the same units.)

- To increase a quantity by, for example, 5%, a quick way is to multiply by 1.05.
- To reduce a quantity by, for example, 12%, a quick way is to multiply by 0.88 (as 100 − 12 = 88).
- To mix in a given ratio both quantities must be multiplied by the same amount.
- To share in a given ratio, first add the parts of the ratio together. Then use the same multiplier for the parts as the total.

Revision exercise 1a

1. Change these fractions to decimals, giving your answer to three decimal places where appropriate.
   (a) $\frac{5}{16}$ (b) $\frac{4}{7}$ (c) $\frac{7}{40}$ (d) $\frac{4}{15}$

2. Change the decimals you found in question 1 into percentages. Give your answers to one decimal place.

3. Find £1.50 as a percentage of £21. Give your answer to the nearest 1%.

4. Find 80 cm as a percentage of 6 m. Give your answer to one decimal place.

5. Karl cuts 20 cm from a piece of wood 1.6 m long. By what percentage has he shortened the piece of wood?

6. The audience for a TV soap increased from 8 million to 10 million. What percentage increase is this?

7. The number of pupils in a school went down from 850 to 799. What percentage reduction is this?

8. In a sale all prices are reduced by 15%. Find the new price of a pair of trainers that originally cost £65.

9. Each year the value of an antique increases by 20% of its value at the beginning of the year. If it was worth £450 on 1 January 1996, what was it worth on 1 January 1997? What was it worth on 1 January 1999?

10. The instructions for making flaky pastry say, 'First mix flour and fat in the ratio 8 parts to 5 parts.' How much fat should I mix with 480 g of flour?

11. Laura and Marie share a flat. They agree to share the rent in the same ratio as their wages. Laura earns £600 per month and Marie earns £800 per month. If the rent is £420 per month, how much do they each pay?

You will need to know:

- how to use simple scales
- how to use compasses and a protractor (or angle measurer).

Isometric drawings

For these drawings, you will need a ruler, a pencil and triangle spotty paper.

Isometric drawing is a way of representing 3D shapes. Triangle spotty paper is ideal for this.

Examiner's tip

Make sure that the spotty paper is the right way round. Draw in pencil, then any mistake can be put right.

Example 1   Look at this shape.

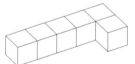

This is an isometric drawing of the same shape.

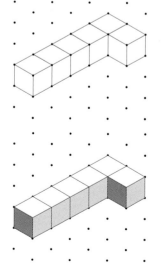

You can make the drawing look more realistic and easier to understand by using different shadings for the sides facing in the three different directions.

1. (a) On triangle spotty paper, make isometric drawings of these shapes.

    (i)    (ii) a cuboid with dimensions   (iii)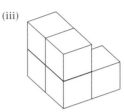
    4 by 3 by 2

    (b) How many cubes make up this shape?

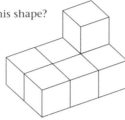

2. Use triangle spotty paper to make isometric drawings of these solid shapes. Use a scale of
   1 centimetre to 2 units.

   (a) 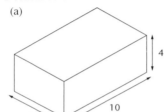   (b) 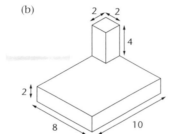   (c)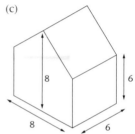

1. This shape is made from five cubes.
   Draw all the other different shapes that can be
   made from just five cubes in one layer.
   (**Hint**: There are eleven more to find.)

2. Use triangle spotty paper to make isometric drawings of these shapes. You will need to choose a suitable
   scale for each drawing.

   (a) 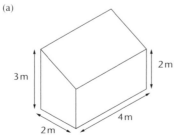   (b) 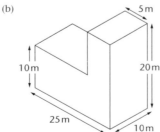   (c)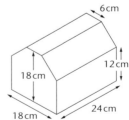

Chapter 2   Drawing

Drawing and constructing

As well as a ruler, a protractor or angle measurer and a pair of compasses, a set-square is useful.

Drawing parallel lines

Parallel lines are always the same distance apart.

Example 2

Draw two parallel lines, 3 cm apart.

Draw the base line. Place one side of the set-square on this line.

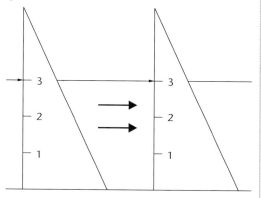

Measure and mark a height of 3 cm from the base line. Slide the set-square along, measure and mark a height of 3 cm again.

Join the points. The two lines will be parallel.

Examiner's tip

When using a set-square, the two perpendicular sides (these two →△) are the important ones.

Finding the distance from a point to a line

The length you need to find is always the **perpendicular** distance from the point onto the line. Again, you can use a set-square to help.

Example 3

Find the distance from point P to the line AB.

Place the set-square with one side on the given line. Make sure the other side touches the given point. Measure the length from the line to the point.

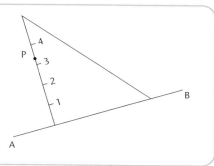

Chapter 2 Drawing and constructing

Drawing triangles

Given two sides and the included angle

The **included** angle is the angle between the two given sides.

Make an accurate drawing of the triangle sketched opposite.

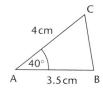

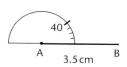

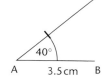

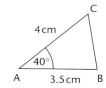

Draw the line AB and measure the 40° angle from A.

Draw a line from A through the point.

Mark the point C on this line, 4 cm from A. Join C to B.

Exercise 2.2a

1. Make accurate full-size drawings of these triangles.

(a)

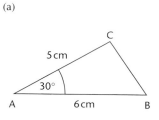

(b)

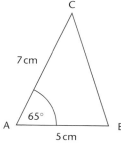

(c)

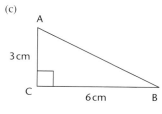

(d)

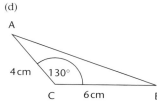

2. For each triangle in question 1, measure the unmarked side and the other two angles on your drawing.

3. For each triangle in question 1, find the perpendicular distance of C from AB on your drawing.

Exercise 2.2b

1. Draw these triangles accurately.

(a)

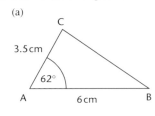

(b)

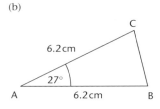

(c)

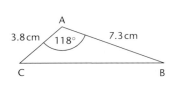

(d)

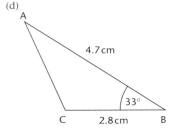

2. For each triangle in question 1, measure the unmarked side and the other two angles on your drawing.

3. For each triangle in question 1, draw a line through C parallel to AB on your drawing.

Given one side and two angles

Example 5

In triangle PQR, PQ = 6 cm, angle RPQ = 35° and angle PQR = 29°.

Make an accurate drawing of triangle PQR.

Draw a sketch.

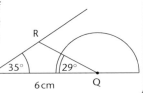

Draw the line PQ. Measure the angle of 35° at P and draw a line, any length but longer than PQ.

Measure the angle of 29° at Q. Draw a line to meet the previous line at R. This completes the triangle.

Examiner's tip

Draw a sketch of the triangle first. When an angle is written as three letters, the middle letter indicates the **vertex** (corner) of the angle.

Examiner's tip

If the two given angles were at R and Q, then angle P could be found by adding the two known angles and subtracting from 180°.

Chapter 2 **Drawing triangles**

Given three sides

Example 6   Make an accurate drawing of triangle ABC where AB = 5 cm, BC = 4 cm and AC = 3 cm.

Draw a sketch.

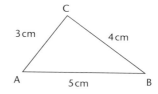

Draw the line AB. From A, with compasses set to a radius of 3 cm, draw an arc above the line.

From B, with compasses set to a radius of 4 cm, draw another arc to intersect the first. The point where the arcs meet is C.

Given two sides and a non-included angle

A **non-included** angle is one which is not between the two known sides.

Example 7   Make an accurate drawing of the triangle sketched opposite.

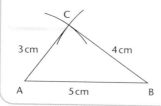

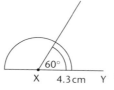

Draw the line XY. Measure the angle of 60° at X and draw a line of any length, but longer than XY.

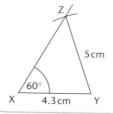

From Y, with compasses set to a radius of 5 cm, draw an arc to intersect the line. The point where the arc cuts the line is Z.

1. Make accurate full-size drawings of these triangles.

(a)

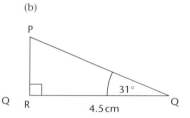

(b)

(c) triangle PQR with PQ = 7 cm, angle RPQ = 41°, angle PQR = 53°
(d) triangle PQR with PR = 3.6 cm, angle PRQ = 126°,
    angle RPQ = 18°

2. In each of the triangles in question 1, measure the unmarked sides on your drawing.

3. Draw these triangles accurately.

(a)

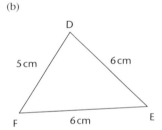

(b)

(c) triangle DEF with DE = 4.3 cm, EF = 7.2 cm, FD = 6.5 cm

4. In each of the triangles in question 3, measure the angles on your drawing.

5. Draw these triangles accurately.

(a)

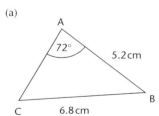

(b)

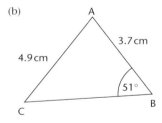

1. Make accurate full-size drawings of these triangles.

(a)

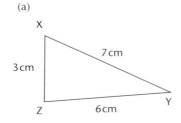

(b)

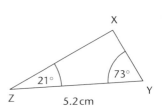

(c)

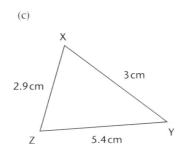

(d)

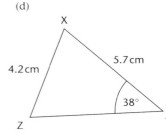

(e) triangle XYZ with XY = 4.9 cm, angle XYZ = 33°, angle YXZ = 66°
(f) triangle XYZ with XY = 8.3 cm, YZ = 4.3 cm, ZX = 5.1 cm

2. In each of the triangles in question 1, measure all the unknown lengths and angles on your drawings.

Plan and elevations

Any one view of a three-dimensional object will not show all of its features. For the full picture, you need views from three different perpendicular directions. The three views are called:
- the **plan**
- the **front elevation**
- the **side elevation**.

Example 8 Sketch the plan and elevations of this solid.

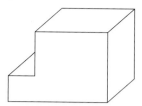

Examiner's tip

Notice that the only indication of the step in the solid is a line on the plan. This view will not tell you how deep the step in the solid is.

The **plan view** of the shape is the view of the surfaces that you would see if you looked at it from directly above. This is sometimes called the **bird's-eye view**.

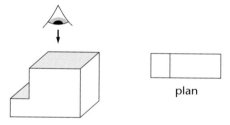

plan

Examiner's tip

Notice that this view will not tell you how wide the solid is.

The **front elevation**, or front view, of the shape is the view of the surfaces that you would see if you looked at it directly from the front.

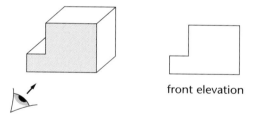

front elevation

Examiner's tip

The step in the solid is hidden from view from this direction. To show it is there, but cannot be seen, use a dotted line. Dotted lines are always used to indicate hidden features of a solid.
The side elevation can be viewed from either side of the solid. From the opposite side, the view would be:

which is the same but with a solid line instead of a dotted line. The step in the solid can be seen from this direction.

The **side elevation**, or side view, of the shape is the view of the surfaces that you would see if you looked at it directly from the side.

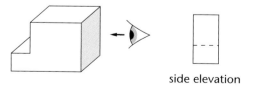

side elevation

Chapter 2 Plan and elevations

25

Notice that the front elevation and the side elevation are the same. The two dotted lines indicate the hidden edges of the inside of the ring. The curved part of the ring appears as a rectangle in the front and side elevations.

Example 9

Draw accurately the plan (P), the front elevation (F) and the side elevation (S) of this ring.

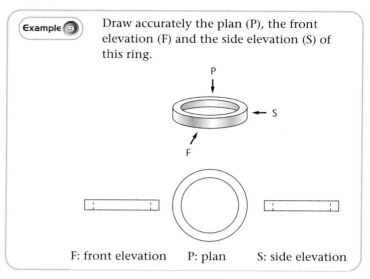

F: front elevation   P: plan   S: side elevation

Example 10

Draw accurately the plan (P), the front elevation (F) and the side elevation (S) of this small wedge.

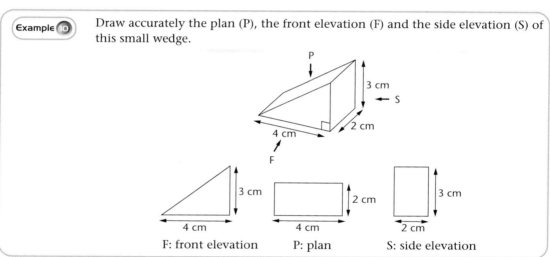

F: front elevation   P: plan   S: side elevation

Sketch the plan and elevations for each of these solid shapes.
The arrows indicate the directions of the plan (P), front
elevation (F) and side elevation (S) in each case.

1.

2.

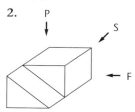

3.

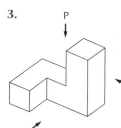

4.

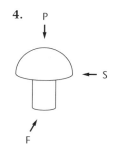

5.

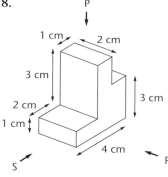

6.

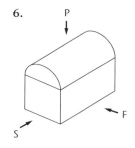

7.

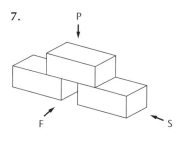

Draw the plan and elevations of each of these shapes accurately.

8.

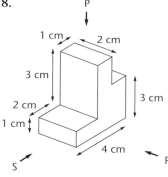

1 cm
2 cm
3 cm
2 cm
1 cm
4 cm
3 cm
S
F
P

9.

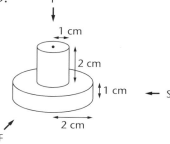

1 cm
2 cm
1 cm
2 cm
S
F
P

Hint: You will need compasses to draw the plan.

10.

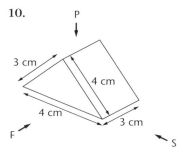

3 cm
4 cm
4 cm
3 cm
F
S
P

Hints: (a) You will need compasses to draw the front elevation.
(b) To draw the side elevation, you will need to measure the height of the triangle in the front elevation.

Exercise 2.4b

Sketch the plan and elevations for each of these solid shapes.
The arrows indicate the directions of the plan (P), front elevation (F) and side elevation (S) in each case.

1.

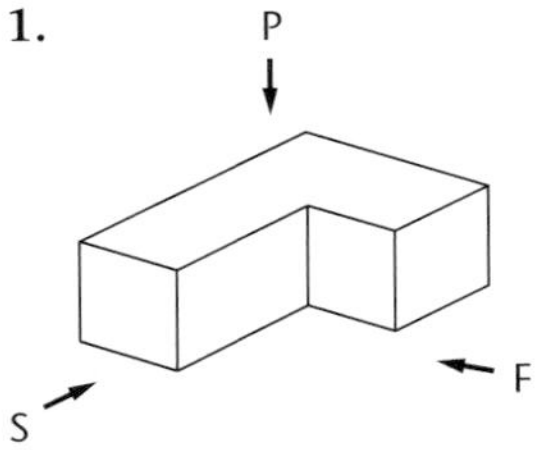

P
S
F

2.

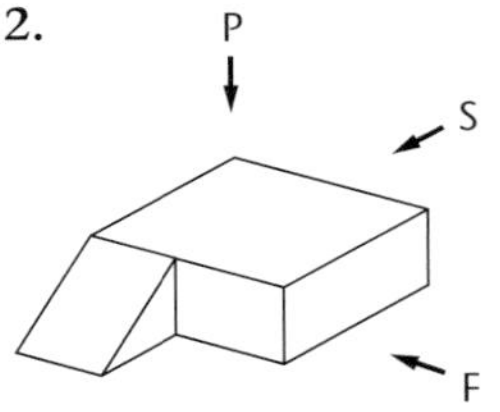

P
S
F

3.

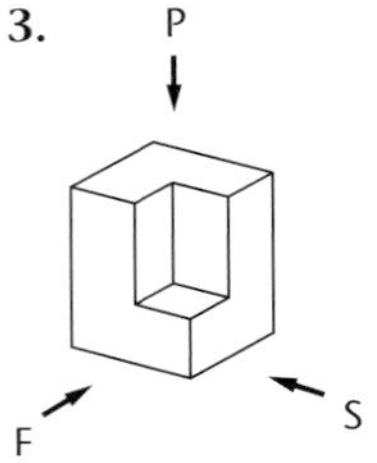

P
F
S

4.

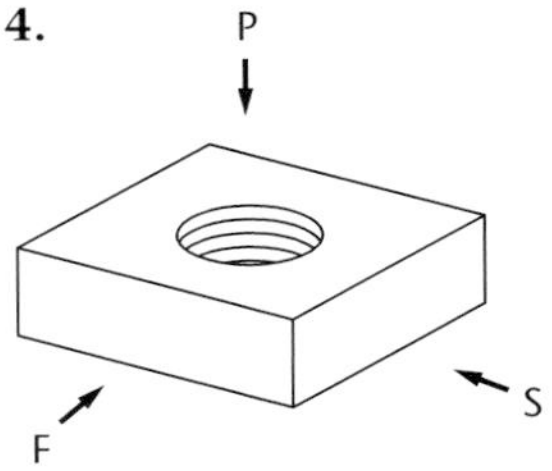

P
F
S

5.

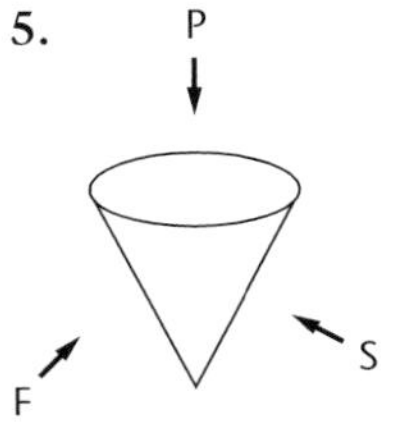

P
F
S

6.

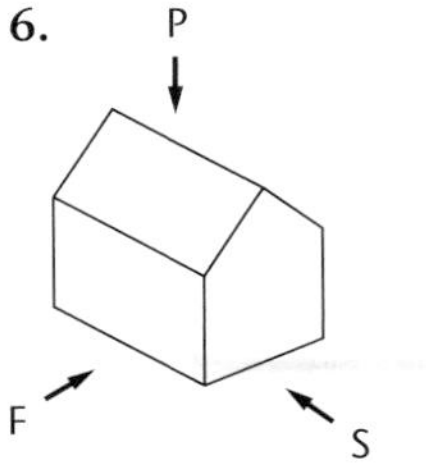

P
F
S

7.

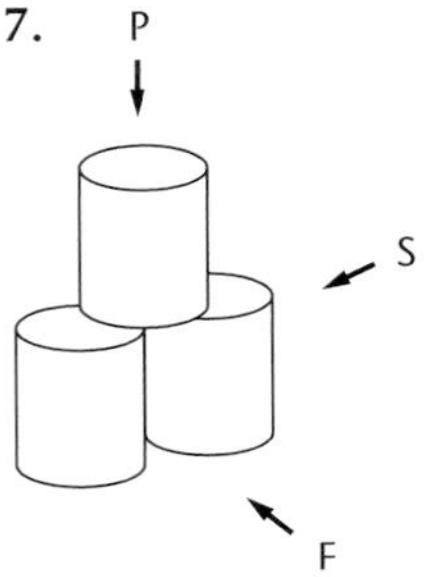

P
S
F

Draw the plan and elevations of each of these shapes accurately.

8.

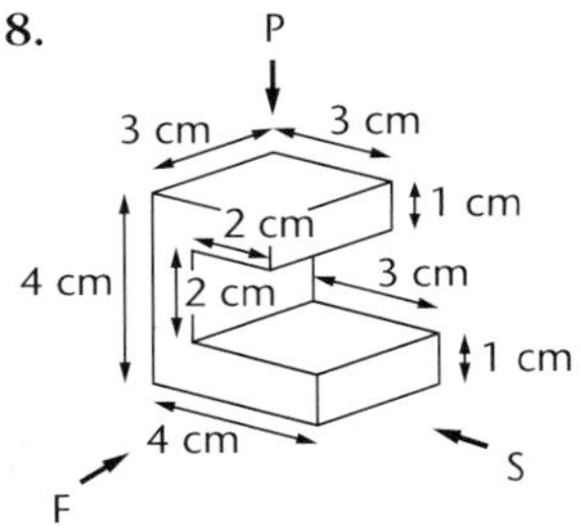

P
3 cm
3 cm
1 cm
2 cm
4 cm
2 cm
3 cm
1 cm
4 cm
F
S

Hint: You will need to draw the side elevation before drawing the front elevation.

9.

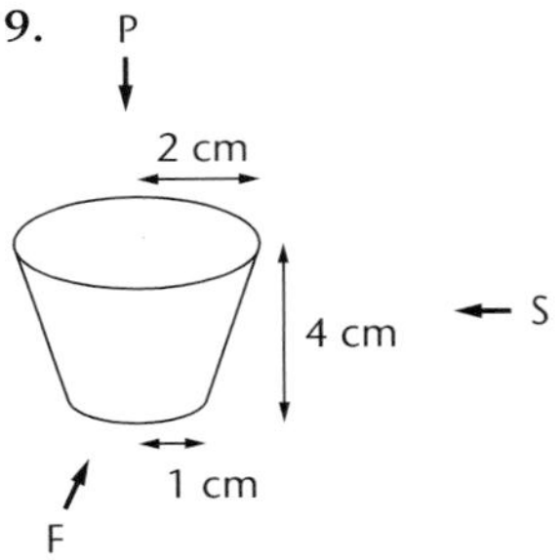

P
2 cm
4 cm
S
1 cm
F

Hint: You will need compasses to draw the plan.

10.

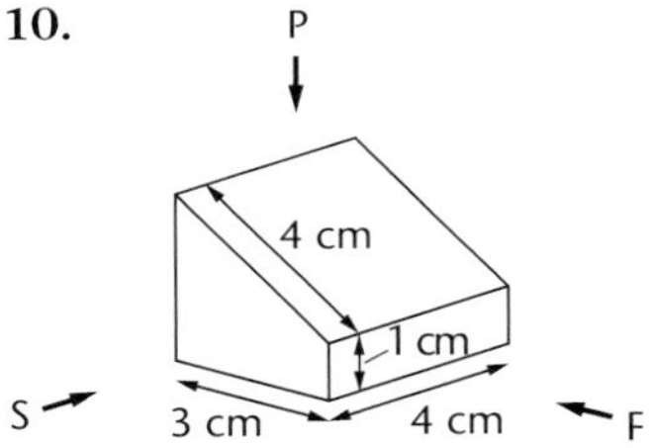

P
4 cm
1 cm
S
3 cm
4 cm
F

Hints: (a) You will need to draw the side elevation before drawing the front elevation.
(b) You will need a compass to draw the side elevation.

Key points

- Accurate drawings must be accurate. Use a sharp pencil and check all measurements.
- Drawing a rough sketch first helps you to draw the sides and angles in the right place.
- Compasses are not just for drawing circles. Use them when you know the distance but not the direction.

1. Make accurate drawings of the following triangles. On your drawings measure the lengths or angles indicated.

   (a) Find BC.

   (b) Find angle SRT.

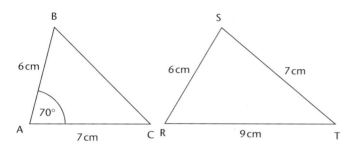

   (c) Find PR.

   (d) Find (i) XY (ii) angle XZY.

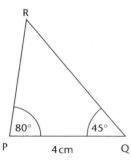

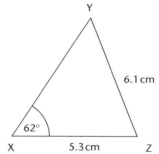

   (e) Find AC.

   (f) Find (i) LM (ii) angle MLN.

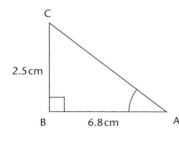

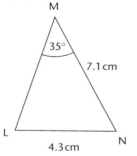

2. Make accurate drawings of these triangles.
   (a) triangle ABC, where AB = 6 cm, AC = 5 cm and angle BAC = 60°
   (b) triangle PQR, where PQ = 6 cm, PR = 4.5 cm and QR = 3.9 cm
   (c) triangle BCD, where BC = 7 cm, angle DBC = 41° and angle BCD = 57°
   (d) triangle RST, where RS = 5.5 cm, RT = 4.8 cm and angle RST = 46°

3. Draw an isosceles triangle with one side measuring 4.6 cm and the other two measuring 6.2 cm. Measure the angles inside this triangle.

4. Use what you learned in the first part of this chapter to make an accurate drawing of each of these shapes.

(a)

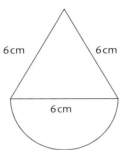

6 cm   6 cm

6 cm

(b) (i) What is the name of this shape?
(ii) The opposite sides should be parallel. How can you check this?

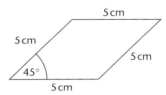

5 cm

5 cm

5 cm

5 cm

45°

(c) This is the net of a solid shape. What is the solid shape?

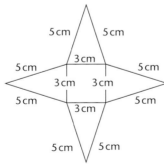

5 cm   5 cm

5 cm   3 cm   5 cm

3 cm   3 cm

5 cm   5 cm

3 cm

5 cm   5 cm

(d)

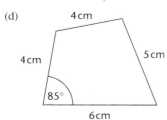

4 cm

4 cm   5 cm

85°

6 cm

5. Draw triangle ABC where BC = 7 cm, angle ABC = 62° and angle ACB = 42°.
(a) Draw a line through A parallel to BC.
(b) What is the distance from A to BC?

6. Sketch the plan and elevations of each of the following shapes.

(a)

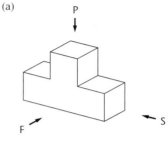

P

F   S

(b)

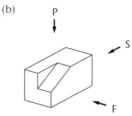

P

S

F

7. Draw accurately the plan and elevations of this shape.

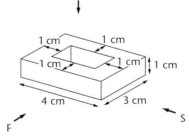

P

1 cm   1 cm

1 cm   1 cm   1 cm

4 cm   3 cm

F   S

8. Sketch the object shown in the plan and elevations.

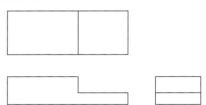

Sequences, symbols and equations

Letters for unknowns

Imagine you had a job where you were paid by the hour. You would receive the same amount for each hour you worked.

How could you work out how much you will earn in a week?

You would need to work it out as:
  the number of hours worked multiplied by the amount you are paid for each hour.

This is a **formula** in words.

If you work 35 hours at £4.50 an hour, it is easy to work out $35 \times £4.50$, but what if the numbers change?

The calculation '$35 \times £4.50$' is only right if you work 35 hours.

Suppose you move to a better job where you are paid more for each hour?

You need a simple formula that always works. You can use symbols to stand for the numbers that can change.

You could use ? or □, but it is less confusing to use letters. Using letters to stand for unknown numbers is called **algebra**.

Start by using algebra to work out your wages.
  Let: the number of hours be N
          the amount you are paid each hour be P
          the amount you earn in a week be W.
  Then $W = N \times P$

 Example ①   Find W when $N = 40$, $P = £5.00$.

$W = 40 \times 5 = £200$

 Example ②   When you make a journey, S is the speed, d is the distance travelled and t is the time taken.

To find the speed, you divide the distance by the time, so the formula for S is:

$S = d \times t$

 **Examiner's tip**

If you are not sure whether to multiply or divide, try an example with numbers first.

 Example 3

The cost (C) of hiring a car is a fixed charge (f), plus the number of days (n) multiplied by the daily rate (d), so:

$$C = f + n \times d$$

Exercise 3.1a

Write these formulae, using the letters given.
1. The cost (C) of x pencils at y pence each.
2. The area (A) of a rectangle m cm long and n cm wide.
3. The height (h) of a stack of n tins each t cm high.
4. The temperature (F) in °F is 32, plus 1.8 times the temperature in °C (C).
5. My gas bill (B) is a charge (s), plus the number (n) of units used multiplied by the cost (u) of each unit.
6. The mileage performance (number of miles per litre), (R), of a car is the number (m) of miles travelled divided by the number (p) of litres of petrol used.
7. The time (T) to cook a turkey is 30 minutes plus 40 minutes for each kilogram (k).
8. The area (A) of a triangle is half the base (b) times the height (h).
9. The number (d) of dollars is 1.65 times the number (p) of pounds.
10. The current in a circuit (i) is the voltage (e) divided by the resistance (r).

Exercise 3.1b

Write these formulae, using the letters given.
1. The cost (c) of petrol is the number (n) of litres multiplied by the price (p) of petrol per litre.
2. The total wages (w) in a factory is the number (n) of workers multiplied by the weekly wage (q).
3. The perimeter (p) of a quadrilateral is the sum of the lengths of its sides (a, b, c, d).
4. The cost (P) of n books at q pounds each.
5. The numbers (N) of books that can fit on a shelf is the length (L) of the shelf divided by the thickness (t) of each book.
6. The time (t) for a journey is the distance (d) divided by the speed (s).
7. The number (F) of French francs is 9 times the number (P) of pounds.
8. The number (Q) of posts for a fence is the length (R) of the fence divided by 2, plus 1.
9. The number (n) of eggs in a box a eggs across, b eggs along and c eggs up.
10. The approximate circumference (c) of a circle is 6 multiplied by its radius (r).

Exercise 3.2a

Use the formulae in Exercise 3.1a to find:
1. C when $x = 15$, $y = 12$
2. A when $m = 7$, $n = 6$
3. h when $n = 20$, $t = 17$
4. F when $C = 40$
5. B when $s = 9.80$, $n = 234$, $u = 0.065$
6. R when $m = 320$, $p = 53.2$
7. T when $k = 9$
8. A when $b = 5$, $h = 6$
9. d when $p = 200$
10. i when $e = 13.6$, $r = 2.5$

Exercise 3.2b

Use the formulae in Exercise 3.1b to find:
1. c when $n = 50$, $p = 70$
2. w when $n = 200$, $q = 150$
3. p when $a = 7$, $b = 5$, $c = 8$, $d = 2$
4. P when $n = 25$, $q = 7$
5. N when $L = 90$, $t = 3$
6. t when $d = 260$, $s = 40$
7. F when $P = 50$
8. Q when $R = 36$
9. n when $a = 12$, $b = 20$, $c = 6$
10. c when $r = 5$

Making it simple

Here is the formula for the perimeter P of this rectangle.

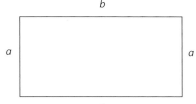

$P = a + a + b + b$

$a + a$ is twice as big as a.

so $a + a = 2 \times a$, which is usually written as $2a$, meaning
'2 times a'.

So $P = 2a + 2b$.

Here is the formula for the area of the rectangle.

$A = a \times b$

The short way of writing $a \times b$ is ab.

Here is the formula for the area of this square.

$A = s \times s$

The short way of writing $s \times s$ is s^2 (said as 's to the power 2' or
's squared').

Example 4

$$3a + 2b - 2a + 5b = 3a - 2a + 2b + 5b$$
$$= a + 7b$$

Example 5

$$p \times q \times p^2 \times q = p \times p^2 \times q \times q$$
$$= p^3 \times q^2$$
$$= p^3q^2 \quad \text{said as '}p\text{ cubed }q\text{ squared'}$$

Exercise 3.3a

Simplify these.

1. $x + x + x + x + x$
2. $y + y + y + z + z$
3. $a \times a \times a$
4. $a \times a + b \times b$
5. $p + p + q + p + q + q + p$
6. $a + 2b + 2a + b$
7. $3m + 2n + n + m$
8. $5x - 3x + 2y - y$
9. $6p + 2q - 5q - 3p$
10. $a^2 + 2a - 3a - 6$
11. $a + a + a \times a$
12. $2b + 3b$
13. $2b \times 3b$
14. $2a \times 3b$
15. $2a \times b + a \times 3b$
16. $2a + 3b - 5a$
17. $pq \times p^2q$
18. $a^2b^2 \times 2ab$
19. $xy^2 + 2xy + 3xy^2$
20. $2a^2 \times 3a^3$

Exercise 3.3b

Simplify these.

1. $x + x + y + y$
2. $a + b + a + b + a$
3. $2p + 3q + 2q + 5p$
4. $2m + 3n - m - n$
5. $x^2 + 3x - 5x - 15$
6. $3p \times 4q$
7. $5x \times 2y + x \times 3y$
8. $xy^2 \times xy$
9. $3x^2 + 2y^2 - 5x^2$
10. $ab^2 \times 3a^2b^2$

Sequences

A group of numbers, written in order, is called a **sequence** if there is a rule to find the next number. The numbers in a sequence are called **terms**.

5   6   7   8   ?

The rule is 'add 1' and the next term is 9.

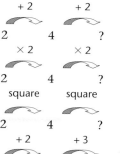

1   2   4   8   ?

The rule is 'multiply by 2' and the next term is 16.

When you are given only a few terms, you can sometimes spot more than one rule.

+ 2   + 2

2   4   ?   The rule is 'add 2', the next term is 6.

× 2   × 2

2   4   ?   The rule is 'multiply by 2', the next term is 8.

square   square

2   4   ?   The rule is 'square', the next term is 16.

+ 2   + 3

2   4   ?   The rule is 'add one more', the next term is 7.

Can you find any more?

Exercise 3.4a

For each sequence, find a rule and write down the next number.

1.	1	2	3	4
2.	2	4	6	8
3.	41	43	45	47
4.	10	20	30	40
5.	2	4	8	16
6.	5	5	5	5
7.	3	9	27	81
8.	27	25	23	21
9.	10	7	4	1
10.	72	36	18	9

Exercise 3.4b ⚠

For each sequence, find a rule and write down the next number.

1.	$\frac{1}{2}$	$\frac{1}{3}$	$\frac{1}{4}$	$\frac{1}{5}$
2.	1	3	6	10
3.	3	9	81	6561
4.	64	32	16	8
5.	1.2	1.5	1.8	2.1
6.	243	81	27	9
7.	3	7	13	21
8.	1	2	6	15
9.	2	8	18	32
10.	1	1	2	3

The *n*th term

It is often possible to find a formula to give the terms in a sequence.

You usually use *n* to stand for the number of a term.

Example 6

If the formula is *n*th term = 2*n* + 1, then:

the first term $= 2 \times 1 + 1 = 3$
the second term $= 2 \times 2 + 1 = 5$
the third term $= 2 \times 3 + 1 = 7$
and so on.

Examiner's tip

This is like Exercise 3.2, putting the numbers 1, 2, 3, 4 in for *n*.

Exercise 3.5a

Each of these is the formula for the *n*th term. Find the first four terms of the sequence.

1. *n* + 1
2. 2*n*
3. 2*n* – 1
4. *n* + 5
5. 3*n*
6. 3*n* + 1
7. 5*n* – 3
8. 10*n*
9. 7*n* – 7
10. 2 – *n*

Exercise 3.5b

Each of these is the formula for the *n*th term. Find the first four terms of the sequence.

1. *n*
2. *n* + 3
3. 4*n*
4. *n* – 1
5. 2*n* + 1
6. 3*n* – 1
7. 6*n* + 5
8. 2*n* – 3
9. 5 – *n*
10. 10 – 2*n*

Notice how this works.
If the formula contains a 2*n*, the terms increase by 2 each time; if it contains 5*n*, the terms increase by 5 each time.
So to find a formula for a given sequence, find how much more (or less) each term is than the one before it.

Example 7

+2   +2   +2

3   5   7   9

The differences between the terms is 2, so the formula will include 2*n*.

When *n* = 1, 2*n* = 2, but the first term is 3, which is 1 more.

The formula will be *n*th term = 2*n* + 1.

Examiner's tip

This will always work if the differences are the same each time.

Example 8

$$\overset{-2}{9 \curvearrowright 7} \quad \overset{-2}{\curvearrowright 5} \quad \overset{-2}{\curvearrowright 3}$$

The differences here are still 2 but they must be **negative 2**, as the terms are getting smaller.

This time the formula will include $-2n$, to make the terms get smaller.

When $n = 1$, $-2n = -2$ but the term is 9, which is 11 more.

The formula for the nth term is $11 - 2n$.

Exercise 3.6a

Find the nth terms for each of these sequences.

1.	1	2	3	4
2.	4	6	8	10
3.	4	8	12	16
4.	0	2	4	6
5.	7	11	15	19
6.	1	7	13	19
7.	11	21	31	41
8.	5	8	11	14
9.	101	201	301	401
10.	25	23	21	19

Exercise 3.6b

Find the nth terms for each of these sequences.

1.	0	1	2	3
2.	2	5	8	11
3.	7	9	11	13
4.	4	9	14	19
5.	15	20	25	30
6.	−1	3	7	11
7.	5	7	9	11
8.	101	102	103	104
9.	4	3	2	1
10.	7	4	1	−2

Examiner's tip

This is called **expanding the brackets**. Remember to multiply the number or letter outside by each term inside the brackets.

Brackets

What is $2 \times 3 + 4$? Is it 14? Is it 10?

The rule is 'do the multiplication first', so the answer is 10.

If you want the answer to be 14, you need to add 3 and 4 first.

Use brackets to show this: $2 \times (3 + 4)$

Notice that this is equal to $2 \times 3 + 2 \times 4$.

It is the same in algebra.

$a(b + c)$ means 'add b and c then multiply by a' and this is the same as 'multiply a by b, multiply b by c, then add the results'.

So $a(b + c) = ab + ac$.

Example 9 | Expand the brackets.

(a) $4(2x - 1) = 4 \times 2x + 4 \times (-1)$
$$= 8x - 4$$

(b) $-x(x - y) = (-x) \times x + (-x) \times (-y)$
$$= -x^2 + xy$$

Examiner's tip

Remember about multiplying with negative numbers!
$(-3) \times (-x) = 3x$
$(-3) \times x = -3x$
and so on.

Exercise 3.7a

Expand these brackets.

1. $2(a + b)$
2. $3(x + 2)$
3. $4(2x + 1)$
4. $a(a + 2)$
5. $y(y - 1)$
6. $2(1 - x)$
7. $5(p - q)$
8. $3(3x - 1)$
9. $2(3x + 2)$
10. $2(2x - 3)$

11. $x(x + y)$
12. $3x(x + 1)$
13. $y(3 - x)$
14. $z(2x + 3y)$
15. $4p(3p - 5q)$
16. $-3(x + y)$
17. $-3(x - y)$
18. $-a(a + b)$
19. $-a(a - b)$
20. $-3p(2p - 4q)$

Exercise 3.7b

Expand these brackets.

1. $3(x + y)$
2. $2(p + 3)$
3. $x(x + 3)$
4. $4(3x - 2)$
5. $a(a + 2b)$
6. $3c(c + d)$
7. $x(2 - x)$
8. $-y(2 + y)$
9. $-z(z - 2)$
10. $-2x(5x - 3y)$

Forming equations

You can use algebra to solve some problems. The first step is to write the problem in a mathematical way.

Example 10 | The price of a carton of milk has gone up 5 pence.

It now costs 59 pence. How much did it cost before?

Let the original price be c pence.

Then the new price is $c + 5$.

But this is equal to 59, so $c + 5 = 59$.

This is an **equation**.

Example 11 | Here is a puzzle. John is 3 years older than his sister. He is also twice as old as his sister. How old is John?

Let John be x years old.

Then his sister is $(x - 3)$ years old.

But John is twice as old as his sister, so he is $2 \times (x - 3) = 2(x - 3)$ years old.

The equation is $x = 2(x - 3)$.

Chapter 3 Forming equations

37

Exercise 3.8a

Form equations, using the given letter for the unknown.

1. A magazine costs £m.
   (a) How much do three magazines cost?
   (b) If three magazines cost £6, form an equation in m.
2. A loaf costs b pence.
   (a) How much do five loaves cost?
   (b) If five loaves cost 240 pence, form an equation in b.
3. My age now is y years.
   (a) What will my age be in six years' time?
   (b) In six years' time, I shall be 21. Form an equation in y.
4. Five years ago, Mary's age was 61. Now Mary's age is x years.
5. The length of a rectangle is 3 cm more than its width. Its perimeter is 14 cm. The width is x cm.
6. The length of a rectangle is 5 metres. Its area is 20 square metres. The width is w metres.
7. In three years' time, Amy will be twice as old as she is now. Amy is a years old.
8. I think of a number, double it, subtract 3 and the result is 14. The number is x.
9. I think of a number, subtract 3 then double the result. This gives 14. The number is y.
10. I think of a number, multiply it by 3, add the original number and subtract 7. This gives 29. The number is z.

Save these equations until Exercise 3.13a.

Exercise 3.8b

Form equations, using the given letter for the unknown.

1. The cost of four cups of coffee is £6. Each one costs c pounds.
2. The cost of five bags of crisps is 85p. Each one costs x pence.
3. My age now is T years. 25 years ago I was 36.
4. The area of a rectangle is 36 square metres. The width is 4 metres. The length is L metres.
5. I think of a number, multiply it by 5 and the result is 30. The number is N.
6. In seven years' time, Sarah will be twice as old as she is now. Her age now is y years.
7. Each of the equal sides of an isosceles triangle is 3 cm longer than the third side. The perimeter is 21 cm. The third side is x cm.
8. I think of a number, add 7, then double the result. This gives 24. The number is k.
9. I think of a number, multiply it by 5, add 8 and then add twice the original number. The result is 1. The number is q.
10. John is five years older than Peter. The total of their ages is 59. Peter is y years old.

Save these equations until Exercise 3.13b.

Solving equations

Look at the equation in Example 10: $c + 5 = 59$.

In this case, the problem can easily be solved without the equation.

If the price of a carton of milk went up by 5 pence to 59 pence, it must have been $(59 - 5)$ pence before, i.e. 54 pence.

The equation has two sides, $c + 5$ and 59. The amounts on either side are equal.

If you subtract 5 from $c + 5$, you are left with c. But then the two sides will no longer have the same value.

To keep them the same, subtract 5 from 59.

Write this as:
$$c + 5 = 59$$
$$[c + 5 - 5 = 59 - 5]$$
$$c = 54$$

With practice, you can leave out the line in square brackets.

Example 12
Solve the equation $2d = 12$.

The d has been multiplied by 2.

To find d, you must halve the $2d$, or divide by 2.

To keep the sides the same, you must also divide the 12 by 2.

Write this as:
$$2d = 12$$
$$[2d \div 2 = 12 \div 2]$$
$$d = 6$$

Example 13
Solve the equation $p - 3 = 7$.

On the left-hand side, 3 has been subtracted from p.

To find p you must add 3.

Write this as:
$$p - 3 = 7$$
$$[p - 3 + 3 = 7 + 3]$$
$$p = 10$$

Examiner's tip

Always perform the same operation to the **whole** of each side.

Exercise 3.9a

Solve these equations.

1. $3a = 12$
2. $4b = 24$
3. $2c = 16$
4. $d + 1 = 3$
5. $e + 6 = 17$
6. $f + 4 = 3$
7. $g - 3 = 6$
8. $h - 5 = 1$
9. $5 + j = 1$
10. $2k = 17$

Exercise 3.9b

Solve these equations.

1. $2x = 14$
2. $5y = 35$
3. $4z = 4$
4. $p + 2 = 5$
5. $q + 5 = 16$
6. $r + 3 = 1$
7. $a - 3 = 1$
8. $3b = -18$
9. $3c = 20$
10. $5d = -22$

Two-step equations

Examiner's tip

Remember, all the operations must be done to all of each side.

Example 14

Solve the equation $3x - 1 = 5$.

This needs two steps, using the same rules as before.

$$3x - 1 = 5$$
$$[3x - 1 + 1 = 5 + 1] \quad \text{Add 1 to each side.}$$
$$3x = 6$$
$$[3x \div 3 = 6 \div 3] \quad \text{Divide each side by 3.}$$
$$x = 2$$

It is more difficult to start with '÷ 3'.

$$[3x \div 3 - 1 \div 3 = 5 \div 3] \quad \text{Divide each side by 3.}$$
$$x - \tfrac{1}{3} = 1\tfrac{2}{3}$$
$$x - \tfrac{1}{3} + \tfrac{1}{3} = 1\tfrac{2}{3} + \tfrac{1}{3} \quad \text{Add } \tfrac{1}{3} \text{ to each side.}$$
$$x = 2$$

But it still works, if you don't forget the extra term!

Exercise 3.10a

Solve these equations.

1. $2x + 1 = 3$
2. $2x - 1 = 3$
3. $2x + 3 = 1$
4. $2x - 3 = 1$
5. $3x + 2 = 8$
6. $5x + 2 = 7$
7. $7x - 3 = 18$
8. $4x + 7 = 3$
9. $2x - 3 = 4$
10. $3x + 2 = 9$

Exercise 3.10b

Solve these equations.

1. $2x + 5 = 9$
2. $2x - 5 = 9$
3. $3x + 7 = 4$
4. $3x - 7 = 8$
5. $4x - 11 = 5$
6. $11x - 4 = 7$
7. $10x + 3 = 18$
8. $2x - 7 = 4$
9. $2x + 7 = 4$
10. $5x - 3 = -5$

Equations with brackets

Sometimes the equation formed to solve a problem will involve brackets.

Examiner's tip

You may find it easier to do it the first way. By expanding the brackets, you will have an equation like those in Exercise 3.10.

Example 15

Solve the equation $2(x + 3) = 6$.

Here are two methods.

Either:
$$[2 \times x + 2 \times 3 = 6] \quad \text{Expand the bracket.}$$
$$2x + 6 = 6$$
$$[2x + 6 - 6 = 6 - 6] \quad \text{Subtract 6 from each side.}$$
$$2x = 0$$
$$[2x \div 2 = 0 \div 2] \quad \text{Divide each side by 2.}$$
$$x = 0$$

Example 15
continued

Or:

$[2(x + 3) \div 2 = 6 \div 2]$ Divide each side by 2.

$$x + 3 = 3$$

$[x + 3 - 3 = 3 - 3]$ Subtract 3 from each side.

$$x = 0$$

Examiner's tip

This method is usually shorter.

Exercise 3.11a

Solve these equations.

1. $2(x + 1) = 10$
2. $3(x + 2) = 9$
3. $4(x - 1) = 12$
4. $5(x + 6) = 20$
5. $2(x - 3) = 7$
6. $3(2x - 1) = 15$
7. $2(2x + 3) = 18$
8. $5(x - 1) = 12$
9. $3(4x - 7) = 24$
10. $2(5 + 2x) = 17$

Exercise 3.11b

Solve these equations.

1. $2(x + 4) = 8$
2. $2(x - 4) = 8$
3. $5(x + 1) = 35$
4. $3(x + 7) = 9$
5. $2(x - 7) = 3$
6. $4(3x - 1) = 20$
7. $7(x + 4) = 21$
8. $3(5x - 13) = 21$
9. $2(4x + 7) = 12$
10. $2(2x - 5) = 11$

Equations with the unknown on both sides

Sometimes the equation formed to solve a problem will have the unknown on both sides.

Example 16

Solve $2x + 1 = x + 5$.

The first step is the same as before.

$[2x + 1 - 1 = x + 5 - 1]$ Subtract 1 from each side.

$$2x = x + 4$$

Now use the same idea for the x-term on the right-hand side.

$[2x - x = x - x + 4]$ Subtract x from each side.

$$x = 4$$

Example 17

Solve $2(3x - 1) = 3(x - 2)$.

Expand the brackets.

$[2 \times 3x + 2 \times (-1) = 3 \times x + 3 \times (-2)]$

$$6x - 2 = 3x - 6$$

Now use the previous method.

$[6x - 2 + 2 = 3x - 6 + 2]$ Add 2 to each side.

$$6x = 3x - 4$$

$[6x - 3x = 3x - 3x - 4]$ Subtract $3x$ from each side.

$$3x = -4$$

$[3x \div 3 = -4 \div 3]$ Divide each side by 3.

$$x = -1\tfrac{1}{3}$$

Chapter 3 Forming equations

Exercise 3.12a

Solve these equations.

1. $2x - 1 = x + 3$
2. $3x + 4 = x + 10$
3. $5x - 6 = 3x$
4. $4x + 1 = x - 8$
5. $2(x + 3) = x + 7$
6. $5(2x - 1) = 3x + 9$
7. $2(5x + 3) = 5x - 1$
8. $3(x - 1) = 2(x + 1)$
9. $3(3x + 2) = 2(2x + 3)$
10. $3(4x - 3) = 10x - 1$

Exercise 3.12b

Solve these equations.

1. $2x + 3 = x + 6$
2. $4x - 1 = 3x + 7$
3. $4x - 3 = x$
4. $5x + 7 = 2x + 16$
5. $2(x - 1) = x + 2$
6. $2(2x + 3) = 3x - 7$
7. $5(3x + 2) = 10x$
8. $3(4x - 1) = 5(2x + 3)$
9. $3(3x + 1) = 5(x - 7)$
10. $7(x - 2) = 3(2x - 7)$

Exercise 3.13a

Solve the equations you formed in Exercise 3.8a.

Exercise 3.13b

Solve the equations you formed in Exercise 3.8b.

Exercise 3.14a

Solve these equations.

1. $3x - 1 = 20$
2. $3(x - 1) = 15$
3. $3(x + 1) = 2(x + 1)$
4. $2(2y - 1) = y + 2$
5. $15x = 210$
6. $4x - 3 = 2x + 7$
7. $5x + 2 = 3(x - 6)$
8. $7x + 5 = 2(x + 6)$
9. $3y = 2(y - 2)$
10. $8z - 3 = 3(2z + 5)$

11. $3(4x + 5) = 4(2x - 1)$
12. $5(x - 3) = 2(x + 4)$
13. $27x + 15 = 13x - 13$
14. $15y - 22 = 4(2y + 5)$
15. $2(5z - 1) = 3(3z + 4)$
16. $4(4x + 3) = 5(2x - 6)$
17. $3(4x + 5) = 5(x - 1) - 1$
18. $\frac{1}{2}x = 2$
19. $3(2z - 1) = 2 - 3z$
20. $2x - 1 = 3(x + 2)$

Exercise 3.14b

Solve these equations.

1. $2x - 5 = 17$
2. $10x = 75$
3. $4x = 3(x - 7)$
4. $5(2x + 5) = 3(3x + 7)$
5. $6x - 7 = 2(x + 3)$
6. $3(4x + 2) = 2(5x - 3)$
7. $2(3x - 1) = 8 - 4x$
8. $\frac{1}{4}x = 4$
9. $7x - 2 = 2(4x + 3)$
10. $8(2x + 1) = 7(3x - 1)$

Inequalities

Sometimes being equal does not help! If the height of the lorry is h metres, then h must be less than 4 metres if the lorry is to go under the bridge. This is an inequality, and is written as:

$h < 4$

Similarly, if the height of the lorry is actually 4.5 metres and it has to go under another bridge of height b metres, then b must be greater than 4 m, 5 metres if the lorry is to go through.

$h > 4.5$

These inequalities can be shown on the number lines.

h < 4   b > 4.5

Show each of these inequalities
on a number line.

1. $x < 3$
2. $y > 3$
3. $z > 0$
4. $2x > 4$
5. $2x < 1$

Show each of these inequalities
on a number line.

1. $a < 5$
2. $b > 4$
3. $c > 0$
4. $5d > 30$
5. $2x < 5$

Key points

- Algebra involves working with letters which stand for unknown numbers. You can always put numbers in when you are not sure.
- The two sides of an equation must always be kept equal. Operations carried out to simplify or solve an equation must always be the same for each side.

- To find the nth term of a sequence, find the differences between terms. If the differences are all the same, this difference, multiplied by n, will be part of the nth term. Put in a value of n to find the number.
- Everything outside a bracket must be multiplied by everything inside the bracket.

Revision exercise 3a

1. Solve these equations.
   (a) $3x = 42$
   (b) $2x - 1 = 5$
   (c) $3(x + 1) = 7$
   (d) $4x - 3 = 3x + 2$
   (e) $2(3x - 1) = 2x + 1$

2. Simplify these expressions.
   (a) $4x - 2y + 3y - 2x$
   (b) $a^2b + 2ab + 3a^2b - ab$
   (c) $8y \times 3z$
   (d) $pq \times p^2q$
   (e) $3x^2 \times 2xy$

3. Expand the brackets.
   (a) $2(3a - b)$
   (b) $-7(4 - 2c)$
   (c) $2x(3x - 5)$
   (d) $5x(2y + 3z)$
   (e) $-3x(3x - 8)$

4. Find the nth term for each sequence.
   (a)   3   5   7   9
   (b)   5   10   15   20
   (c)   11   12   13   14
   (d)   −2   −4   −6   −8
   (e)   17   14   11   8

5. $P = a^2 + b^2$
   (a) Find P when $a = 4$, $b = 7$
   (b) Find P when $a = 3$, $b = 4$
   (c) Find P when $a = -3$, $b = -4$

6. (a) Write this problem as an equation.
       I think of a number (x)
          subtract 7
       multiply the result by 3
          add the original number.
       The result is −5.
   (b) Solve the equation to find the number.

7. Solve these equations.
   (a) $2(2x - 1) = 3(x - 2)$
   (b) $3(5x + 2) = 7(2x + 1)$
   (c) $4x + 7 = 3(4 + x)$
   (d) $5(2x - 3) = 15$
   (e) $3(3x - 2) = 2(2x - 3)$

8. Show these inequalities on number lines.
   (a) $4x < 20$ (b) $2y > 13$

9.* Solve these equations.
   (a) $\frac{1}{2}x + 1 = 4$
   (b) $3 - x = 2x - 3$
   (c) $2(3x - 2) = 4(2x + 3) - x$
   (d) $3(7 - x) = 4(1 - 2x)$
   (e) $\frac{2}{3}x = \frac{3}{2}$

4 Calculating and illustrating data

Grouping data

When you have to analyse a lot of data it is sometimes helpful to group or sort the data into bands, or **intervals**, of equal width. This makes it easier to find the frequencies, especially if you draw up a tally chart.

Example

Here are the prices of irons from the catalogue of an electrical superstore.

£9.60	£12.95	£13.90	£13.95	£14.25
£16.75	£16.90	£17.75	£17.90	£19.50
£19.50	£21.75	£22.40	£22.40	£24.50
£24.90	£26.00	£26.75	£26.75	£27.50
£29.50	£29.50	£29.50	£32.25	£34.25
£34.25	£34.50	£35.25	£35.75	£38.75
£39.00	£39.50	£47.00	£49.50	

The prices can be grouped into bands that are £5 wide.

Price (£)	Tally	Frequency
5.00–9.99	/	1
10.00–14.99	////	4
15.00–19.99	ŦŦŦ /	6
20.00–24.99	ŦŦŦ	5
25.00–29.99	ŦŦŦ //	7
30.00–34.99	////	4
35.00–39.99	ŦŦŦ	5
40.00–44.99		0
45.00–49.99	//	2
	Total	34

Example 2

The table shows the marks gained by the 26 children in class 6, in test A of the Key Stage 2 SATs.

28	36	6	17	24	21	24	12	19
27	30	13	19	9	18	35	12	26
9	27	13	35	15	8	25	14	

Their teacher groups the marks into intervals of 10.

Mark	Tally	Frequency
0–9	////	4
10–19	//// ////	10
20–29	//// ///	8
30–39	////	4
	Total	26

Examiner's tip

Unless you are told what size of bands or intervals to use, you should normally try to divide the values up into no more than about eight equally-sized bands.

The frequencies of all the sets of data can be plotted in a diagram like a bar chart, which is called a **frequency chart** or a **frequency diagram**.

Graph A
Frequency diagram showing prices of irons

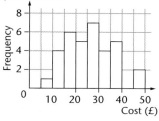

Graph B
Frequency diagram showing class 6's test A scores

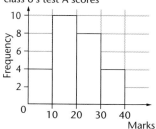

Frequency polygon

A frequency polygon is formed by joining, with straight lines, the midpoints of the tops of the bars in a frequency chart.

Here are the frequency polygons drawn onto the two bar charts shown earlier.

Graph C
Frequency polygon for the prices of irons

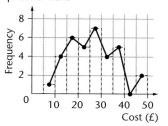

Graph D
Frequency polygon for class 6's test A scores

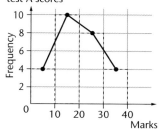

Chapter 4 Grouping data

NB   • The frequency polygon starts at the midpoint of the first bar
         and ends at the midpoint of the last bar and not at the origin
         or zero.
      • As can be seen in graph C a zero value in between other
         values must be shown.
      • The bars are not usually shown.

Exercise 4.1a

For each of the following sets of data, construct frequency tables using
appropriate intervals. Then draw the frequency polygons.

1. The scores for test B for class 6

32	37	5	18	31	29	27	11	21	37	28
22	29	6	33	36	17	33	11	31	22	37
23	12	37	24							

2. The lengths of 36 pea pods

53	81	66	83	75	61	73	85	62	79	83
76	56	77	71	85	54	89	84	58	75	87
71	88	92	75	93	65	94	82	63	79	91
72	69	70								

3. The lifetime, in hours, of 30 batteries

32	47	20	52	48	82	63	40	36	57
40	70	56	54	46	53	47	51	48	69
69	39	55	58	58	63	57	45	61	74

Exercise 4.1b

For each of the following sets of data, construct frequency tables using
appropriate intervals. Then draw the frequency polygons.

1. The number of items in the pencil cases of students in class 8

1	2	2	3	4	4	5	5	5	6	6
8	8	8	9	9	10	10	11	13	14	14
14	14	15	15	18	19	25	26	32		

2. Thirty students in a year 7 class were asked to keep a record of how
    much television they watched in a week. The results are listed
    below as times in hours.

1.5	21	12.5	0	2.5	15	23	19	4	14
8	16	13.5	16.5	6	4.5	9	18	5	10.5
8.5	6	3	9	11.5	3.5	19.5	13	10	9

3. Thirty pupils estimated a length of 30 cm. Here are their estimates,
    in cm.

29.0	30.3	30.9	29.0	26.7	28.3	30.0	28.9	28.6	23.4
30.0	26.2	27.2	27.5	29.4	30.5	25.0	24.6	26.5	27.5
20.0	19.4	23.0	20.0	22.0	20.2	23.8	34.0	22.5	20.5

Pie charts

Pie charts give a useful, clear picture and allow you to make comparisons between different types of data. They are often seen in newspapers.

Pie charts can be drawn quite easily. You just need to remember that there are 360° round a point. The next example explains how to draw a pie chart.

Example 3 Julie does a survey of the 30 students in her maths class to find out which topic of mathematics they prefer.

Number work, percentages etc.	10
Algebra, solving equations, drawing graphs	4
Geometry	7
Handling data and probability	9

The total of 30 students are represented by 360°.

1 student is represented by $\dfrac{360°}{30} = 12°$.

Number	Angle at centre =	$10 \times 12 = 120°$
Algebra	Angle at centre =	$4 \times 12 = 48°$
Geometry	Angle at centre =	$7 \times 12 = 84°$
Handling data	Angle at centre =	$9 \times 12 = 108°$

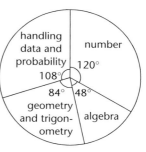

Pie charts and grouped data

Pie charts can be drawn for grouped data. This pie chart shows the data on class 6's scores in test A.

There are 26 students so 1 student is represented by $360 \div 26 = 13.8°$.

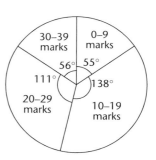

Mark	Frequency	Angle
0–9	4	$13.8 \times 4 = 55.4°$
10–19	10	$13.8 \times 10 = 138.4°$
20–29	8	$13.8 \times 8 = 110.8°$
30–39	4	$13.8 \times 4 = 55.4°$

NB It is impossible to draw angles to the calculated degree of accuracy so the values of the angles are rounded to the nearest degree.

Exercise 4.2a

1. Draw a pie chart showing the lengths of the pea pods given in Exercise 4.1a question 2, using the grouping that you chose.

2. Students in class 10 were asked to estimate a height of 2 metres up a wall. Here are their estimates, in centimetres.

194	130	212	196	142	186	184	150	198	220
156	220	156	158	162	192	218	200	170	210
214	160	198	204	188	184	182	190	206	160

   Group this data into appropriate intervals and show it in a pie chart.

3. Tom earns £15 each Saturday, washing cars at the local garage. He divides his earnings up like this.

| Savings | £2.50 | CDs etc. | £3.00 |
| Clothes | £6.00 | Going out | £3.50 |

   Draw a pie chart to show this.

4. The nutritional information on a packet of a well-known breakfast cereal states:

typical value per 100 g			
protein	6 g	sodium	1.1 g
sugar	10 g	fibre	1.5 g
starch	75 g	other	5.1 g
fat	1.3 g		

   Draw a pie chart to show this.

Exercise 4.2b

1. The table shows the approximate 1998 population figures (in millions), for six African countries. Show this information in a pie chart.

Country	Population (millions)
Rwanda	8.0
Somalia	10.6
Uganda	21.0
Tanzania	30.4
Nigeria	121.6
Ethiopia	58.4
Total	250.0

2. Look in a newspaper or look around the class you are in. Take 15 male names and 15 female names. Draw pie charts to show the frequency of the names of the males and the females. Compare the two pie charts.

3. A survey of the colours of 300 cars in a car park gave the following information. Show this in a pie chart.

Colour	Number
red	102
blue	35
green	63
silver	45
other	55

4. A hospital listed the ages of 200 women who had babies in its maternity ward during the first two months of a year. The details are shown below. Draw a pie chart to show this information.

Age range	Number of women
15 ≤ 20	15
21 ≤ 25	49
26 ≤ 30	71
31 ≤ 35	44
36 ≤ 40	16
41 ≤ 45	3
46 ≤ 50	2

Reading pie charts

You must be able to read the information in a pie chart that someone else has drawn.

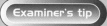

Examiner's tip

Check that the individual amounts add up to the total.

Example 4

Harry earns £18 000 a year. The pie chart shows how he uses this money.

1° represents £18 000 ÷ 360 = £50

Measuring the angles for each sector gives these results.

food	60°	£3000
rent	70°	£3500
savings	50°	£2500
travel	60°	£3000
entertainment	40°	£2000
clothing	40°	£2000
heating and lighting	30°	£1500
other	10°	£500
Total		£18 000

1. 1000 people were interviewed about their holiday destination. The pie chart shows the information they gave. Measure the angles and calculate the number of people going to each destination.

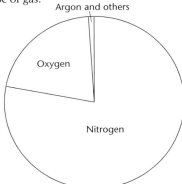

2. The amounts of different types of plants growing on a 10 km² section of moorland are shown in the pie chart. Measure the angles and calculate the area covered by each type of plant.

3. The composition of air is shown in the pie chart. Measure the angles and work out the percentage of each type of gas.

Exercise 4.3b

1. The age-groups of the first 100 people through the doors of a new supermarket are shown in the pie chart. Measure the angles and calculate how many of each age-group there were.

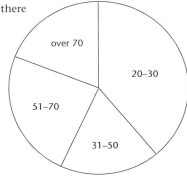

2. A recent analysis, by country of birth, of 60 adults attending an English language class produced this pie chart. Analyse the pie chart and work out the numbers from each country.

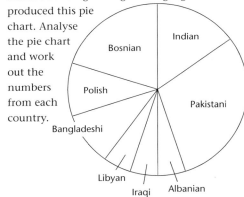

3. An analysis of greenhouse gases in the atmosphere is shown in the pie chart. Measure the angles and calculate the percentage of each gas present in the atmosphere.

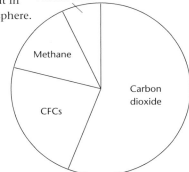

Scatter diagrams

Scatter diagrams (also called scatter graphs) are used to investigate any possible link or relationship between two features or variables. Values of the two features are plotted as points on a graph. If these points tend to lie in a straight line then there is a relationship or **correlation** between the two features.

 Example 5 The table below shows the mean annual temperature for 12 cities which lie north of the equator.

City	Latitude (degrees)	Mean temperature (°C)
Bombay	19	31
Casablanca	34	22
Dublin	53	13
Hong Kong	22	25
Istanbul	41	18
St Petersburg	60	8
Manila	15	32
Oslo	60	10
Paris	49	15
London	51	12
New Orleans	30	22
Calcutta	22	26

When these data are plotted, as shown below, there appears to be a relationship between temperature and latitude: the farther north the city the lower the temperature will tend to be.

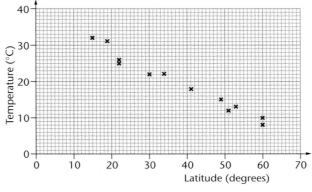

Annual temperature compared with latitude

Example 6 This table shows the percentage marks gained in a maths test and an English test by the same 14 students.

Maths	62	53	23	61	25	46	48	49	60	61	61	69	85	48
English	52	53	45	57	48	49	53	53	56	58	59	54	62	53

The scatter graph showing these data suggests that there is a relationship between the results of the two subjects: the higher the maths mark, the higher the English mark.

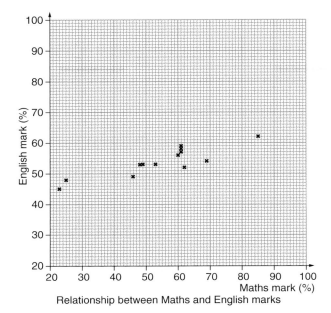

Relationship between Maths and English marks

A 15th student took the maths test but missed the English test. Her maths score was 40%. The scatter graph can be used to estimate her English score as around 50%. Check that you agree.

Exercise 4.4a

1. The marks of ten students in the two papers of a French exam are shown in the table.

Paper 1	20	32	40	45	60	67	71	80	85	91
Paper 2	15	25	40	40	50	60	64	75	76	84

Plot these marks on a scatter diagram.
A student scored 53 marks on paper 1. What would you guess his likely mark to be on paper 2?

2. In the Key Stage 2 maths SATs the pupils take two written tests, test A and test B, and a mental mathematics test. The marks for the 26 pupils in class 6 are shown in the table.

Test A	Test B	Mental mathematics	Test A	Test B	Mental mathematics
28	32	16	9	6	5
36	37	19	18	33	12
6	5	3	35	36	18
17	18	5	12	17	6
24	31	16	26	33	15
21	29	17	9	11	4
24	27	14	27	31	19
12	11	9	13	22	9
19	21	17	35	37	18
27	37	16	15	23	12
30	28	16	8	12	3
13	22	6	25	37	19
19	29	17	14	24	19

Draw scatter diagrams to compare the marks between test A and the mental test score, and between test A and test B.

3. The table below gives the height and the resting pulse rate for 18 people.

   Draw a scatter diagram and investigate if there is a relationship between height and pulse rate.

Height (cm)	Pulse rate (per minute)
160	68
162	64
180	80
173	92
170	80
163	80
148	82
160	84
180	90
165	84
172	116
163	95
168	90
182	76
170	84
155	80
175	104
180	68

Can you estimate from your graph the pulse of someone 185 cm tall? or someone 140 cm tall?

Chapter 4 Scatter diagrams

Exercise 4.4b

1. A biologist studies honey bees. She collects data on the size of the wings for European honey bees and African honey bees.

European bees										
wing length (mm)	9.5	9.28	9.41	9.2	9.53	8.88	9.15	9.53	9.17	8.98
wing width (mm)	3.27	3.27	3.24	3.11	3.23	3.06	3.18	3.27	3.16	3.15
African bees										
wing length (mm)	8.81	8.67	8.67	8.93	8.68	8.60	8.76	8.60	8.68	8.77
wing width (mm)	3.02	3.03	3.05	3.05	3.06	2.99	2.91	3.01	2.97	3.09

(a) Draw a scatter diagram to see if there is any relationship between the length and width of the wings of European bees.

(b) Draw a second scatter diagram to investigate the wings of the African bees.

What do you notice if you compare the two sets of data?

2. This table shows the Olympic gold medal high jump heights for men from 1900 until 1988.

Year	Height	Year	Height	Year	Height
1900	1.9	1932	1.97	1968	2.24
1904	1.8	1936	2.03	1972	2.23
1908	1.91	1948	1.98	1976	2.25
1912	1.93	1952	2.04	1980	2.36
1920	1.94	1956	2.12	1984	2.35
1924	1.98	1960	2.16	1988	2.38
1928	1.94	1964	2.18		

Plot a scatter diagram, with the year plotted along the horizontal axis, to show this data. From your diagram try to estimate the heights for the more recent Olympic games in 1992 and 1996. Try to find the actual heights to see how close your estimates were.

Exercise 4.5a

Look back at your answers to Exercise 4.1a questions 1, 2 and 3, and write down the modal class for each set of data.

Exercise 4.5b

Look back at your answers to Exercise 4.1b questions 1, 2 and 3, and write down the modal class for each set of data.

Mean, median, mode and range of data

You should already know how to find the **mode**, which is the most common value in a set of data or numbers in a list of numbers.

When dealing with grouped data you will be expected to find the **modal class** rather than a single value. This will be the group with the greatest frequency. On a frequency chart it is the group with the highest bar.

If you look back to page 45, in the graph for the prices of irons, you will see that the modal class is £25–29.99 because there are more irons for sale within that price range. The graph for the pupils' marks shows a modal class of 10–19, with 10 pupils.

Calculating mean, median, mode and range

You should be able to calculate the mean, median, mode and range for a set of data.

Example 7

The label of a matchbox is marked 'Average contents 50 matches'. A survey of ten matchboxes produced the following data.

Box	A	B	C	D	E	F	G	H	I	J
Number of matches	45	51	48	47	46	47	49	50	52	47

Find the range, mean, median and mode of the contents.

Writing the numbers in size order gives:

45   46   47   47   47   48   49   50   51   52

Mode (the number that occurs most often) = 47

$$\text{Mean} = \frac{45 + 46 + 47 + 47 + 47 + 48 + 49 + 50 + 51 + 52}{10} = \frac{482}{10} = 48.2$$

Median (the middle number, because there are ten numbers in the data the median lies halfway between the middle two numbers) $= \dfrac{47 + 48}{2} = 47.5$

Range (the difference between the smallest and the largest values) = 52 − 45 = 7

If data are presented in a frequency table, you can work out the range, mean, median and mode without needing to work on all the data.

 Example 8

One day in September, a scientist recorded the highest temperature at 48 coastal towns.

Temperature (°C)	14	15	16	18	19	20	21	22	23	24	25	27
Number of towns (frequency)	1	1	2	2	8	12	10	4	3	2	2	1

- The mode is 20°C because this is the temperature with the greatest frequency.
- The median is midway between the 24th and 25th numbers. You could write out the temperatures: 14, 15, 15, 16, 18, 18, 19, 19, 19
  but it is quicker to add up the frequencies until you reach the place where the next frequency added would be greater than the middle value, here 25.

  So adding up the first five frequencies gives: 1 + 1 + 2 + 2 + 8 = 16.

  Adding on the next frequency, 12, would give a total of 28 so the 24th and 25th values must occur in the band where the frequency is 12 and thus the median temperature is 20°C.

- The mean is worked out by multiplying each temperature by its frequency:
$$\frac{\begin{array}{l}14 \times 1 + 15 \times 1 + 16 \times 2 + 18 \times 2 + 19 \times 8 + 20 \times 12 + \\ 21 \times 10 + 22 \times 4 + 23 \times 3 + 24 \times 2 + 25 \times 2 + 27 \times 1\end{array}}{48} = 20.4°C$$

- The range = 27°C − 14°C = 13°C

Exercise 4.6a

1. A football team played 26 matches during a season. The number of goals they scored were as follows.

Number of goals	0	1	2	3	4	5	6
Number of matches (frequency)	3	5	6	7	2	2	1

Find the mean, median, mode and range of the number of goals per match.

Examiner's tip

Remember to put the data in size order when you are finding the median.

2. The team decides to measure the fitness of its players. One test is to run 800 metres. These are the times, in seconds, that the players recorded.

136.4   186.0   152.6   178.0   157.8   150.8   151.6   162.4
166.0   152.4   167.2   138.2   149.8   176.0   146.4

Calculate the mean, median, mode and range of the times.

Exercise 4.6b

1. The table shows the scores of 40 players after the first round of a golf tournament.

70   68   71   67   74   69   69   71   68   70   71   70   72   69   69   68   71   70   70   72   72   69
68   70   68   69   67   71   69   70   68   67   70   70   73   69   71   67   69   68

Make a frequency table and calculate the mean, median, mode and range of the scores.

2. Thirty pupils estimated a 30 cm length. Here are their estimates:

29.0   30.3   30.9   29.0   26.7   28.3   30.0   28.9   28.6   23.4   30.0   26.2   27.2   27.5   29.4   30.5
25.0   24.6   26.5   27.5   20.0   19.4   23.0   20.0   22.0   20.2   23.8   34.0   22.5   20.5

Calculate the mean, median, mode and range of the estimates.

Comparing data

Example 9

The children in a swimming club record the time it takes them to swim one length of the pool, backstroke.

The results are:

Time (seconds)

Boys	45	46	48	60	42	53	47	51	54	54	49	48	47	53	48	45
Girls	45	47	47	55	46	53	54	63	48	50	46	51	48	48		

Compare the distributions of the times for boys and girls.

Boys: mean = 53   Girls: mean = 54
       median = 48   median = 48
       mode = 48   mode = 48
       range = 60 – 42 = 18   range = 63 – 45 = 18

For both boys and girls the median, mode and range are the same but the mean for the girls is slightly higher so you could deduce that the girls are slightly slower than the boys.

1. Here are the English and maths test results for two groups of students. (You have already done some work with the results for group A in the section on scatter diagrams.)

Group A

Maths	62	53	23	61	25	46	48	49	60	61	61	69	85	48
English	52	53	45	57	48	49	53	53	56	58	59	54	62	53

Group B

Maths	63	36	69	38	43	47	60	43	86	95	45
English	52	53	45	57	48	49	53	53	56	58	59

Calculate the mean, median, mode and range for each set of data and compare the two groups' performances in maths and English.

2. The temperatures of two towns were recorded for 12 days and the results were recorded.

Day	1	2	3	4	5	6	7	8	9	10	11	12
Town A	11	13	12	11	14	15	17	15	16	15	20	18
Town B	10	12	15	13	16	12	15	16	14	16	17	21

Compare the temperatures for the two towns, using mean, median, mode and range.

3. A shoe manufacturer surveys the shoe sizes of 100 people in two towns. The tables show the findings.

Town A

Shoe size (x)	Frequency (f)	x times f (xf)
4	3	
5	6	
6	15	
7	20	
8	28	
9	22	
10	5	
11	1	
Totals	100	

Town B

Shoe size (x)	Frequency (f)	x times f (xf)
4	8	
5	9	
6	16	
7	18	
8	24	
9	19	
10	3	
11	3	
Totals	100	

Calculate the mean, median and mode for both sets of data and compare the shoe sizes in both towns.

1. A market gardener grew some tomatoes using two different fertilisers to see which gave the better results. The table gives the weights, in grams, of tomatoes from 25 plants using each fertiliser.

Fertiliser A

79 91 48 86 44 67 54 36 83 55 79 26 82 68 77 80 18 42 61 76 24 20
56 73 69

Fertiliser B

92 34 54 79 71 89 40 80 54 88 93 61 25 48 90 99 56 28 78 91 41 53
51 73 78

Compare the two sets of data using the mean, median, mode and range. Which fertiliser do you think is better?

Exercise 4.7b continued

2. Forms 7B and 7M collected coupons from crisp packets. The table shows numbers of coupons they collected.

Form 7B

48   21   36   37   82   29   42   67   56   87   86   55   83   70   68   50   61   54   73   52   44   74
55   80   77   74   76   71   69   76

Form 7M

34   36   40   23   61   18   56   47   53   56   54   50   80   26   25   24   28   36   57   57   79   50
54   60   48   37   78   81   73   72

Calculate the mean, median, mode and range for each set. Which form do you think did better?

Stem and leaf tables

Here are the marks gained by 30 students in an examination.

63	58	61	52	59	65	69	75	70	54
57	63	76	81	64	68	59	40	65	74
80	44	47	53	70	81	68	49	57	61

A different way of showing these data is to make a **stem and leaf** table like this.

First: Write the tens figures in the left-hand column of a table. These are the stems.

4	
5	
6	
7	
8	

Next: Go through the marks in turn and put in the units figures of each mark in the proper row. These are the leaves.

first 63

4	
5	
6	3
7	
8	

then 58

4	
5	8
6	3
7	
8	

then 61

4	
5	8
6	3 1
7	
8	

When all the marks are entered, the diagram will look like this.

4	0 4 7 9
5	8 2 9 4 7 9 3 7
6	3 1 5 9 3 4 8 5 8 1
7	5 0 6 4 0
8	1 0 1

Finally, rewrite the diagram so the units figures in each row are in size order, with the smallest first.

4	0 4 7 9
5	2 3 4 7 7 8 9 9
6	1 1 3 3 4 5 5 8 8 9
7	0 0 4 5 6
8	0 1 1

The finished stem and leaf diagram is like a frequency chart. You can read off information:

- the modal group (the one with the highest frequency) is the 60–69 group

- there are 30 results so the median is midway between the 15th and 16th results. (Starting at the first result, 40, and counting on 15 results gives 63, the 16th result is also 63 so the median 63.)

Exercise 4.8a

1. These are the weights, in kilograms, of 25 newborn babies.

2.6	2.9	3.2	2.5	3.1	1.9
3.5	3.9	4.0	2.8	4.1	1.7
3.8	2.6	3.1	2.4	4.1	2.6
4.2	3.6	2.9	2.8	2.7	3.3
3.8					

   (a) Copy this stem and leaf diagram and complete it.

   ```
   1.
   2.
   3.
   4.
   ```

   (b) Use your table to find the median weight.

2. A group of pupils took two mathematics tests. Here are the stem and leaf diagrams of the marks for the tests.

   Test 1
   ```
   2 | 3 4 5
   3 | 1 3 6 8
   4 | 0 2 5 6 9
   5 | 2 3 5 7
   6 | 1 3 5 5 6 9
   7 | 0 1 3 4 4 6 8
   8 | 0 2 3 3 6 6 7
   9 | 1 2 2 4 5
   ```

   Test 2
   ```
   2 | 0 1 2 6
   3 | 1 2 2 4 5 7 8
   4 | 2 2 5 5 6 8 8 9
   5 | 0 2 2 6 7 8
   6 | 1 3 4 4 9 9
   7 | 0 2 3 5
   8 | 3 7 9
   9 | 1 5
   ```

   (a) – Which test appears to have been harder?

   (b) – Find the median mark for each test.

Exercise 4.8b

1. A scientist measured the lengths of a type of fish caught in two different rivers. Here are their lengths, in cm.

   River A

   38 47 43 51 45 33 62 57
   36 40 49 66 55 49 45 31
   40 44 57 58 35 52 73 39
   38 69 46 55

   River B

   48 52 54 37 42 65 70 49
   61 50 54 45 61 72 74 64
   56 38 65 69 71 67 71 70
   68

   (a) Make a stem and leaf diagram to show the lengths of fish in each river.

   (b) Find the median length for each group of fish.

2. The following table gives the weights, in grams, of 30 tomatoes.

   68 34 59 34 30 33 37 49 53 49
   65 47 36 44 32 31 58 41 41 28
   39 45 27 55 37 34 30 40 30 41

   (a) Make a stem and leaf diagram to show the weights.

   (b) Use your table to find the median weight.

Key points

- A frequency polygon is constructed by joining the midpoints of the tops of the bars, in order, in a frequency diagram.
- A pie chart is constructed by dividing up the 360° at the centre of a circle in proportion to the sizes of the data being considered.
- A scatter diagram shows if there is a relationship between two sets of data.
- A stem and leaf diagram is a type of frequency table.

In a set of data:

- The mode is the number or value that occurs most often.
- The mean is the usual average and is found by adding up all the values and dividing by the number of values.
- The median is the middle value or, if there is an even number of values, it is halfway between the middle two values.
- The range is the difference between the smallest and the largest values.

1. The table below shows the heights and weights of a group of eight people.

Height (cm)	140	132	164	169	157	185	176	143
Weight (kg)	38	29	60	57	55	73	68	41

   Draw a scatter graph and see if there is any relationship between height and weight.
   What would be the likely weight of someone 160 cm tall?
   What would be the likely height of someone weighing 50 kg?

2. During a year a grain merchant sells the following numbers of sacks of grain.
   Barley   12 000   Maize   9000
   Corn   15 000   Wheat   24 000
   Show this information on a pie chart.

3. Tariq did a survey about the lunch-time eating arrangements of his year group. The pie chart
   shows his findings.
   If 97 children have the hot school meal, calculate the numbers for the other choices.

4. Two machines pack paper clips into boxes. Each box should hold 200 paper clips. A sample of 100 boxes

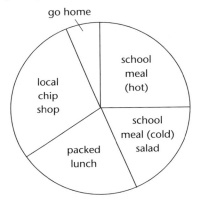

   was checked and the results are given in the table.

Machine A

Number of paper clips	197	198	199	200	201	202	203	
Frequency	1	8	28	27	22	10	4	

Machine B

Number of paper clips	197	198	199	200	201	202	203	204
Frequency	2	4	8	23	29	25	6	3

   For each machine calculate the mean, median, mode and range.
   Comment on your findings.

5. The following table gives the lengths, in cm, of 16 leaves from a tree.
   4.4 5.3 4.9 5.8 5.2 5.9 6.2 6.4
   6.3 6.4 7.6 7.2 7.6 8.1 9.3 9.2
   (a) Make a stem and leaf diagram to show these lengths.
   (b) Use your table to find the median length.

Solving problems and checking results I 5

In this chapter you will look at ways of applying the techniques you have learned in number work to solving problems.

Getting the right answer is especially important in practical problems, so you will also look at ways of checking whether an answer is correct.

Solving ratio problems

Most people think 'value for money' is very important. Although some supermarkets have realised this too, and give comparative prices, you often need to work out for yourself which item gives you better value for your money.

As an example, look at this special offer.

It is easy to spot that the large pack contains three times as much as the small one.

Cost of 1.5 kg = £4.99

Cost of three small packs (3 × 500 g) = 3 × £1.70 = £5.10

The large pack is better value.

If you can't see a quick method, here are two other methods you can use to make the comparison.

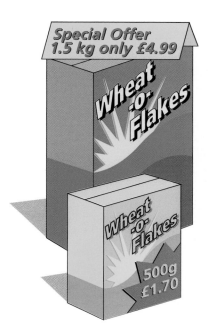

(a) Price per unit

In this example, this is the price per kg.

	Large	Small
Price per kg	$\dfrac{£4.99}{1.5}$ = £3.33	£1.70 × 2 = £3.40

The unit price is lower for the large pack so this is better value.

Examiner's tip

There are different methods you can use for comparing prices. Take care with the units of what you have found and write them down clearly. This will help you to see whether a lower number is better or worse value!
A question may also ask you to consider other factors that may influence a decision. For example a large pack may not be better value if the contents are not used by the 'best before' date.

(b) Number of units per pence or pound (£)

	Large	Small
Number of grams for 1p	$\frac{1500}{499}$ = 3.01 to 2 d.p.	$\frac{500}{170}$ = 2.94 to 2 d.p.

The number of grams you buy for 1p is greater for the large pack, so this is better value.

Other types of ratio problem that you may need to solve include questions about sharing. These worked examples cover two of them.

Example 1

Two families share a meal and the total cost is £77.91. They agree to pay in the ratio 3 to 4. How much should each family pay?

Number of shares = 3 + 4 = 7

Cost per share = £77.91 ÷ 7 = £11.13

One family pays £11.13 × 3 = £33.39

The other family pays £11.13 × 4 = £44.52

Check: £33.39 + 44.52 = £77.91 = total cost

Example 2

Jenny and Mike share some prize money in the ratio 3 to 2. Jenny gets £12.60. How much does Mike get?

3 shares = £12.60

1 share = £12.60 ÷ 3 = £4.20

2 shares = £4.20 × 2 = £8.40

Mike gets £8.40.

1. A 400 ml can of drink costs 38p. How many pence is this for 100 ml?
2. A 420 g bag of Mars bars costs £1.59. How many pence per gram is this?
3. A 325 g bag of Mars bars costs £1.09. How many grams for 1p is this?
4. A 400 g tin of tomatoes costs 28p. How many pence per gram is this?
5. Three litres of sunflower oil cost £2.19. How much is this per millilitre?
6. Show which is better value, 5 litres of spa water for £1.29 or 2 litres for 87p.
7. Here are the prices of some bottles of cola. Which size gives the best value?
   3 litre £1.99   2 litre £1.35   1 litre 57p   500 ml 65p
   Give a reason why you might want to buy another size rather than the best-value one.
8. Here are the prices of some packs of cans of cola. Which pack gives the most cola for your money?
   12 × 150 ml cans for £2.95
     6 × 330 ml cans for £1.59
   12 × 330 ml cans for £2.99
9. Three families share the cost of a holiday in the ratio 4 to 3 to 2. The total cost of the holiday is £2790. How much does each family pay?
10. Petra and Sam took part in a sponsored swim. The amounts they raised were in the ratio 5 to 3. Petra raised £75.50. How much did Sam raise?

1. A 100 ml tube of toothpaste costs £1.79. How much does 1 ml cost?
2. A tin of baked beans and sausages costs 75p for the 420 g size. How many grams is this for 1p?
3. A 680 g pack of mild cheese costs £2.59. Find the cost per gram.
4. A 500 g pack of mild cheese costs £1.99. Find the cost per gram.
5. A brand of shaving gel costs £1.19 for the 75 ml bottle and £2.89 for the 200 ml bottle. Show which is the better value.
6. Here are some comparisons for a brand of toothpaste.

Cost of tube	£1.19	£2.05
Amount	50 ml	100 ml
Cost of 1 ml	2.38p	2.05p

   (a) Which size is the better buy?
   (b) For each size, find how many millilitres you get for 1p.
7. Here are some special offers at two record stores.

Explain which is the better offer if you want to buy:
(a) three CDs   (b) four CDs.

8. Jo and Karen want to go out for a meal on a Wednesday. Their two local restaurants have special offers.

Which offer would be cheaper for them?

9. Jeff and his mother find that their ages are in the ratio 3 to 7. Jeff is 15 years old. How old is his mother?

10. The prices of adults' and children's tickets for a theatre are in the ratio 5 to 2. The total price of one adult's and one child's ticket is £17.50. How much does each of these tickets cost?

Checking your work

When you have solved a problem, how do you know if your answer is right? Sometimes, accuracy is vital. In the news recently was the story of a doctor treating a baby. She put the decimal point in the wrong place in a calculation and prescribed 100 times as much drug as she intended.

Using common sense

When checking anything, the first technique to use is common sense! Does the answer sound sensible in the context of the question? When solving practical problems, your own experience often gives you an idea of the size of the answer you expect. For example, in the case of a shopping bill, you would probably react and check if it came to more than you expected!

Look at the result to this calculation.

$752 \div 24 = 18\,048$

When 752 is divided by a number that is greater than 1, the result should be less than 752.

Instead, it is more. It looks as if the × button was pressed by mistake, instead of the ÷ button. Checking whether the answer is sensible can help to spot errors like this.

Using estimates

Look at this problem.

Kate has £25 birthday money to spend. She sees CDs at £7.99.

How many of them can she buy?

Unless you take a calculator shopping, in this situation the easy way is to do a quick estimate.

Use £8 instead of £7.99.

$25 \div 8 = 3$ 'and a bit' ($3 \times 8 = 24$)

So Kate can buy three CDs.

Most people do this sort of mental check when they are shopping. You can extend it to your maths lessons, and any other subjects where you use calculations.

Using inverse operations

If neither of the above checks is easy to do, try doing the calculation another way. For instance, for a complex calculation you might use brackets instead of the memory. For a simpler one, using inverse operations is often useful.

For example: $6.9 \div 750 = 0.0092$

Check: $0.0092 \times 750 = 6.9$

You will need to use these methods in the rest of the chapter, but don't let it stop there.

Examiner's tip

Develop the habit of always checking whether your answer is about the right size.

Exercise 5.2a

Look at the calculations in questions 1–5. The answers are all wrong. For each one, show how you can quickly tell this without using a calculator to work it out. For some, your method may be different from the one shown in the answers!

1. $-6.2 \div -2 = -3.1$
2. $12.4 \times 0.7 = 86.8$
3. $31.2 \times 40 = 124.8$
4. $\sqrt{72} = 9.49$ to 2 d.p.
5. $0.3^2 = 0.9$

In questions 6–10, use estimates to calculate rough values.

6. The cost of seven packs of crisps at 22p each.
7. The cost of nine CDs at £13.25 each.
8. $65.4 \div 3.9$
9. $\dfrac{194.4 + 16.7}{27.3}$
10. $\dfrac{49.7}{4.1 \times 7.9}$

Exercise 5.2b

Look at the calculations in questions 1–5. The answers are all wrong. For each one, show how you can quickly tell this without using a calculator to work it out. For some, your method may be different from the one shown in the answers!

1. $6.4 \times -4 = 25.6$
2. $24.7 + 6.2 = 30.8$
3. $76 \div 0.5 = 38$
4. $(-0.9)^2 = -0.81$
5. $\sqrt{1000} = 10$

In questions 6–10, use estimates to calculate rough values.

6. The cost of 39 theatre tickets at £7.20 each.
7. The cost of five CDs at £5.99 and two tapes at £1.99.
8. The cost of three meals at £5.70 and two drinks at 99p.
9. 3.1×14.9
10. $47 \times (21.7 + 39.2)$

Solving money problems

This section includes more problems on everyday finance. Income tax may not affect you yet, unless you have lots of money invested in a bank, but sooner or later you will probably have to pay it. It is sensible to learn about it now, so you know what to expect. If you take holidays abroad, you may already be familiar with problems about foreign currency and exchange rates.

Taxation

Unless you are on very low wages, the government does not let you keep everything you earn! Your total pay is called your **gross pay**, and pension contributions are deducted from this. What is left is called your **gross taxable pay**, and this figure is used to calculate how much tax you must pay. Tax and National Insurance contributions are deducted and what is left is your **net pay**, sometimes called your 'take home' pay – the amount you actually keep. £12 000 p.a. means £12 000 per year, as p.a. means *per annum*.

Everyone gets an income tax-free allowance, called their **personal allowance**. Each year in the Budget, the Chancellor of the Exchequer announces how much this will be for the following tax year, and what the rates of income tax will be. There are additional tax-free allowances for some groups of people, such as the elderly whose income is restricted. The income tax year starts on 6 April one year and ends on 5 April the next year.

Here are some tax rates and allowances for the tax year 1999–2000.

Personal allowance		£4335
Bands of taxable income:		
		(£ per year)
Starting rate	10%	0–1500
Basic rate	23%	1501–28 000
Higher rate	40%	over 28 000

Examiner's tip

Spreadsheets can be a useful tool when it comes to calculations like this. See Example 4.

Example 3

Jim's gross pay, after pension contributions, for 1999–2000 is £16 800. What tax does he pay for this tax year?

Personal allowance = £4335

Income tax is due on his taxable income of £16 800 – £4335 = £12 465.

Starting rate tax = 10% of £1500 = £150

Basic rate tax = 23% of (£12 465 – 1500) = 23% of £10 965 = £2521.95.

So total tax = £2671.95

Example 4

Juanita's gross annual pay for 1999–2000 was £38 000. She paid pension contributions of 9% of her pay. How much tax should have been deducted from her pay each month?

	£	£	£
Gross pay		38 000	
Pension contributions @ 9%		3420	
Taxable pay	34 580		
Personal allowance	4335		
Tax due on	30 245		
Tax @ 10% on £1500	1500	150	
Tax @ 23% on £26 500	26 500	6095	
Tax @ 40% on £2245	2245	898	
Total tax due for the year			7143
Tax per month			595.25

Foreign currency exchange rates

In a hypermarket in Calais in April 99, Joy saw a T-shirt for FF72.40 (French francs). Was it good value? She used a rough rule of FF10 ≈ £1 to estimate that it was, and bought it. When her Visa bill arrived it showed:

72.40 FRENCH FRANC RATE 9.55 £7.58

Decisions like this about cost and value are vital, both when you plan the trip and when you are travelling abroad. You will need to use your estimating and checking skills from earlier in this chapter as well as the ability to do accurate calculations. Knowing when to divide and when to multiply is essential when you are converting from one currency to another.

Exchange rates also vary, depending on whether you are buying or selling. You will get a different rate when you buy francs, for example, to go on holiday than when you come back and change your remaining francs back to pounds. The Euro should make travelling within Europe much easier when it is fully implemented.

As well as calculations, you could use **conversion graphs**.

Example 5

A cup of coffee in Germany cost DM4.80. The exchange rate was DM3.14 to £1. How much did the coffee cost in £?

$$4.80 \div 3.14 = £1.53$$

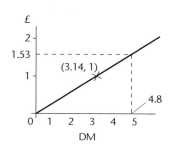

Example **6**

(a) Margaret changed £200 into pesetas at a rate of Pta245.01 to £1. How many pesetas did she get?

$200 \times 245.01 = 49\,002$

(b) Her holiday had to be cancelled and she received a rate of Pta267.51 to £1 when she changed the money back. How much did she receive?

$49\,002 \div 267.51 = £183.18$

Exercise 5.3a

Use the 1999–2000 tax rates on page 66 to answer questions 1–5.

1. Pia's gross pay is £12 400 per annum. She just gets the personal allowance. On what income is tax due?
2. Peter's gross salary is £23 500 p.a. He pays 9% in pension contributions. In addition to his personal allowance, he gets another £242 allowance for professional expenses. On what income is tax due?
3. Tax is due on £14 216 of Pali's income. How much tax does he have to pay?
4. Jim has to pay tax on £29 834 of his income. What is his tax for the year?
5. Sean's annual salary is £15 800. His pension contributions are 6% and he has other allowances totalling £325 per year in addition to his personal allowance. How much tax does he pay each month?

For questions 6–9 use the following exchange rates, which are all equivalent to £1.

Israeli shekel	6.83	Japanese yen	197.15
French franc	9.66	South African rand	9.49
Greek drachma	4.78	US dollar	1.572

6. In Jerusalem, Israel, a cup of tea cost 9 shekels. What was this in £?
7. Jenny changed £100 to go to Japan. How many yen did she receive?
8. Car hire in Greece cost 200 drachma per day. What was this in £?
9. Which is the cheaper drink of cola: one at FF5.50 or one at $0.78?
10. Tim went to the USA for six months.
    (a) He took £1500 in traveller's cheques at an exchange rate of $2.15 to £1. How many dollars' worth of cheques did he receive?
    (b) When he came back, he had $450 left. The exchange rate was then $1.69 to £1. How much did he get back when he changed this money?

Exercise 5.3b

Use the 1999–2000 tax rates on page 66 to answer questions 1–5.

1. Sarah's gross pay is £16 400 per year. She pays 6% of her salary in pension contributions. On what income is tax due?
2. Paul has a part-time job and earns £5200 in a year. What tax does he have to pay?
3. Mary earns £7925 p.a. How much tax does she pay in a year?

4. Deena earns £19 800 and pays 10% pension contributions. She only gets the personal allowance. How much tax does she pay in the year?
5. Mr Archibald earns £45 000. He pays 12% to his pension fund and has allowances of £1283 in addition to his personal allowance. He also pays no tax on his £1800 gift aid to charities.
    (a) On what income is tax due?
    (b) Calculate his annual income tax.

For questions 6–9 use the following exchange rates, which are all equivalent to £1.

Israeli shekel   6.83   Japanese yen   197.15
Franch franc   9.66   South African rand   9.49
Greek drachma   4.78   US dollar   1.572

6. The entry fee to a museum in Paris was FF45. How much was this in £?

7. Pat changed £350 into rands to go to South Africa. How many rands did he get?

8. In New York, Dilip bought a pair of jeans costing $33. A similar pair in London would cost £27.99. How much cheaper, in pounds, were they in New York?

9. Yoko saw a dress priced at £43.75. How many yen was this?

10. Mary wanted 4000 Canadian dollars for her holiday. The bank charged her 2% commission for changing her money, and gave her an exchange rate of 2.303 Canadian dollars to £1. How much did she have to pay?

Selecting suitable methods

When you are faced with a numerical problem to solve, where do you start? Earlier in this chapter you saw several different common types of problem. In this section you will consider problems that need more than one step to solve them.

Example 7

Chris bought 33.7 litres of petrol at 78.9p per litre. The price at another garage was 76.4p. How much more petrol could Chris buy for the same money if she uses this garage?

First step: Chris spent 33.7 × 78.9p = 2659p or £26.59

Second step: Number of litres of petrol she could have bought at the second garage for this money = 2659p ÷ 76.4p = 34.8

Third step: The extra amount of petrol Chris could buy = 34.8 – 33.7 = 1.1 litre

Examiner's tip

Give yourself plenty of practice with the sorts of problem covered in this chapter. Then you are much more likely to recognise the type of method required to solve a given problem.

To solve problems like this, you need to find a way of breaking the problem down into steps you can work out. Think what information you have and what you can find out from it straight away. Then work towards what you are asked to find.

Trial and improvement methods

If you cannot see a way of calculating the answer to a problem straight away, you may need a trial and improvement method.

Example 8

A cuboid has a square base and its height is 2 cm more than a side of the square base.

Its volume is 500 cm³. Find the height, to two decimal places.

Length (cm)	Width (cm)	Height (cm)	Volume (cm³)	Comment
5	5	7	175	First try too small
10	10	12	1200	Too large, try in between
7	7	9	441	Getting nearer
8	8	10	640	The height must be between 9 and 10 cm
7.5	7.5	9.5	534.375	Too large, try something smaller
7.4	7.4	9.4	514.744	Still too large
7.3	7.3	9.3	495.597	The height is between 9.3 and 9.4 cm
7.31	7.31	9.31	497.490 091	
7.32	7.32	9.32	499.387 968	
7.33	7.33	9.33	501.290 637	The height is between 9.32 and 9.325 cm
7.325	7.325	9.325	500.338 703	The height is 9.32 to 2 d.p.

 **Examiner's tip**

Don't use trial and improvement and multiplying when a straightforward division would do the job more efficiently!

A spreadsheet has been used here for the calculations, but you could use a calculator.

The comments indicate what each trial has shown, and how to decide what to try next.

Trial and improvement was a sensible method to use for this problem, but it isn't for a problem like 'Find the number that, when multiplied by 5, gives the answer 73.' The quick method here is to use inverse operations. The answer is 73 ÷ 5 = 14.6.

Giving answers to a specified number of decimal places

When solving practical problems you may have to round the answer on your calculator display.

Example 9

Round 10.45723... to two decimal places (2 d.p.).
This number is between 10.45 and 10.46.

Look at the figure in the third decimal place (the one after the place to which you are rounding). If it is 5 or more, round the answer up. If it is 4 or less, round down. Here it is 7, so the answer is 10.46 to two decimal places.

Exercise 5.4a

1. Round these numbers to two decimal places.
   (a) 207.424 (b) 485.346 (c) 27.899

2. Round these numbers to one decimal place.
   (a) 47.938 (b) 6.78 (c) 90.55

3. Use trial and improvement to find, to one decimal place, a number for which:
   number × (number + 1) × (number + 2) = 900

4. Two numbers add up to 50. Use trial and improvement to find the greatest product they can have.

5. A 250 g packet of chocolate biscuits costs 69p. One month the packets were on offer: '20% more for the same price'. How many grams for 1p were there in the special offer packets? Give your answer to one decimal place.

6. Kate travelled for $1\frac{1}{2}$ hours at 70 miles per hour and for 20 minutes at 30 miles per hour. How far did she go altogether?

7. Jean bought 2 kg of cheese at £4.38 per kg. Her friend saw some cheaper cheese at £3.92 per kg. How much more of this cheese could Jean have bought for the same price? Give your answer to the nearest gram.

8. A football stadium has 18 500 seats. At one game there were 11 345 spectators. What percentage of the seats were empty?

9. Four friends went out to a restaurant and agreed to share the cost of their meal equally.

Chicken Korma	£4.50	Chicken Tikka	£5.30
Lamb Madras	£4.10	Vegetable Biriani	£5.50
Prawn Biriani	£6.70	Pilau rice	£1.50
Onion Bhaji	£1.80	Naan bread	£1.40

   From this menu they had three Lamb Madras, one Chicken Tikka, two Vegetable Biriani, one Onion Bhaji, three Pilau rice and two Naan bread. Their drinks cost £6.00. How much did they each pay?

10. A Youth Club outing cost £266. There were 56 members of the Youth Club, but only $\frac{3}{4}$ of them went on the outing. How much did they each pay?

Exercise 5.4b

1. Round these numbers to two decimal places.
   (a) 4.851 (b) 72.3047 (c) 840.916 83

2. Round these numbers to three decimal places.
   (a) 49.8346 (b) 572.6901 (c) 3.141 59

3. Use trial and improvement to find, to two decimal places, a number for which:
   number × (number + 3) × (number + 5) = 900

4. When two numbers are multiplied together the answer is 50. Use trial and improvement to find, to two decimal places, the least sum they can have.

5. On average, the Jones family eats four bananas a day. One week they ate 37 bananas. How many extra bananas did they eat that week?

6. In the local elections, 42% of those eligible to vote did so. 3570 people voted. How many people were eligible to vote?

7. In a sale there was a reduction of 15% off the normal prices. On the last day there was a further reduction of $\frac{1}{3}$ off the sale price. How much did Mary pay on the last day for a jacket originally priced at £45?

8. Ali travelled 100 miles at 70 miles per hour. How long did he take? Give your answer to the nearest minute.

9. Students from year 10 are going to an art gallery. Seven staff and 105 students are going.
   (a) The coaches seat 45 people. How many coaches are needed?
   (b) Each coach costs £140 to hire, and the cost is shared among the students. The entrance fee to the gallery is £1.50 each for the students. How much each will the trip cost for the students?

10. A mobile phone company had two tariffs. Tariff A had a monthly charge of £17.50 and a call charge of 18 pence a minute. Tariff B had a monthly charge of £25 and a call charge of 12 pence a minute.
    (a) Which tariff is cheaper for making 100 minutes of calls a month?
    (b) For how many minutes a month would the charge for the two tariffs be the same?

Key points

Solving ratio problems
- Best buy – compare the price per unit or the number of units you get for £1 or 1p. Sometimes, you may be able to see a quicker method.
- Ratios – think how many shares there are. Often, you need to find the value of one share.

Checking your work
- Does the answer sound sensible in the context of the question?
- Do a quick estimate.
- Do the calculation another way to check.
- Work backwards from your answer.

Solving money problems
- Taxation – find the taxable pay, deduct the allowances, then work out the tax due on the remaining income.

- Foreign currency – knowing when to divide and when to multiply is vital when converting currencies. Think whether your answer is sensible and do estimates to check.

Selecting suitable methods
- In multi-step problems, find a way of breaking the problem down into steps you can work out. Think what information you have and what you can find out straight away from it. Then work towards what you are asked to find. Write down enough words to convince yourself and the examiner that you know what information a calculation has told you!
- Trial and improvement – show your trials and the result of each trial calculation clearly. Think after each trial what you have found out and what is a good idea to try next.

Revision exercise 5a

1. A 800 g white 'bloomer' loaf of bread costs 69p. How many grams do you buy for 1p?
2. A 500 ml pack of milk costs 27p. A 2 litre container of milk costs 75p. How much less do I pay if I buy 2 litres of milk in a container rather than in 500 ml packs?
3. Two friends shared the cost of a holiday in the ratio 3 : 2. The total cost of the holiday was £1062. How much did they each pay?
4. The distances travelled by Ravi and Sue to a party were in the ratio 5 : 2. Sue travelled 6.4 miles. How far did Raji travel?
5. Estimate the cost of travelling 48 miles by car if the cost per mile is 31p.
6. Stephen has £25. Show a rough calculation to check whether he has enough money to buy six CDs at £3.98.
7. Suppose the income tax personal allowance is £4875 and the basic rate of tax is 22%. How much tax would Kerry pay with an income of £14 500 and no other tax allowances?
8. Maria bought a sunhat in Greece. It cost 12 drachma. The exchange rate was 4.82 drachma to £1. How much did the hat cost in £?
9. The sum of two numbers is 6. Their product is 7. Use trial and improvement to find the larger of the two numbers, correct to one decimal place.
10. Four friends had a meal. The bill came to £35.50 plus a service charge of 12%. They shared the cost equally. How much did they each pay?

You will need to know:

- language about shapes, such as 'polygon', 'vertex'
- how to identify lines of symmetry
- how to identify rotational symmetry.

Regular and irregular polygons

Look at these pentagons.

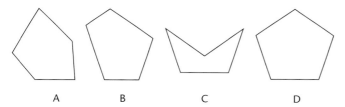

In B, all the angles are the same. In C, all the sides are the same length.

In D, the sides are the same and the angles are the same. This is a **regular** pentagon. All the others are irregular pentagons.

> A regular polygon has equal angles and equal sides.

Symmetry properties of regular polygons

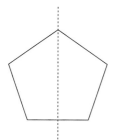 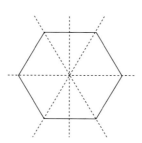 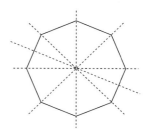

Some of the **lines of symmetry** of these regular polygons have been drawn. Can you see how many lines of symmetry there will be on each diagram if they are all drawn?

The pentagon has five lines of symmetry altogether. Each one goes through a vertex and the middle of the opposite side.

The hexagon has six lines of symmetry altogether. Each one goes through opposite vertices, or through the midpoints of opposite sides.

The octagon has eight lines of symmetry altogether. Each one goes through opposite vertices, or through the midpoints of opposite sides.

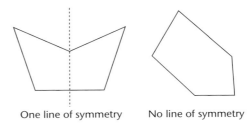

A regular polygon has the same number of lines of symmetry as it has sides.

An irregular polygon may have fewer lines of symmetry than this, or no lines of symmetry at all.

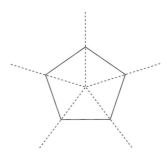

One line of symmetry   No line of symmetry

Some shapes have **rotational symmetry**, which means they look the same when they are rotated about their central point. You may already have used rotational symmetry without realising it, to draw a regular polygon in a circle.

The regular pentagon has rotational symmetry of order 5.

Any regular pentagon has the same order of rotational symmetry as it has sides.

Symmetry properties of some 3D shapes

Here are two examples of regular **polyhedra**, which are also known as 3D shapes.

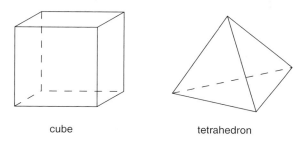

cube   tetrahedron

Each of their faces is a regular polygon.

They have rotational symmetry, for example about these axes.

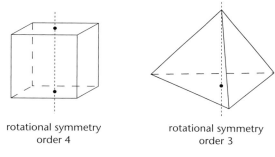

rotational symmetry   rotational symmetry
order 4   order 3

The cube also has its diagonals as axes of rotational symmetry.

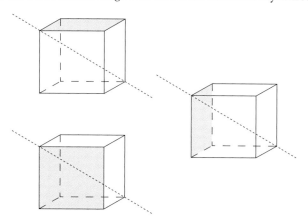

Using these axes of rotational symmetry, the order of rotational symmetry is only 3. Try it with a real cube to check.

Examiner's tip

To find the order of rotational symmetry, think how many positions one corner could take as the shape is rotated about the axis.

To check for planes of symmetry, imagine cutting the shape into halves which are reflections of each other.

The cube and tetrahedron also have planes of symmetry. Here are some examples, can you spot any more?

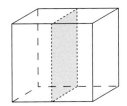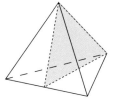

The cube has nine planes of symmetry – can you find them all?

Two more shapes which also have symmetry are the cuboid and sphere.

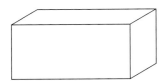

If all its faces are rectangles (not squares), a cuboid has:
3 planes of symmetry
3 axes of rotational symmetry.

A sphere has infinitely many planes of symmetry and axes of rotational symmetry.

Exercise 6.1a

1. How many lines of symmetry does a square have?
2. What is the order of rotational symmetry of a nonagon?
3. Draw a hexagon which has rotational symmetry of order 3.
4. When a sphere is cut in half by a plane of symmetry, what shape is the cross-section?
5. How many planes of symmetry does a tetrahedron have?
6. This octagonal floor tile is not regular.

   (a) How many lines of symmetry does it have?
   (b) What is its order of rotational symmetry?

7. A biscuit tin is an octagonal prism.
   (a) How many faces does it have?
   (b) If its base is a regular octagon, in how many ways will the lid fit on?
8. Here is a design for some fabric. Describe its symmetry.

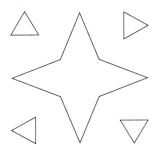

9. Regular pentagons do not tessellate. Draw a pentagon which will tessellate.
10. Draw a pattern on a regular octagon so that the completed design has two lines of symmetry *and* rotational symmetry of order 2.

1. How many lines of symmetry does a rectangle have?
2. Draw a hexagon with just one line of symmetry.
3. Draw a pentagon with rotational symmetry of order 1.
4. When a tetrahedron is cut in half by a plane of symmetry, what shape is the cross-section?
5. What is the order of rotational symmetry of these polygonal stars?

(a) 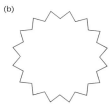　(b)

6. A jewellery box is a prism, with a regular hexagon as its cross-section. It has a separate lid. In how many different positions can the lid be put on?
7. This quadrilateral has no symmetry but it does tessellate. Draw enough copies of this quadrilateral pattern to show how it will tessellate.

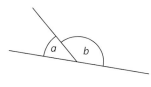

8. John is packing boxes into a carton. Each box is a cuboid with square ends. It does not matter which way up the box is packed.

How many ways can John fit a box into one section of the carton?
9. A patchwork quilt is made of regular hexagons cut from three different fabrics. Show an arrangement of hexagons so that the overall design has rotational symmetry of order 3.
10. Describe the symmetry of this dodecagon.

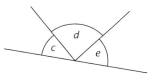

Some basic angle facts

You may have already met and used many of the angle facts listed below. Check that you know them all.

1. **The sum of angles on a straight line is 180°**

$a + b = 180°$　　$c + d + e = 180°$

2. **The sum of angles round a point is 360°**

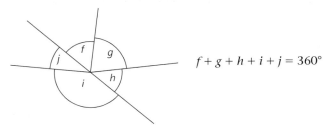

$$f + g + h + i + j = 360°$$

3. **When two straight lines cross, the opposite angles are equal**

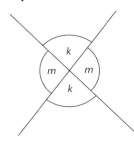

The angles *k* are equal.
The angles *m* are equal.
They are **opposite angles at a point**. Angles like these are also called **vertically opposite angles**.
Also, $k + m = 180°$ as they are angles on a straight line.

4. **The sum of the angles in a triangle is 180°**

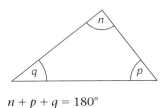

$$n + p + q = 180°$$

5. **The sum of the angles in a quadrilateral is 360°**

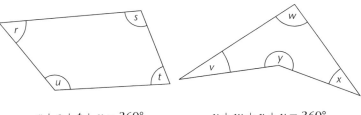

$$r + s + t + u = 360° \qquad v + w + x + y = 360°$$

The sum in the second quadrilateral is still 360°, even though it is re-entrant rather than convex in shape. This is easy to see if you split the quadrilateral into two triangles. Each triangle has an angle sum of 180°, so the sum of the angles of the quadrilateral is $2 \times 180° = 360°$.

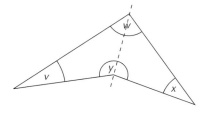

Example ① Find the missing angles in these diagrams.

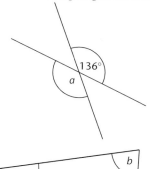

$a = 136°$ opposite angles at a point are equal

$55 + 97 + 130 = 282$
So $b + 282° = 360°$
$b = 78°$ angle sum of a quadrilateral is 360°

Example ② Find the value of a in this isosceles triangle.

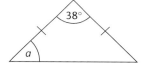

Since the triangle is isosceles, the unmarked angle is also a.

$2a + 38° = 180°$

So $2a = 180° - 38°$ angle sum of a triangle is 180°

$2a = 142°$

$a = 71°$

Exercise 6.2a

1. Two of the angles of a triangle are 30° and 72°. What is the size of the third angle of the triangle?
2. Find the sizes of these lettered angles.

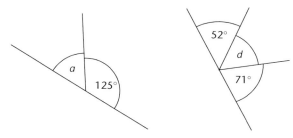

3. Two lines cross, making an angle of 35°. What are the sizes of the other three angles?
4. Two angles make a straight line. One of the angles is 87°. What is the other?
5. Three angles together make a straight line. Two of them are 52°. What is the other?

6. Find the sizes of the lettered angles, if AB and CD are straight lines.

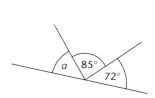

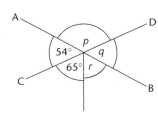

7. Look at these angles, then copy the table below and complete it.

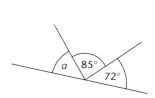

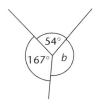

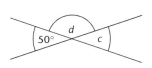

Angle	Size	Reason
a	____°	angles on a straight line add to ____°
b	____°	angles at a point add to ____°
c	____°	… angles at a point are equal
d	____°	angles on a straight line …

8. Three angles of a quadrilateral are 100°. What is the size of the fourth angle?
9. A quadrilateral has two pairs of equal angles. One pair are each 55°. What size is each of the other two angles?
10. Draw a circle, then draw a quadrilateral with all its vertices on the edge of the circle. Measure the opposite angles of the quadrilateral. If your drawing is accurate, the opposite pairs should add to 180°.

Exercise 6.2b

1. Draw an isosceles triangle and its line of symmetry. Show that the line of symmetry splits the isosceles triangle into two congruent right-angled triangles.

2. One of the equal angles of an isosceles triangle is 35°. Find the other two angles.

3. One of the angles of an isosceles triangle is 126°. Find the other two angles.

4. One of the angles of an isosceles triangle is 50°. Find the other two angles. (**Hint:** There are two sets of possible answers.)

5. A circle is divided into six equal sectors. What is the angle at the centre of the circle, for each sector?

6. A circle is divided into nine equal sectors. What is the angle at the centre of the circle, for each sector?

7. Three lines cross at a point. Two of the angles formed are equal, as shown in the diagram. Find the value of a and the sizes of the other angles in the diagram.

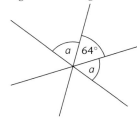

8. The two unmarked angles in this kite are equal. Find their size.

9. A kite has two angles of 125° and one angle of 75°. Find the size of its other angle.

10. The two unmarked angles in this arrowhead are equal. Find their size.

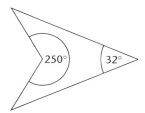

Angles with parallel lines

When a line crosses a pair of parallel lines, sets of equal angles are formed. The line cutting the parallels is called a **transversal**.

Learn to recognise each type of pair so that you are able to give reasons for angles being equal.

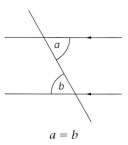

$a = b$

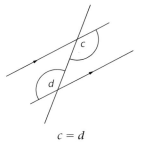

$c = d$

The diagrams above show pairs of equal **alternate angles**. These are on opposite sides of the transversal.

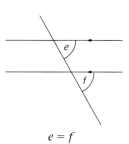

$e = f$

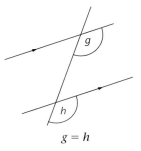

$g = h$

The diagrams above show pairs of equal **corresponding angles**. These are in the same position between the transversal and one of the parallel lines.

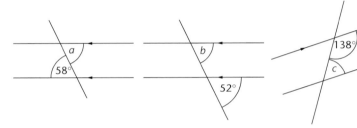

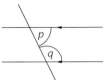

Examiner's tip

Remembering a ⊏ shape or using facts about angles on a straight line may help you to remember that allied angles add up to 180°.

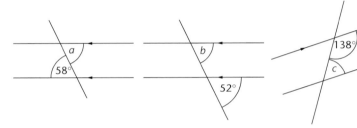

$p + q = 180°$

$r + s = 180°$

These pairs of angles are not equal. Instead, they add to 180°. They are called **allied angles** or **co-interior angles**.

Example 3

Find the sizes of the lettered angles in these diagrams, giving your reasons.

$a = 58°$ alternate angles are equal
$b = 52°$ corresponding angles are equal
$c = 42°$ allied angles add to 180°

Notation for lines and angles

You often use single letters for angles and lines, as in this chapter so far. Sometimes, the ends of a line are labelled with letters. Then you can use these letters to name the angles.

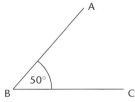

The line AB is the line joining A and B.

The angle at B is the angle made between the lines AB and BC. It may be written as angle ABC.

In this diagram, angle ABC = 50°.

It is essential to use three letters when there is more than one angle at a point.

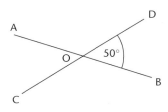

In this diagram, there are four angles at O, and they are two different sizes.

Angle AOC = angle BOD = 50°

Angle BOC = angle AOD = 130°

1. Find the value of x, giving your reason.

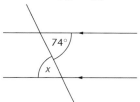

2. Find the value of z, giving your reason.

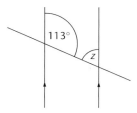

3. Draw accurately a pair of parallel lines and a line crossing them. Mark on your drawing a pair of acute corresponding angles. Measure them and check that they are equal.

4. Draw accurately a pair of parallel lines and a line crossing them. Mark on your drawing a pair of obtuse alternate angles. Measure them and check that they are equal.

5. Find the sizes of the lettered angles in this diagram.

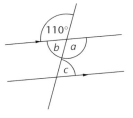

6. Find the sizes of the lettered angles in this diagram.

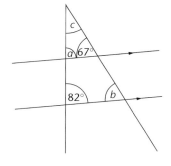

7. In a parallelogram ABCD, angle ABC = 75°. Draw a sketch and mark this angle and the sizes of the other three angles.

8. One of the angles of an isosceles trapezium is 64°. Draw a sketch and write the sizes of all the angles in suitable positions on your sketch.

9. Sketch an isosceles trapezium and write down all you can about its angles and symmetry.

10. Find the sizes of the lettered angles in this diagram.

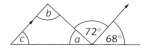

1. Find the value of y, giving your reason.

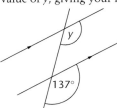

2. Draw accurately a pair of parallel lines and a line crossing them. Mark on your drawing a pair of acute alternate angles. Measure them and check that they are equal.

3. Draw accurately a pair of parallel lines and a line crossing them. Mark on your drawing a pair of corresponding obtuse angles. Measure them and check that they are equal.

4. Draw accurately a pair of parallel lines and a line crossing them. Mark on your drawing a pair of allied angles. Measure them and check that they add up to 180°.

5. Find the sizes of the lettered angles in the diagram.

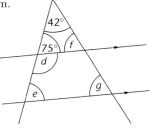

6. In a parallelogram, one angle is 126°. Make a sketch of the parallelogram and mark this angle. Calculate the other three angles and label them on your sketch.

7. In an isosceles trapezium ABCD, angle BCD is 127°. Make a sketch of the trapezium and mark this angle. Calculate the other three angles and label them on your sketch.

8. Sketch a parallelogram and write down all you can about its angles and its symmetry.

9. Find the sizes of the lettered angles in this diagram.

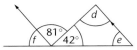

10. Use what you know about triangles to find the sizes of angles ABE and CDE in this diagram. Show why BE and CD are parallel.

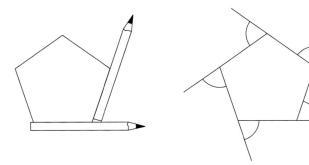

Angles in polygons

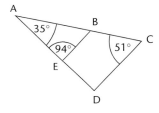

Put a pencil along the base of the pentagon, as shown, then slide it to the right until its end reaches the end of the base. Now carefully rotate the pencil about its end until it is along the next side of the pentagon. Slide it up until its end reaches the top of that side. Continue in this way until you reach the base again, as shown. When you get back to the beginning, the pencil will have turned through 360°.

It will have turned through the angles shown in the diagram on the right, next to the first diagram. These angles are called **exterior angles**. You have shown that the sum of the exterior angles of the pentagon is 360°.

This method could have been used for any convex polygon.

> The sum of the exterior angles of any convex polygon is 360°.

At each vertex, the interior and exterior angles make a straight line.

> For any convex polygon, at any vertex:
> interior angle + exterior angle = 180°

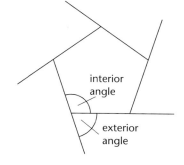

interior angle

exterior angle

Example **4**

The two unlabelled exterior angles of this pentagon are equal. Find their size.

First, find the sum of the angles that are given.

72° + 80° + 86° = 238°

The sum of all the exterior angles is 360°.

The sum of remaining two angles
= 360° − 238° = 122°.

So each angle is 122 ÷ 2 = 61°.

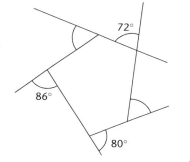

72°

86°

80°

In many problems about polygon angles, the easiest way to solve them is to use the fact that the sum of the exterior angles is 360°.

Sometimes, however, it is useful to find the sum of the interior angles.

At each vertex of a convex polygon:
    interior angle + exterior angle = 180°

> Sum of (interior + exterior angles) for the polygon
> = 180° × number of angles = 180° × number of sides

But the sum of the exterior angles is 360°.

So the sum of the interior angles of the polygon
= 180° × number of sides − 360°.

Putting this algebraically, for an *n*-sided convex polygon:

the sum of the interior angles = $(180n - 360)°$

Example 5

Find the value of the missing angle in this pentagon.

For a pentagon:
the sum of the interior angles
= $(180 \times 5 - 360)° = 540°$

The sum of the four angles given = 437°

So the missing angle = 540° − 437° = 103°

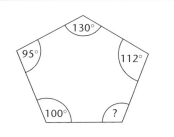

Finding angles in regular polygons

For regular polygons, each interior angle is the same, each exterior angle is the same. This is an easy way to work out the angles.

Another useful fact about regular polygons is that they may be divided, from their centre, into congruent isosceles triangles. This is often useful when you are asked to draw regular polygons accurately. For example, for a regular pentagon, the angle at the centre is 360° ÷ 5 = 72°.

The base angles of the triangle together equal 180° − 72° = 108°. So each base angle is 54°.

From the diagram, these base angles are each half of an interior angle, so this gives another way of working out interior angles, too.

Each exterior angle
$$= \frac{360°}{\text{number of sides}}$$
then use:
interior angle
= 180° − exterior angle

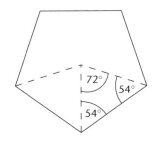

Example 6

Find the interior angle of a regular octagon.

For a regular octagon, the exterior angle
= 360° ÷ 8 = 45°.

So the interior angle = 180° − 45° = 135°.

Exercise 6.4a

1. Three of the exterior angles of a quadrilateral are 90°, 52° and 87°. Find the size of the other exterior angle.

2. Four of the exterior angles of a pentagon are 70°, 59°, 83° and 90°. Find the size of the other exterior angle.

3. Five of the exterior angles of a hexagon are 54°, 48°, 65°, 35° and 80°. Find the size of the other exterior angle.

4. Find the sizes of the interior angles of the pentagon in question 2.

5. Find the sizes of the interior angles of the hexagon in question 3.

6. A regular polygon has nine sides. Find the sizes of its exterior and interior angles.

7. Find the interior angle of a regular dodecagon (12 sides).

8. A regular polygon has an exterior angle of 24°. How many sides does it have?

9. What is the sum of the interior angles of:
   (a) a hexagon (b) a decagon?

10. Six of the angles of a heptagon are 122°, 141°, 137°, 103°, 164° and 126°. Calculate the size of the remaining angle.

Exercise 6.4b

1. Three of the exterior angles of a quadrilateral are 110°, 61° and 74°. Find the size of the other exterior angle.

2. Four of the exterior angles of a pentagon are 68°, 49°, 82° and 77°. Find the size of the other exterior angle.

3. Four of the exterior angles of a hexagon are 67°, 43°, 91° and 37°. Find the size of the other exterior angles, given that they are equal.

4. Find the sizes of the interior angles of the pentagon in question 2.

5. Find the sizes of the interior angles of the hexagon in question 3.

6. A regular polygon has 15 sides. Find the sizes of its exterior and interior angles.

7. Find the interior angle of a regular 20-sided polygon.

8. A regular polygon has an exterior angle of 30°. How many sides does it have?

9. What is the sum of the interior angles of:
   (a) an octagon (b) a nonagon?

10. A polygon has 11 sides. Ten of its interior angles add up to 1490°. Find the size of the remaining angle.

Key points

Regular and irregular polygons
- A regular polygon has equal angles and equal sides.

Symmetry properties of regular polygons
- A regular polygon has the same number of lines of symmetry as it has sides.
- Any regular polygon has the same order of rotational symmetry as its number of sides.

Some basic angle facts
- Angles on a straight line add up to 180°.
- Angles round a point add up to 360°.
- When two straight lines cross, the opposite angles are equal.

- Angles in a triangle add up to 180°.
- Angles in a quadrilateral add up to 360°.

Angles with parallel lines
- Thinking of a Z shape may help to remember alternate angles are equal – or looking at the diagrams upside-down and seeing the same shapes.
- Thinking of an F shape or a translation may help to remember that corresponding angles are equal.
- Remembering a ⊏ shape or using facts about angles on a straight line may help to remember that allied angles add up to 180°.

Angles in polygons
- The sum of the exterior angles of a convex polygon is 360°.
- Interior angle + exterior angle = 180°.
- For an n-sided polygon:
  sum of the interior angles = $(180n - 360)°$.

Finding angles in regular polygons
- Each exterior angle = $\dfrac{360°}{\text{number of sides}}$.
- Then use:
  interior angle = 180° − exterior angle.

Revision exercise 6a

1. How many lines of symmetry does a regular octagon have?

2. What is the order of rotational symmetry of this shape?

3. Sketch an isosceles triangle and show any symmetry it has.

4. An isosceles triangle has an angle of 40°. Find the sizes of the other angles in the triangle. There are two possible sets of answers – find both sets.

5. Three angles of a quadrilateral are 62°, 128° and 97°. Find the size of the other angle.

6. Find the sizes of the lettered angles, giving your reasons.

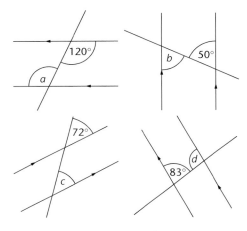

7. Four of the exterior angles of a pentagon are 85°, 66°, 54° and 97°. Find the size of the other exterior angle.

8. Calculate the sum of the interior angles of a heptagon.

9. A regular polygon has ten sides. Find the size of its: (a) exterior (b) interior angles.

10. An irregular polygon has eight sides. Seven of its interior angles add up to 940°. Calculate the size of its other interior angle.

Coordinates, graphs and bearings

Points in all four quadrants

You can plot points in any of the four quadrants of the Cartesian plane, using positive and negative coordinates.

Example ❶ The four points on the grid are A(2, 1), B(2, –2), C(−2, −3), D(−4, 3).

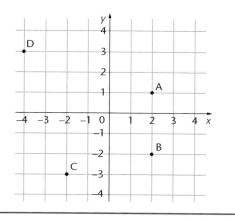

Examiner's tip

Always label the axes clearly.

Make sure you get the sign correct.

Straight-line graphs

The simplest equations that give straight-line graphs are of the type $y = 3$ or $x = 2$.

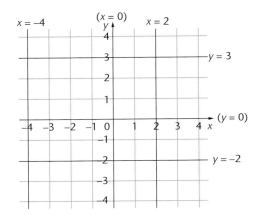

Every point on the line $y = 3$ has y-coordinate 3.

Every point on the line $x = 2$ has x-coordinate 2.

The x-axis has the equation $y = 0$ and the y-axis has the equation $x = 0$.

So $y = 3$ is the equation of the line through 3 on the y-axis and is parallel to $y = 0$, and $x = 2$ is the equation of the line through 2 on the x-axis and is parallel to $x = 0$.

These are both drawn on the graph. Can you see the lines $y = -2$ and $x = -4$, which are also drawn on the graph?

Lines of the type $y = mx + c$

The most common straight-line graphs you need to draw have equations of the form $y = mx + c$, for example $y = 3x + 4$, $y = 2x - 5$ and $y = -3x + 2$.

You need only two points to draw a straight line, but you should work out three points to check. It is best to use two values that are as far apart as possible, and 0 if it is in the range.

Example 2

Draw the graph of $y = 2x + 3$ for values of x from -3 to $+3$.

You can use any three values of x from -3 to $+3$. In this example the best values to use are -3, 0 and $+3$.

When $x = -3$, $y = 2 \times -3 + 3$
$\qquad = -6 + 3 = -3$.

When $x = 0$, $y = 2 \times 0 + 3 = 0 + 3 = 3$.

When $x = +3$, $y = 2 \times 3 + 3 = 6 + 3 = 9$.

The graph needs to include x-values from -3 to $+3$ and y-values from -3 to $+9$.

Plot the points $(-3, -3)$, $(0, 3)$ and $(3, 9)$ and join them with a straight line. Label the line.

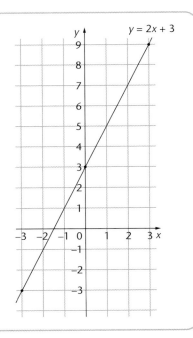

Example **E** Draw the graph of $y = -2x + 1$, for values of x from -4 to $+2$.

When $x = -4$, $y = -2 \times -4 + 1 = +8 + 1 = 9$.

When $x = 0$, $y = -2 \times 0 + 1 = 0 + 1 = 1$.

When $x = 2$, $y = -2 \times 2 + 1 = -4 + 1 = -3$.

In this case the y-values are from -3 to 9.

Plot the points $(-4, 9)$, $(0, 1)$ and $(2, -3)$ and join them with a straight line.

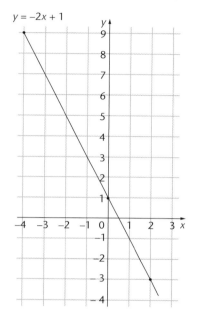

Notice that this line slopes the opposite way.

Examiner's tip

Any line with equation $y = ax + b$ will cross the y-axis at $y = b$.

Exercise 7.1a

1. On a grid, draw and label axes from -6 to $+6$ for both x and y.
   Plot and label the points A(1, 3), B(-2, -6), C(-4, 3), D(-6, 3), E(2, -5), F(-2, 0), G(0, 5).

2. On a grid, draw and label axes from -6 to $+6$ for both x and y. Then draw the line for each of these equations.
   $x = 3$, $x = -3$, $y = 4$, $y = -5$

3. Draw the graph of $y = 4x$, for $x = -3$ to $+3$.

4. Draw the graph of $y = x + 3$, for $x = -3$ to $+3$.

5. Draw the graph of $y = 3x + 2$, for $x = -4$ to $+2$.

6. Draw the graph of $y = x - 3$, for $x = -2$ to $+4$.

7. Draw the graph of $y = 3x - 4$, for $x = -2$ to $+4$.

8. Draw the graph of $y = 4x - 2$, for $x = -3$ to $+3$.

9. Draw the graph of $y = -2x + 5$, for $x = -2$ to $+4$.

10. Draw the graph of $y = -3x - 4$, for $x = -4$ to $+2$.

Chapter 7 Points in all four quadrants

Exercise 7.1b

1. On a grid, draw and label axes from -6 to $+6$ for both x and y. Then draw the line for each of these equations.
   $x = 5$, $x = -2$, $y = 2$, $y = -3$
2. Draw the graph of $y = 3x$, for $x = -3$ to $+3$.
3. Draw the graph of $y = x + 6$, for $x = -4$ to $+2$.
4. Draw the graph of $y = 4x + 2$, for $x = -3$ to $+3$.
5. Draw the graph of $y = 3x + 5$, for $x = -4$ to $+2$.
6. Draw the graph of $y = x - 5$, for $x = -1$ to $+6$.
7. Draw the graph of $y = 2x - 5$, for $x = -1$ to $+5$.
8. Draw the graph of $y = -x + 1$, for $x = -3$ to $+3$.
9. Draw the graph of $y = -3x - 2$, for $x = -4$ to $+2$.
10. Draw the graph of $y = -2x - 4$, for $x = -4$ to $+2$.

Harder straight-line equations

Sometimes you could be asked to draw a straight line with an equation of the form $3x + 2y = 12$, where both the x-term and the y-term are on the same side of the equation. In this case, it is easier to work out the value of x when y is 0, and the value of y when x is 0.

Draw the line joining these two points. Then check that the coordinates of another point on the graph fit the equation.

Example 4

Draw the graph of $4x + 3y = 12$.

When $x = 0$, $3y = 12$, $y = 4$.

When $y = 0$, $4x = 12$, $x = 3$.

The axes on the graph need to be labelled for x from 0 to 3, and for y from 0 to 4.

Plot the points $(0, 4)$ and $(3, 0)$ and join them with a straight line.

Check with a point on the line, such as $x = 1.5$, $y = 2$.

$4x + 3y = 4 \times 1.5 + 3 \times 2 = 6 + 6 = 12$, which is correct.

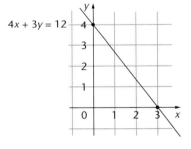

Other types of equation include $2y = 3x + 5$, where you first need to find $2y$, and then divide by 2.

You may also be asked to draw graphs of equations involving fractions.

1. Draw the graph of $3x + 5y = 15$.
2. Draw the graph of $7x + 2y = 14$.
3. Draw the graph of $2y = 5x + 3$, for $x = -3$ to $+3$.
4. Draw the graph of $2y = 3x - 5$, for $x = -2$ to $+4$.
5. Draw the graph of $y = \frac{1}{2}x + 3$, for $x = -6$ to $+4$.
6. Draw the graph of $y = \frac{1}{3}x - 4$, for $x = -3$ to $+6$.
7. (a) On the same grid, draw the graphs of:
     (i) $y = 8$ (ii) $y = 4x + 2$
     for $x = -3$ to $+3$.
   (b) Write down the coordinates of the point where the two lines cross.
8. (a) On the same grid, draw the graphs of $y = 2x + 3$ and $2x + y = 7$, for $x = 0$ to 5.
   (b) Write down the coordinates of the point where the two lines cross.

1. Draw the graph of $2x + 5y = 10$.
2. Draw the graph of $3x + 2y = 15$.
3. Draw the graph of $3y = 2x + 6$, for $x = -3$ to $+3$.
4. Draw the graph of $2y = 5x - 8$, for $x = -2$ to $+4$.
5. Draw the graph of $y = \frac{1}{2}x + 9$, for $x = -6$ to $+4$.
6. Draw the graph of $y = \frac{3}{5}x - 4$, for $x = -5$ to $+10$.
7. (a) On the same grid, draw the graphs of:
     (i) $y = x + 3$
     (ii) $y = 4x - 3$
        for $x = -2$ to $+3$.
   (b) Write down the coordinates of the point where the two lines cross.
8. (a) On the same grid, draw the graphs of $y = 3x - 2$ and $4x + y = 12$, for $x = 0$ to 5.
   (b) Write down the coordinates of the point where the two lines cross.

Quadratic graphs

Quadratic graphs are graphs of equations of the form $y = ax^2 + bx + c$, where a, b, c are constants and b and c may be zero, for example, $y = x^2 + 3$, $y = x^2 - 4x + 3$. When the graph is drawn, it produces a curve called a **parabola**.

In an examination you may be expected to complete a table of values to work out the points. Even if you are not given a table, it is still best to use one.

Example 5

Draw the graph of $y = x^2 + 4$, for values of x from -3 to $+3$.

First draw a table of values.

x	-3	-2	-1	0	1	2	3
x^2	9	4	1	0	1	4	9
$y = x^2 + 4$	13	8	5	4	5	8	13

Now label the axes. The values of x go from -3 to $+3$ and the values of y go from 4 to 13. It is better to include 0 in the values of y, so let them go from 0 to 15.

Always include the x-axis even when the y-values are all positive.

On this graph the scale used is 1 square to 1 unit on the x-axis and 1 square to 5 units on the y-axis.

Plot the points and join them with a smooth curve. Label the curve.

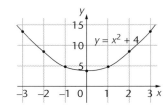

Example 6

Draw the graph of $y = x^2 + 3x$, for values of x from -5 to $+2$.

Label the axes from -5 to $+2$ for x and from -5 to 10 for y.

x	-5	-4	-3	-2	-1	0	1	2
x^2	25	16	9	4	1	0	1	4
$3x$	-15	-12	-9	-6	-3	0	3	6
$y = x^2 + 3x$	10	4	0	-2	-2	0	4	10

Add the numbers in rows 2 and 3 to give the value of y.

Here you can see that $x = -1$ and $x = -2$ both give $y = -2$, and the lowest value (called the **minimum**) will be when x is between -1 and -2.

It is useful to work out the value of y when $x = -1.5$. To do this, add some more values to your table:

$x = -1.5$, $x^2 = 2.25$, $3x = -4.5$, $y = -2.25$.

Plot $(-1.5, -2.25)$, which is the lowest point of the graph.

Examiner's tip

A common error is to include the values in the x-row when adding to find y.

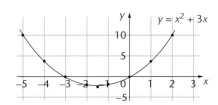

Example 7

(a) Draw the graph of $y = x^2 - 2x - 3$, for values of x from -2 to $+4$.

Label the axes from -2 to $+4$ for x, and from -5 to $+5$ for y.

(b) Solve the equation $x^2 - 2x - 3 = 0$ from your graph.

(a)

x	-2	-1	0	1	2	3	4
x^2	4	1	0	1	4	9	16
$-2x$	4	2	0	-2	-4	-6	-8
-3	-3	-3	-3	-3	-3	-3	-3
$y = x^2 - 2x - 3$	5	0	-3	-4	-3	0	5

To find the value of y, add together the values in rows 2, 3 and 4.

Use a scale of 1 square to 1 unit on the x-axis and 2 squares to 5 units on the y-axis.

Now plot the points $(-2, 5)$, $(-1, 0)$ and so on. Join them with a smooth curve. Label the curve.

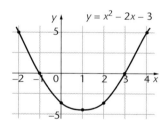

(b) $x^2 - 2x - 3 = 0$ when $y = 0$, which is when the curve crosses the x-axis. This is when $x = -1$ or $x = +3$.

 Examiner's tip

If you find that you have two equal lowest (or highest) values for y, the curve will go below (or above) that point. You will need to find the y-value between the two equal values. To do this, find the value of x halfway between the two points and substitute, to find the corresponding value of y.

 Examiner's tip

The y-scale for Examples 5 and 6 could be 1 cm to 5 units. This is satisfactory and does not take a lot of space, but if you do have enough space it will look better and be easier to plot if you use a scale of 2 cm to 5 units, as in Example 7.

Sometimes questions will be put in context. They will be about a real-life situation, rather than just being about a graph in terms of x and y.

Chapter 7 Quadratic graphs

Example 8

The cost C, in pounds, of circular plates is given by the formula $C = \dfrac{x^2}{10} + 2$, where x is the radius of the plate in centimetres.

(a) Draw up a table of values and complete it.

(b) (i) Draw the graph of C against x, for values of x from 0 to 20, and y from 0 to 45.

(ii) From your graph find the size of plate that would cost £16.40.

(a)

x	0	5	8	10	15	20
x^2	0	25	64	100	225	400
$\dfrac{x^2}{10}$	0	2.5	6.4	10	22.5	40
$C = \dfrac{x^2}{10} + 2$	2	4.5	8.4	12	24.5	42

(b) (i)

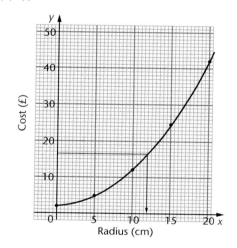

(ii) You can buy a plate with radius 12 cm for £16.40.

Use 2 mm graph paper for these questions.

1. Copy the table for $y = x^2 + 5$ and complete it. Do not draw the graph.

x	-3	-2	-1	0	1	2	3
x^2	9						
5	5						
$y = x^2 + 5$	14						

2. Copy the table of values for $y = x^2 + 3x - 7$ and complete it. Do not draw the graph. Notice the extra value at $x = -1.5$.

x	-4	-3	-2	-1	0	1	2	-1.5
x^2			4					
$3x$			-6					
-7			-7					
$y = x^2 + 3x - 7$			-9					

3. Copy the table of values for $y = -x^2 - 5x + 6$ and complete it. Do not draw the graph. Remember $-x^2$ is always negative. Notice the extra value at -2.5.

x	-6	-5	-4	-3	-2	-1	0	1	2	-2.5
$-x^2$			-16						-4	
$-5x$			20						-10	
6			6						6	
$y = -x^2 - 5x + 6$			10						-8	

4. Draw up a table of values for $y = x^2 - 3x + 1$, for $x = -2$ to $+4$. Do not draw the graph.

5. (a) Copy the table of values for $y = x^2 - 2$ and complete it.

x	-3	-2	-1	0	1	2	3
x^2							
-2							
$y = x^2 - 2$							

   (b) Draw the graph of $y = x^2 - 2$ for $x = -3$ to $+3$. Label the x-axis from -3 to $+3$ and the y-axis from -5 to $+10$. Use a scale of 1 cm to 1 unit on the x-axis and 2 cm to 5 units on the y-axis.

6. Draw the graph of $y = -x^2 + 4$ for $x = -3$ to $+3$. Label the x-axis from -3 to $+3$ and the y-axis from 0 to 15. Use a scale of 1 cm to 1 unit on the x-axis and 2 cm to 5 units on the y-axis.

7. (a) Copy this table of values for $y = x^2 - 3x$ and complete it.

x	-1	0	1	2	3	4	5	1.5
x^2								
$-3x$								
$y = x^2 - 3x$								

   (b) Draw the graph of $y = x^2 - 3x$, for $x = -1$ to $+5$. Label the x-axis from -1 to $+5$ and the y-axis from -5 to $+10$.

   Use a scale of 1 cm to 1 unit on the x-axis and 2 cm to 5 units on the y-axis.

   (c) Use the graph to solve the equation $x^2 - 3x = 0$.

8. (a) Draw the graph of $y = x^2 - 4x + 3$, for $x = -1$ to $+5$.
   (b) Use the graph to solve $x^2 - 4x + 3 = 0$.
9. (a) Draw the graph of $y = x^2 - 5x + 2$, for $x = -1$ to $+6$.
   (b) Use the graph to solve $x^2 - 5x + 2 = 0$. Give your answers correct to one decimal place.
10. When a stone is dropped from the edge of a cliff, the distance, d metres, it falls is given by the formula $d = 5t^2$, where t is the time in seconds.
    (a) Work out the values of d for $t = 0$ to 5.
    (b) Draw the graph for $t = 0$ to 5.
    (c) The cliff is 65 metres high. How long does it take the stone to reach the bottom of the cliff? Give the answer correct to one decimal place.

Exercise 7.3b

1. Copy the table of values for $y = x^2 + 6$ and complete it. Do not draw the graph.

x	-3	-2	-1	0	1	2	3
x^2		4					
6		6					
$y = x^2 + 6$		10					

2. Copy the table of values for $y = 2x^2 - 8$ and complete it. Do not draw the graph.

x	-3	-2	-1	0	1	2	3
x^2	9						
$2x^2$	18						
-8	-8						
$y = 2x^2 - 8$	10						

> To find the value of y rows 3 and 4 are added. Some people do not include the x^2 row but it can help.

3. Draw up a table of values for $y = x^2 - 5x + 8$, for $x = -2$ to $+4$. Do not draw the graph.
4. Draw the graph of $y = x^2 + 4x$, for $x = -6$ to $+2$. Label the x-axis from -6 to $+2$ and the y-axis from -5 to $+15$.
   Use a scale of 1 cm to 1 unit on the x-axis and 1 cm to 5 units on the y-axis.
5. Draw the graph of $y = -x^2 + 2x + 6$, for $x = -2$ to $+4$. Label the x-axis from -2 to $+4$ and the y-axis from -5 to 10.

> An extra point at $x = 2.5$ might be useful.

   Use a scale of 1 cm to 1 unit on the x-axis and 1 cm to 5 units on the y-axis.
6. (a) Draw the graph of $y = x^2 - 6x + 5$, for $x = -1$ to $+6$.
   (b) Use the graph to solve $x^2 - 6x + 5 = 0$.
7. (a) Draw the graph of $y = -x^2 + 4x - 3$, for $x = -1$ to $+5$.
   (b) Use the graph to solve $x^2 - 4x + 3 = 0$.

8. (a) Draw the graph of $y = 2x^2 - 5x + 1$, for $x = -2$ to $+4$.

> Note that in this case the values of y are not symmetrical.

   (b) Use the graph to solve $2x^2 - 5x + 1 = 0$.
        Give your answers correct to one decimal place.
9. (a) Draw the graph of $y = 2x^2 - 12x$ for values of x from -1 to $+7$.
    (b) Write down the values of x where the curve crosses $y = 5$.
10. The surface area (S) of a cube is given by the formula $S = 6x^2$, where x is the length of an edge of the cube.
    (a) Copy this table of values and complete it to find S.

x	0	1	2	3	4	5	6
x^2				9			
$S = 6x^2$				54			

   (b) Draw the graph of $S = 6x^2$ for values of x from 0 to 6.
   (c) From your graph, find the length of the edge of a cube with surface area 140 cm².

Harder graphs

You may be asked to draw graphs of equations such as
$y = x^3 + 3x^2$, $y = \dfrac{a}{x}$ or $y = \dfrac{k}{x^2}$, but these are not set very often at
intermediate level.

Example 9

(a) Draw the graph of $y = x^3 - 2x$ for values of x from -2 to $+2$. Label the x-axis from -2 to $+2$ and the y-axis from -5 to $+5$.

(b) Use the graph to solve the equation $x^3 - 2x = 0$. Give your answers correct to one decimal place.

(a)
x	-2	-1	0	1	2	-0.5	0.5
x^3	-8	-1	0	1	8	-0.125	0.125
$-2x$	4	2	0	-2	-4	1	-1
$y = x^3 - 2x$	-4	1	0	-1	4	0.875	-0.875

> It helps to see more clearly where the curve is highest and lowest if you work out the values of y for $x = -0.5$ and $x = 0.5$.

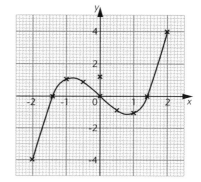

(b) To solve $x^3 - 2x = 0$, you need to find where $y = x^3 - 2x$ crosses $y = 0$, which is the x-axis.

From the graph, the answers are $x = -1.4$ or 0 or $x = 1.4$.

Example 10 Draw the graph of $y = \dfrac{4}{x}$.

x	-4	-3	-2	-1	1	2	3	4
$y = \dfrac{4}{x}$	-1	-1.3	-2	-4	4	2	1.3	1

You cannot use 0 as a point in this type of graph, since you cannot divide 4 (or any other number) by 0. Again it is helpful to work out extra points to give a better curve. In this case you might work out the value of y when $x = -1.5$ and $+1.5$, giving y-values of -2.7 and 2.7 to one decimal place.

It is also useful to have the same scale for both x and y in this case.

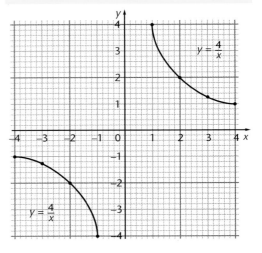

Examiner's tip

When the scale for the axes is not given, always use a scale that is easy to read, such as 1 cm to 1, 5 or 10 units or 2 cm to 5 units.

Sometimes you may be asked to draw graphs of more than one equation on one grid. These equations may not be particularly difficult but you could be asked to solve one or more equations using these graphs. These are not always immediately obvious but, with care, you can work them out.

(a) Draw the graphs of $y = x^2 - 3x - 1$ and $y = 1 - 2x$ on the same grid, for $x = -2$ to $+5$.

(b) Use your graphs to find solutions to these equations.
   (i) $x^2 - 3x - 1 = 0$
   (ii) $x^2 - 3x - 4 = 0$
   (iii) $x^2 - x - 2 = 0$

(a) Working out the table, in the usual way, for $y = x^2 - 3x - 1$ gives the points:
   $(-2, 9), (-1, 3), (0, -1), (1, -3),$
   $(2, -3), (3, -1), (4, 3), (5, 9),$
   $(1.5, -3.25).$
   For $y = 1 - 2x$, three points are $(-2, 5), (0, 1)$ and $(5, -9)$.

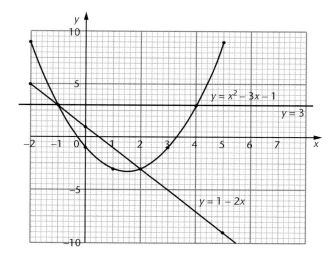

(b) (i) $x^2 - 3x - 1 = 0$
   Here $y = 0$ so find where the curve $y = x^2 - 3x - 1$ meets the line $y = 0$.
   From the graph, this is when $x = -0.3$ or 3.3.

   (ii) $x^2 - 3x - 4 = 0$
   Rewrite the equation so that the left-hand side is $x^2 - 3x - 1$.
   This gives $x^2 - 3x - 1 = 3$, so it is where $y = x^2 - 3x - 1$ cuts the line $y = 3$.
   At these points, $x = -1$ or 4.

   (iii) $x^2 - x - 2 = 0$
   This can be rearranged to give
   $x^2 - 3x - 1 = 1 - 2x$.
   So it is where the curve $y = x^2 - 3x - 1$ and the line $y = 1 - 2x$ cross.
   At these points, $x = -1$ or 2.

1. Draw up a table of values for $y = x^3 - 12x + 2$, for $x = -3$ to $+4$. Do not draw the graph.

2. Draw up a table of values for $y = x^3 - x^2 + 5$, for $x = -2$ to $+4$. Do not draw the graph.

3. Draw up a table of values for $y = \dfrac{8}{x}$ for $x = -8$, -4, -2, -1, 1, 2, 4, 8. Do not draw the graph.

4. (a) On the same grid, draw the graphs of $y = x^2 - 2x - 3$ and $y = x + 1$, for $x = -2$ to $+4$.
   (b) Find the points where the two graphs cross.

5. (a) Draw the graph of $y = x^3 - 3x$, for $x = -3$ to $+3$.
   (b) Solve the equation $x^3 - 3x = 0$.

6. (a) Draw the graph of $y = x^3 - 12x + 2$, for $x = -3$ to $+4$.
   (b) Solve the equation $x^3 - 12x + 2 = 0$.

7. Draw the graph of $y = \dfrac{8}{x}$ for values of x from -8 to $+8$.
   Use a scale of $2\,cm$ to 5 units on both axes.

8. (a) Draw up a table of values for $y = \dfrac{5}{x}$ for $x = -5$, -4, -2.5, -2, -1, 1, 2, 2.5, 4, 5.
   (b) Draw the graph of $y = \dfrac{5}{x}$. Use a scale of $1\,cm$ to 1 unit on both axes.
   (c) On the same grid, draw the graph of $y = x$.
   (d) Use your graph to solve $x^2 = 5$, giving the answers to one decimal place.

9. (a) Draw the graph of $y = x^3 - 4x$ for values of x from -3 to $+3$.
   (b) Use the graph to solve the equation $x^3 - 4x - 2 = 0$.

10. (a) Draw the graphs of $y = 4x^2 - 5$ and $y = 3x + 2$ for $x = -3$ to $+3$.
    (b) Use the graphs to solve these equations.
        (i) $4x^2 - 5 = 20$ (ii) $4x^2 - 3x - 7 = 0$

1. Draw up a table of values for $y = x^3 - 3x + 4$, for $x = -3$ to $+4$. Do not draw the graph.

2. Draw up a table of values for $y = x^3 + 2x^2 - 3$, for $x = -4$ to $+2$. Do not draw the graph.

3. Draw up a table of values for $y = \dfrac{36}{x^2}$ for $x = -6$, -4, -3, -2, -1, 1, 2, 3, 4, 6. Do not draw the graph.

4. (a) On the same grid, draw the graphs of $y = x^2 - 4$ and $y = 9 - x^2$, for $x = -4$ to $+4$.
   (b) Find the coordinates of the point where the two graphs cross.

5. (a) Draw up a table of values for $y = x^3 - x^2 - 6x$, for $x = -3$ to $+4$. Draw the graph.
   (b) Use the graph to solve the equation $x^3 - x^2 - 6x = 0$.

6. (a) Draw up a table of values for $y = \dfrac{12}{x}$, for $x = -12$, -8, -6, -4, -3, -2, -1, 1, 2, 3, 4, 6, 8, 12.
   (b) Draw the graph.

7. (a) Draw the graph of $y = -2x^2 + 3x + 8$, for values of x from -2 to $+4$.
   (b) Use the graph to solve these equations.
        (i) $2x^2 - 3x - 8 = 0$ (ii) $2x^2 - 3x = 0$

8. (a) On the same grid, draw the graphs of $y = x^3$ and $y = 5x$, for $x = -3$ to $+3$.
   (b) (i) Show that when the two curves cross the equation is $x^3 - 5x = 0$.
       (ii) Find the solution to the equation $x^3 - 5x = 0$, giving the answer correct to one decimal place.

9. (a) Draw the graph of $y = x^3 - 8x + 12$ for values of x from -3 to $+3$.
   (b) Use the graph to solve the equation $x^3 - 8x = 0$, giving the answers correct to one decimal place.

10. (a) On the same grid, draw the graphs of $y = x^2 + 2x - 5$, $y = x + 1$ and $y = x - 1$, for values of x from -4 to $+2$.
    (b) Use the graphs to solve these equations.
        (i) $x^2 + 2x - 5 = 0$ (ii) $x^2 + x - 6 = 0$

Bearings

If you want to describe the direction in which something is moving, relative to your own position, you might say, 'Due north' or 'north-east'. If you need to describe the direction accurately you should use **bearings**. These describe a direction as an angle measured clockwise from north.

The most common type of bearing is a **three-figure bearing**. The three-figure bearing for 'due east' is 090°. The zero is put in front of the 90, to make up the three figures. A bearing of 'due south' is 180° and a bearing of 'due west' is 270°.

In the diagram AB is on a bearing of 060° from A.

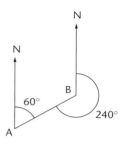

> A is the fixed reference point for the bearing. Bearings are always given from some fixed point, such as a lighthouse.

As the bearing is measured from A, the protractor is placed with its centre over A and its 0 on the north line.

You can write this as 'the bearing of B **from A** is 060°' or
'the bearing **from A** to B is 060°'.

You can also say 'the line BA is on a bearing of 240° **from B**' or
'the bearing of A **from B** is 240°' or
'the bearing **from B** of A is 240°'.

Examiner's tip

Bearings must have three figures so if the angle is less than 100° a zero must be put in front of the figures. Notice that the bearing of A from B = the bearing of B from A + 180° (or −180° if the first bearing is more than 180°).

Example 12 Use a protractor to measure and write down the bearings of:

(a) A from O (b) B from O
(c) C from O.

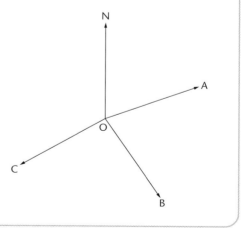

(a) 070°   (b) 146°
(c) 240°

> If you have a circular protractor you can measure all of these directly.
> If you only have a semicircular protractor then for (c) you need to measure the obtuse angle and subtract it from 360°.

Example 13

Three towns are Ayesford (A), Bagshot (B) and Carhill (C).

B is 20 km from A, on a bearing of 085° from A.

C is 15 km from B, on a bearing of 150° from B.

(a) Make a scale drawing showing the three towns.
Use a scale of 1 cm to 5 km.

(b) (i) How far is Ayesford from Carhill?
(ii) What is the bearing of Ayesford from Carhill?

(a) Make a sketch and label the angles and lengths for the final diagram.
Then draw the diagram, starting far enough down the page to make sure it will all fit in.

Sketch

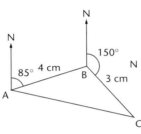

Scale drawing (half-size)

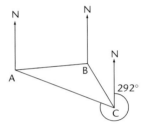

(b) (i) CA measures 5.9 cm, so the distance is 5.9 × 5 = 29.5 km.
(ii) The bearing is 292°.

Example 14

This is a sketch, not to scale, of three points in a field.

Work out the bearings of:

(a) B from A   (b) C from B   (c) A from C.

Sketch

(a) 040°.

(b) The bearing of C from B is labelled y and
the bearing of A from C is labelled q.
From the diagram, the bearing of A from B is $y + 106°$.
By calculation, the bearing of A from B is $180° + 40° = 220°$.
So $y + 106° = 220°$
$y = 220° - 106° = 114°$.
The bearing of C from B = 114°.

(c) From the diagram, the bearing of B from C is $q + 47°$.
By calculation, the bearing of B from C = $180° + 114° = 294°$.
So $q + 47° = 294°$.
$q = 294° - 47° = 247°$
The bearing of A from C is 247°.

1. Measure the bearings of A, B, C and D from O.

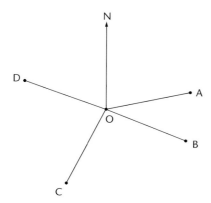

2. A boat is sailing due west. On what bearing is it sailing?

3. Tadmouth is on a bearing of 124° from Easton. What is the bearing of Easton from Tadmouth?

4. The road AB goes exactly north-east. On what bearing does it lie?

5. The map shows a lighthouse L and two boats A and B.

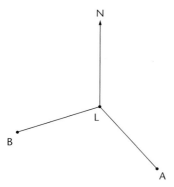

   Measure the bearing from the lighthouse of:
   (a) boat A   (b) boat B.

6. Dorton (D) is 8 miles from Brenwood (B), on a bearing of 058°. Make a sketch showing these two towns. Mark the distance and the angle clearly. Do not draw an accurate plan.

7. A boat leaves port and sails for 10 miles on a bearing of 285°. Make a sketch showing the direction in which the boat is sailing. Mark the distance and the angle clearly. Do not draw an accurate plan.

8. This sketch map shows the position of three oil platforms.

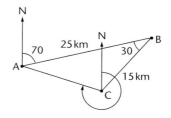

   (a) Make an accurate drawing of the sketch. Use a scale of 1 cm to 5 km.
   (b) Measure the bearing of A from C.

9. In an orienteering competition, competitors have to go from point A to point B to point C.
   A to B is 1 km on a bearing of 125°.
   B to C is 1.5 km on a bearing of 250°.
   (a) Draw an accurate plan of the route. Use a scale of 5 cm to 1 km.
   (b) Measure:
      (i) the bearing of C from A
      (ii) the distance from C to A.

10. This sketch map shows the positions of three hills P, Q and R.

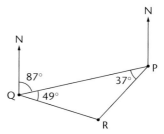

   Work out the bearing of:
   (a) R from Q
   (b) R from P
   (c) P from R.
   Do not make a scale drawing.

1. Mark a point O. Draw a north line through O. Mark points A, B and C, all 5 cm from O on these bearings.
   OA 055°, OB 189°, OC 295°

2. Gravesend is on a bearing of 290° from Gillingham.
   What is the bearing of Gillingham from Gravesend?

3. An aeroplane is flying directly south-east. On what bearing is it flying?

4. On this drawing measure the bearing of:
   (a) P from Q   (b) Q from R.

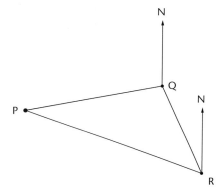

5. The sketch show the position of three buoys in a harbour. Work out the bearing of:
   (a) A from B   (b) C from A.
   Do not make a scale drawing.

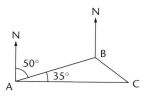

6. Kings Lynn is 60 km from Norwich, on a bearing from Norwich of 277°.
   Use a scale of 1 cm to 10 km to make an accurate drawing of the positions of these two towns.

7. This is an accurate drawing of the course of a yacht race.

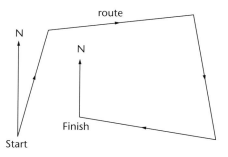

   (a) On what bearing do the yachts go on the first leg?
   (b) What is the bearing of the start from the finish?

8. David and Phil both leave point A and walk across the moors.
   After an hour David is 4 km from A, on a bearing of 075°, and Phil is 5.5 km from A, on a bearing of 230°.
   (a) Make an accurate drawing of their routes. Use a scale of 1 cm to 1 km.
   (b) Find the bearing and distance of David from Phil.
   Remember to do a sketch first.

9. A glider leaves Sutton Bank airfield and flies 15 miles north-east. It then flies 22 miles east.
   (a) Make a scale drawing of its journey. Use a scale of 1 cm to 5 miles.
   (b) How far must it fly back to Sutton Bank, and on what bearing?

10. On the opposite sides of the entrance to a harbour are two radar beacons. The distance between them is 3.5 km and Beacon A is on a bearing of 110° from Beacon B. A ship is on a bearing of 320° from Beacon A and a bearing of 080° from Beacon B.
    Draw a map to a scale of 2 cm to 1 km, showing the two beacons and the ship.

- When an equation contains only x, y and number terms the graph will be a straight line. Work out three points to draw the line.
- When an equation contains an x^2 term but no higher power, it will give a parabola.

  Work out a table, plot the points and join with a smooth curve.

- The x-coordinate(s) of the point(s) where two graphs meet will be the solution to an equation. The equation can be found by writing the equations of both graphs in the form '$y = \ldots$' and putting one right-hand side equal to the other.
- The value of x^2 is always positive and the value of x^3 has the same sign as x.
- Bearings are measured clockwise from north and always have three figures.

Revision exercise 7a

1. On a piece of squared paper, draw a set of x- and y-axes, both labelled from -6 to $+6$.
   - (a) Plot and label the points A$(-3, 2)$, B$(0, -2)$, C$(2, -5)$, D$(5, 0)$.
   - (b) Draw the lines $x = 3$, $x = -5$, $y = -3$, $y = 4$.

2. Draw a set of axes, labelling the x-axis from -3 to $+3$ and the y-axis from -10 to $+12$.
   - (a) Draw the graph of $y = 3x$, for $x = -3$ to $+3$.
   - (b) Draw the graph of $y = -2x + 5$.
   - (c) Write down the coordinates of the point where the two lines cross.

3. (a) Draw the graph of $y = 2x - 4$, for $x = -2$ to $+5$.
   - (b) On the same grid, draw the graph of $5y + 2x = 10$.
   - (c) Write down the coordinates of the point where the two lines cross.

4. (a) Draw the graph of $y = x^2 - 6x + 3$, for $x = -1$ to $+7$.
   Label the x-axis from -1 to $+7$ and the y-axis from -10 to $+10$.
   - (b) Use the graph to find solutions to the equation $x^2 - 6x + 3 = 0$.

5. (a) Draw the graph of $y = -x^2 - x + 12$, for $x = -5$ to $+4$.
   Label the x-axis from -5 to $+4$ and the y-axis from -10 to 15.
   - (b) Use the graph to solve the equation $x^2 + x - 12 = 0$.

6. The height h, in metres, of a ball thrown upwards at 40 metres per second is given by the formula:
   $h = 40t - 5t^2$, where t is the time in seconds.
   - (a) Copy this table of values and complete it.

t	0	1	2	3	4	5	6	7	8
t^2									
$40t$									
$-5t^2$									
$h = 40t - 5t^2$									

   - (b) Draw the graph of $h = 40t - 5t^2$, for values of t from 0 to 8.
   - (c) Find the times when the ball is 70 metres above the ground. Give your answers correct to one decimal place.

7. Mick, the builder works out his charges C, in pounds, for tiling by using the formula $C = \dfrac{n}{4} + 20$, where n is the number of tiles.
   - (a) (i) Copy the table of values and complete it.

n	0	40	80	120	160	200
$C = \dfrac{n}{4} + 20$						

   - (ii) Draw the graph of $C = \dfrac{n}{4} + 20$, for values of n from 0 to 200.

   Dave, another builder, makes a basic charge of £25 and then 20 pence per tile.
   - (b) Draw another line on the graph to show Dave's charges for up to 200 tiles.

(c) Margaret wants to have her bathroom tiled and it will need 150 tiles.
Which builder is cheaper and how much does he charge?

8. This is a sketch of four trees in a field, with some angles marked.

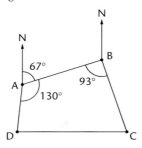

Work out the bearing of:
(a) A from B   (b) C from B   (c) A from D.

9. This is a sketch of three buoys in the sea.

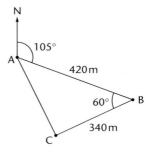

Make a scale drawing and find the distance and bearing of A from C.
Use a scale of 1 cm to 50 m.

10. Theresa is taking part in a yacht race. She sails 15 miles on a bearing of 140°, then 12 miles on a bearing of 025°. Use a scale of 1 cm to 2 miles to make a scale drawing of her route. Find how far she is from the start and on what bearing.

You will need to know:

- the meanings of line and rotational symmetry
- how to plot and read coordinates of a point
- how to reflect a simple shape in a vertical (up the page) line or horizontal (across the page) line
- how to find the interior angle of a regular polygon.

Transformations

Reflections

You may find tracing paper helpful with some of these transformations.

You have already looked at reflections in a vertical line, but you need to be able to reflect a shape in any mirror line.

When a shape is **reflected** it is 'turned over'. The reflection or image is exactly the same size and shape as the original, but in reverse. The shape and its image are **congruent**. For example the reflection of this ⊢ shape would look like this ⊣.

Corresponding points are the same distance from the mirror line but on the opposite side.

Reflections can be drawn on plain paper, but you are often asked to draw them on squared paper.

In this diagram the shape has been reflected in the line PQ.

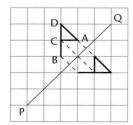

This means:

- the image of point A is half a square diagonally on the other side of the line PQ
- the image of B is half a square diagonally on the other side
- the image of C is a full square diagonally on the other side
- the image of D is one and a half squares diagonally on the other side.

Instead of counting squares you could trace the shape and turn the tracing paper over on the mirror line.

Example 1   Reflect this shape in the line AB.

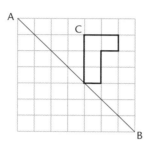 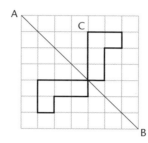

One point is on the line, so its image stays there. The images of the other points are each the same distance from the line as the original point, but on the opposite side. For example, point C and its image are both one and a half squares from the mirror line.

Rotations

When a shape is **rotated** it is turned relative to a fixed point, to a different angle. The shape is still the same way round, and the original shape and its image are congruent. For example

this shape  could turn to any of these positions:

To rotate a shape you need to know three things:

- the angle of rotation
- the direction of rotation
- the centre of rotation.

Remember that rotations are anticlockwise, unless you are told otherwise.

This chapter covers:

- quarter-turns (90° anticlockwise)
- half-turns (180°)
- three-quarter-turns (90° clockwise)

about the centre of the shape or the origin.

In this diagram the shape P has been rotated through a quarter-turn (90° anticlockwise) about point O.

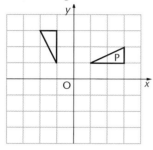

You can do this by counting squares, or by tracing the shape and turning it, or by using a pair of compasses with the point at O and drawing quarter circles from each point.

Examiner's tip

When you have drawn the rotation, turn the page through the correct angle to check it looks like the original.

Example 2

Rotate the shape through a through a three-quarter-turn about O.

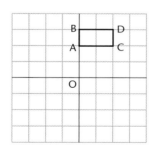 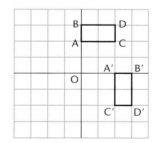

You can think of a three-quarter-turn (anticlockwise) as a quarter-turn clockwise.

You can draw the rotation by counting squares.

A is 2 squares above O, so its image A′ is 2 squares to the right.

B is 3 squares above O, so its image B′ is 3 squares to the right.

C is 2 squares above and 2 to the right of O, so its image C′ is 2 to the right and 2 below O.

D is 3 squares above and 2 to the right of O, so its image D′ is 3 to the right and 2 below O.

Again, you could do it by tracing.

Translations

When a shape is **translated** all its points move the same distance in the same direction. Again, the image and the original shape are congruent. For example this shape [⚑] could translate to this [⚑] .

To describe a translation, you need to say how far it moves across, left or right, and how far it moves up or down.

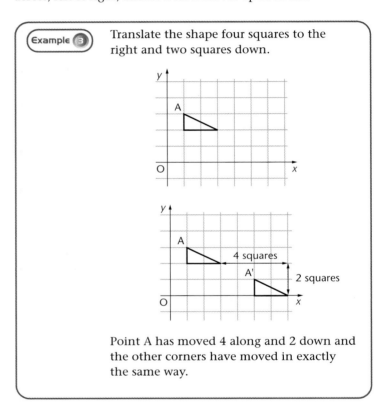

Example 3

Translate the shape four squares to the right and two squares down.

Point A has moved 4 along and 2 down and the other corners have moved in exactly the same way.

Describing transformations

As well as carrying out transformations you also need to be able to describe a transformation that has been done.

If it is a reflection you need to describe the mirror line.

If it is a translation you need to say how far the shape has moved, and in what direction.

Example ④ For each of these diagrams describe fully the transformation of T to T'.

(a)

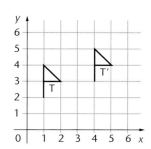

(b)

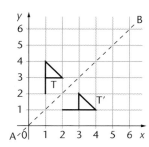

(a) The image is the same way round and the same way up as the original shape, so it is a translation.

It has moved 3 squares to the right and 1 up. | Check by counting squares.

This is a translation of three squares to the right and one up.

(b) The image is the opposite way round from the original shape, so it is a reflection.

To find the mirror line, draw lines between matching points in the original shape and the image.

Then draw a line through the midpoints of these joining lines.

Then label the line or give its equation.
This is a reflection in the line AB or $y = x$.

Creating patterns

Successive transformations can be made on a shape, either in a line or around a point, to create patterns.

Example ⑤ Reflect the triangle in the line $y = 0$ then in $y = -x$, $x = 0$, $y = x$ and so on to make the full pattern.

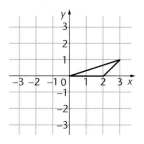

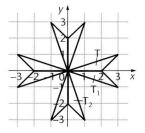

T_1 is the reflection of T in $y = 0$, T_2 is the reflection of T_1 in $y = -x$ and so on.

Use squared paper to answer these questions.

1. Reflect this shape in the line AB.

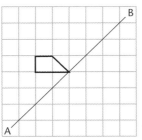

2. Rotate this shape through a half-turn about the origin.

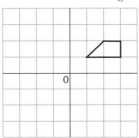

3. Translate this shape two squares to the left and one square down.

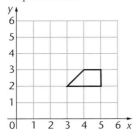

4. Reflect this shape in the line AB.

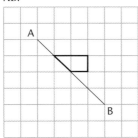

5. Rotate this shape through a three-quarter-turn anti-clockwise about the origin.

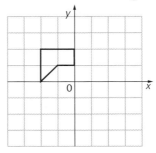

6. Translate this shape three squares to the right and two squares up.

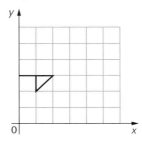

7. Rotate this shape through a quarter-turn anticlockwise about its centre A.

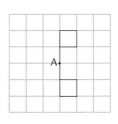

8. Reflect this shape in the line AB.

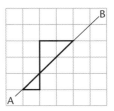

9. In each of these diagrams, describe fully the transformation that maps A onto B.

(a)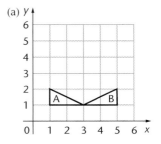

(b)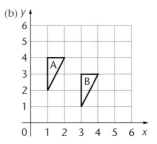

10. Reflect the shape in the line AB, then in a vertical line two squares to the right of AB, and so on to give a strip pattern. Show at least eight shapes.

Use squared paper to answer
these questions.

1. Reflect this shape in the line
   AB.

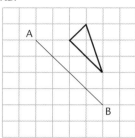

2. Rotate this shape through a
   quarter-turn anticlockwise
   about A.

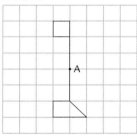

3. Translate this shape five
   squares to the right.

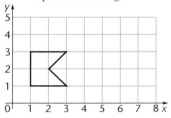

4. Reflect the shape in the line
   AB.

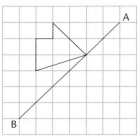

5. Describe the transformation
   that maps A onto B.

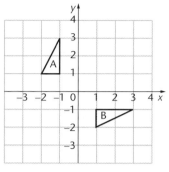

6. Rotate this shape through a
   quarter-turn anticlockwise
   about the origin.

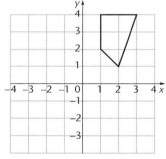

7. Translate this shape two
   squares to the left and four
   squares up.

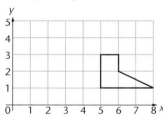

8. Reflect this shape in the line
   AB.

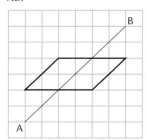

9. Draw a set of axes for x and
   y and label them both from
   -4 to $+4$.
   (a) Plot the points $(-2, 1)$,
       $(-1, 1)$, $(-1, 3)$, then
       join them up to make a
       triangle. Label it A.
   (b) Rotate A through a
       half-turn about the
       origin and label the
       image B.
   (c) Reflect A in the y-axis
       and label the image C.
   (d) Describe fully the
       transformation that
       maps B onto C.

10. Draw a set of axes for x and
    y and label them both from
    -4 to $+4$.
    (a) Plot the points $(0, 0)$,
        $(1, 3)$, $(1, 1)$, then join
        them to make a
        triangle.
    (b) Reflect the triangle in
        the line $y = x$.
    (c) Rotate the shape made
        by the two triangles by
        a quarter-turn
        anticlockwise about the
        origin.
    (d) Reflect all the shapes
        you now have in the
        x-axis to complete the
        pattern.

Chapter 8 Transformations

115

Enlargements

If you draw an enlargement of a shape, the image is the same shape as the original, but it is larger or smaller. You may just be asked to draw a '3 times enlargement' of a simple shape, without being given any other information. In this case copy the shape, making each side three times as long as the original one.

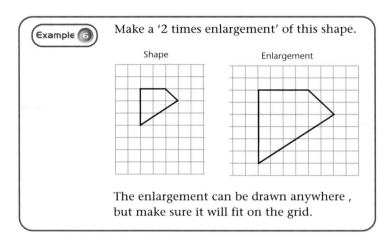

Example 6

Make a '2 times enlargement' of this shape.

Shape

Enlargement

The enlargement can be drawn anywhere , but make sure it will fit on the grid.

In the example, each side in the image is twice the length of the corresponding side in the original. The enlargement and the original are not congruent because they are different sizes, but they are **similar**.

More usually, you will be asked to draw an enlargement from a given point. In this case the enlargement has to be the correct size and also be in the correct position. The given point is the **centre of enlargement**.

For a '2 times enlargement', each point in the image must be twice as far from the centre as the corresponding point in the original is.

If the centre is on the original shape, that point will not move and the enlargement will contain the original shape.

The number used to multiply the lengths for the enlargement is called the **scale factor** and is sometimes called k.

You may be asked to describe an enlargement. To do this you must give the scale factor and the centre of enlargement.

Example 7

Enlarge this shape by scale factor 3 with the origin as centre.

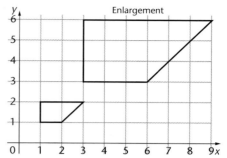

Enlargement

Scale factor 3 means this is a '3 times enlargement', so each point in the image is three times as far from the origin as the corresponding point in the original shape is.

The image of the point (1, 1) is at (3, 3). The image of the point (2, 1) is at (6, 3) and so on.

Symmetry in shapes

This shape has one line of symmetry (vertical). The two halves are reflections of one another.

This shape has rotational symmetry of order 2, but no line symmetry. This means if it is rotated through a half-turn about its centre its image will fit over the shape itself.

Example 8 Describe the symmetry of each of these shapes.

(a)   (b)

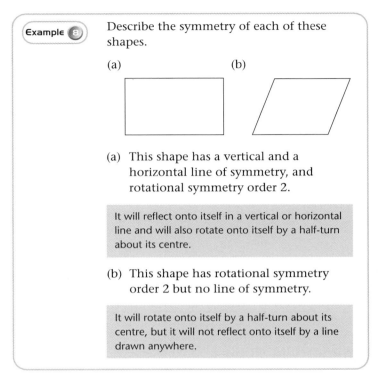

(a) This shape has a vertical and a horizontal line of symmetry, and rotational symmetry order 2.

> It will reflect onto itself in a vertical or horizontal line and will also rotate onto itself by a half-turn about its centre.

(b) This shape has rotational symmetry order 2 but no line of symmetry.

> It will rotate onto itself by a half-turn about its centre, but it will not reflect onto itself by a line drawn anywhere.

Tessellations

If a shape will tessellate, it will fill the full page without any gaps.

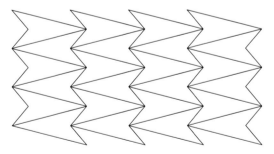

Squares and rectangles will obviously tessellate.

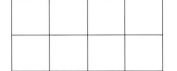 

Circles will not tessellate.

Many other shapes tessellate, but you may need to draw quite a lot of them to show whether they do.

Regular polygons tessellate if their corners meet at a point with no gaps. They will do this if their interior angles will divide exactly into 360°.

Example 9

Show that this shape will tessellate.

It is clear that this shape will tessellate as two fit together to make a rectangle.

Sometimes you will need to draw more shapes before it becomes clear.

Example 10

Without trying to draw the tessellation, show whether a regular pentagon will tessellate.

Find the interior angle of a pentagon.

Exterior angle = 360° ÷ 5 = 72°

Interior angle = 180° − 72° = 108°

Check if 108 divides into 360. It goes 3 times with a remainder of 36.

So a regular pentagon will not tessellate.

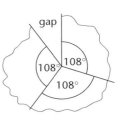

Use squared paper to answer these questions.

1. Draw a '4 times enlargement' of this shape.

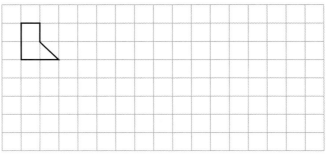

2. Draw a '3 times enlargement' of this shape.

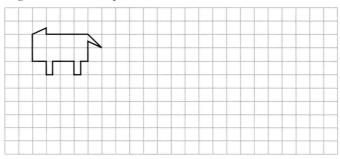

3. Enlarge this shape by scale factor 2, centre the origin.

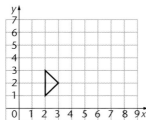

4. Enlarge this shape by a scale factor 3, centre the point A.
   Note that the enlargement includes the first triangle and A does not move.

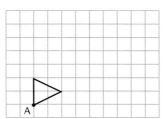

5. Enlarge this shape by scale factor 2, centre the point A.

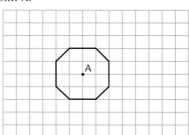

6. Enlarge this shape by scale factor 2, centre the point (2, 1).

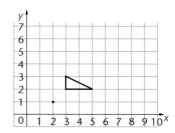

7. Describe the symmetry of these letters.

(a) **B**   (b) **S**

8. Show whether this shape will tessellate.

9. Does a regular hexagon tessellate? Do not try to draw the tessellation. Justify your answer.

10. Describe fully the transformation that will map A onto B.

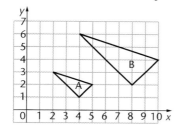

Exercise 8.2b

Use squared paper to answer these questions.

1. Draw a '3 times enlargement' of this shape.

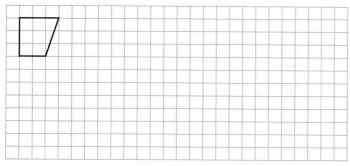

2. Draw a '2 times enlargement' of this shape.

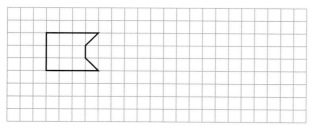

3. Enlarge this shape by scale factor 3, centre the origin.

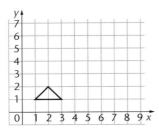

4. Enlarge this shape by scale factor 2, centre the point A.

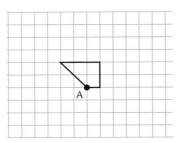

5. Enlarge this shape by scale factor 3, centre the point A.

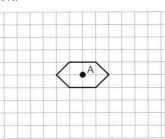

6. Enlarge this shape by scale factor 2, centre the point (2, 2).

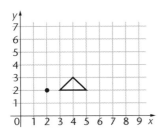

7. Describe the symmetry of these shapes.

(a) 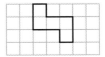   (b)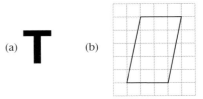

8. Show whether this shape will tessellate.

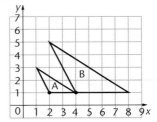

9. Describe fully the transformation that maps A onto B.

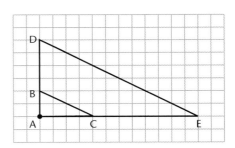

10. Describe fully the transformation that maps ABC onto ADE.

- When a shape is reflected, each point on the shape and the corresponding point on its image are the same distance from the mirror line, but on opposite sides.
- When describing a reflection, you need to define the mirror line.
- A translation moves every point on the shape the same distance, in the same direction.
- In a rotation, all the points move through the same angle about the same centre.
- When describing a rotation, you need to identify the centre, the angle and the direction (anticlockwise is positive).
- A translation is defined by movement to the left or right and up or down.

- When a shape is reflected, rotated or translated, it stays exactly the same shape. The shape and its image are congruent.
- When a shape is enlarged by scale factor k, from a given centre, each point on the enlargement will be k times as far from the centre as the corresponding point on the original shape is. The shape and its enlargement are similar.
- When describing the symmetry of a shape, both line symmetry and rotational symmetry need to be considered.
- A shape will tessellate if a number of copies of the shape will fit together without any gaps.

Revision exercise 8a

Use squared paper to answer these questions.

1. Reflect this shape in the line AB.

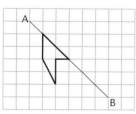

2. Translate this shape four units to the right and three units down.

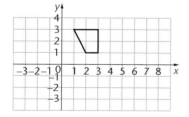

3. Rotate this shape through a three-quarter-turn anticlockwise about the origin.

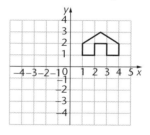

4. Enlarge this shape by scale factor 2, centre the origin.

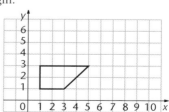

5. Describe the symmetry of these letters.

(a) **H**   (b) **R**

6. Prove that a regular octagon will not tessellate. Do not try to draw the tessellation.

7. Show whether this shape will tessellate. Draw sufficient shapes to make it clear.

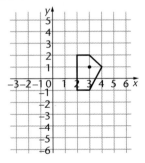

8. Enlarge this shape by scale factor 3, centre the point (3, 1).

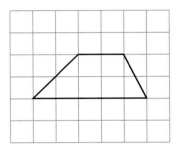

9. Describe the symmetry of the following shapes.

(a)   (b)   (c)

10. Describe fully the transformation that maps:

(a) A onto B   (e) F onto B
(b) B onto D   (f) A onto G
(c) B onto C   (g) G onto B.
(d) D onto E

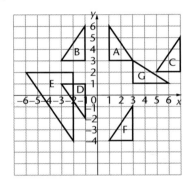

You will need to know:

the meaning of each of these words:

- digit
- factor
- multiple
- prime number
- consecutive

- square number
- square root
- cube
- cube root
- divisor

This is the first of several chapters in the book which are written especially to give you an opportunity to develop the skills and techniques you need in order to answer the extended questions in the timed tasks at GCSE.

Advice about these skills and techniques is given in the Introduction and you are recommended to look back at these pages and the hints given in the example.

Activity 1

Consecutive numbers

This investigation into consecutive numbers is given in two versions.

Version A is set as a structured task, version B is an unstructured task.

The solutions and comments for both versions are also given below.

Work through the solution and comments. Ask your teacher if you are not sure of anything.

Version A

It is claimed that it is possible to make any number by adding consecutive numbers.

For example:

$2 + 3 = 5$
$2 + 3 + 4 = 9$
$2 + 3 + 4 + 5 = 14$

1. Can all odd numbers be written as the sum of two consecutive numbers?

2. Can even numbers be written as the sum of two consecutive numbers?

3. Investigate any relationship between the number of consecutive numbers added together, and their sum.

4. Investigate any other patterns you notice.

Version B

$15 = 7 + 8 = 4 + 5 + 6$
$\quad = 1 + 2 + 3 + 4 + 5$

Investigate consecutive numbers.

Solution and comments for Version A

Comments in *italics* are taken from the table in the Introduction.

Examiner's tip

Start with the smallest odd number and try to match each odd number to an addition sum. Work systematically, write the results in order.

In the answers to these two questions you are showing that you can:
- *organise work, check results*
- *find out necessary information*
- *discuss work*
- *show you understand the task*
- *find examples which fit a statement*
- *search for a pattern, using at least three results*
- *explain your reasoning and make statements about the results you found.*

Question 1

    1
    3 = 1 + 2
    5 = 2 + 3
    7 = 3 + 4
    9 = 4 + 5
   11 = 5 + 6

This suggests that all odd numbers can be written as the sum of two consecutive numbers.

Test: $63 = 31 + 32$

Proof: If the first number is n then the consecutive number is $n + 1$. The sum is $n + (n + 1) = 2n + 1$. Doubling a number gives an even number, adding 1 then will always make the answer odd. Therefore adding two consecutive numbers will always give an odd answer.

Question 2

Even numbers are 2, 4, 6, 8, ...

Adding two consecutive numbers always gives an odd number (see the notes above), so it is not possible to make even numbers in this way.

Proof: As stated in question 1 above, $n + (n + 1)$ will always be odd.

Question 3

Experiment with three consecutive numbers:

$1 + 2 + 3 = 6$
$2 + 3 + 4 = 9$
$3 + 4 + 5 = 12$

Work systematically, write the results clearly. These results suggest the numbers go up in 3s, and are in the 3 times table.

With four consecutive numbers:

$1 + 2 + 3 + 4 = 10$
$2 + 3 + 4 + 5 = 14$
$3 + 4 + 5 + 6 = 18$

These results suggest the numbers go up in 4s.

So with five consecutive numbers will the total increase by 5 each time?

Check:
$1 + 2 + 3 + 4 + 5 = 15$
$2 + 3 + 4 + 5 + 6 = 20$
$3 + 4 + 5 + 6 + 7 = 25$

The prediction is correct and the numbers are in the 5 times table.

Question 4
(a) With three consecutive numbers, the sum is 3 times the middle number.
$3 + 4 + 5 = 12 = 3 \times 4$
$5 + 6 + 7 = 18 = 3 \times 6$

explaining your reasoning and making a statement about the results found

Proof: If the three numbers are $n, n + 1, n + 2$ then:
$n + n + 1 + n + 2 = 3n + 3$
$\qquad\qquad\qquad\quad = 3(n + 1)$
i.e. three times the middle number.
Thus the numbers are in the 3 times table, 3 is a factor of the totals.

beginning to justify solutions, explaining why the results occur

To find three consecutive numbers that make a given total, divide the total by 3, then add 1 and subtract 1 to find all three numbers.
Example
Which three consecutive numbers make 72?
$72 \div 3 = 24$, so the numbers are
$24 - 1 = 23$, 24 and $24 + 1 = 25$.

(b) For four consecutive numbers, the sum is twice the sum of the middle two numbers.
$1 + 2 + 3 + 4 = 10 = 2 \times (2 + 3)$
$3 + 4 + 5 + 6 = 18 = 2 \times (4 + 5)$

explaining your reasoning and making a statement about the results found

Proof:
$n + n + 1 + n + 2 + n + 3$
$\quad = n + (n + 1 + n + 2) + n + 3$
$\quad = n + (2n + 3) + n + 3$
$\quad = (2n + 3) + (2n + 3)$
$\quad = 2(2n + 3)$

beginning to justify solutions, explaining why the results occur

(c) For five consecutive numbers the totals will be in the 5 times table and will be 5 times the middle number.

Can you think of a reason for this?
Hint: Take n as the middle number, then write down the numbers on either side of n (i.e. $n - 1$ and $n + 1$).

(d) The following numbers cannot be given as the sum of consecutive numbers:
2, 4, 8, 16, …

> explaining your reasoning and making a statement about the results found
> Can you think of a reason why this happens?
> Think about the factors of, for example,
> 16: 1, 2, 4, 8, 16
> and of 20: 1, 2, 4, 5, 10, 20
> What do you notice?

Each number is double the previous number – they are powers of 2:

$$4 = 2 \times 2 = 2^2$$
$$8 = 2 \times 2 \times 2 = 2^3$$
$$16 = 2 \times 2 \times 2 \times 2 = 2^4$$

The next number will be 32 which is 2^5.

Comments for Version B

Because this task is set in an unstructured form you will need to think carefully and systematically about how you will work through it and show your results.

There are two possible approaches:
1. start with the totals, i.e. 1, 2, 3, 4, … and try to make them with sums of consecutive numbers
2. start with the actual sums, i.e. $1 + 2$, $1 + 2 + 3$, … and see what totals you can make.

Approach 1 is probably easier, and you are more likely to be systematic in your working than with approach 2. Approach 1 also allows you to see patterns more clearly, especially if you show the results in a table.

This table, for the numbers up to 30, is not complete.

Copy it and and complete it to show that you can see all the patterns.

1				
2				
3	1 + 2			
4				
5	2 + 3			
6		1 + 2 + 3		
7	3 + 4			
8				
9	4 + 5	2 + 3 + 4		
10			1 + 2 + 3 + 4	
11	5 + 6			
12		3 + 4 + 5		
13	6 + 7			
14			2 + 3 + 4 + 5	
15	7 + 8	4 + 5 + 6		1 + 2 + 3 + 4 + 5
16				
17	8 + 9			
18		5 + 6 + 7	3 + 4 + 5 + 6	
19				
20				2 + 3 + 4 + 5 + 6
21				
22			4 + 5 + 6 + 7	
23				
24				
25				3 + 4 + 5 + 6 + 7
26				
27				
28				
29				
30				

However a task is presented it is a good idea to try to extend it by asking some questions of your own.

Here are three possible extension questions to ask if you are attempting version B.

Work through them and/or any ideas or questions of your own.

Activity 2

Adding consecutive numbers

(a) What happens if you add two consecutive odd numbers i.e. $3 + 5$, $5 + 7$?

(b) What happens if you add three consecutive odd numbers?

(c) What happens if you add four consecutive odd numbers?

Look back at the table on the previous page. What do you notice about those numbers that can only be made by adding two consecutive numbers?

Activity 3

Differences between squares

What do you notice if:

(a) you write down two consecutive numbers, square both these numbers and find the difference between the squares

(b) you write down three consecutive numbers, square the middle one, multiply the first and the third numbers together and find the difference between the answers?

Can you prove either or both of the results?

Activity 4

An investigation into squares

(a) How many squares are there in each diagram?

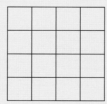

....

    A   B   C

(b) How many squares will there be in the sixth diagram? And how many in the tenth?

Activity 5

3s and 5s

Some multiples of 3 are:

$9 = 3 \times 3$ or $3(3)$

$12 = 4 \times 3$ or $4(3)$

$30 = 10(3)$

Similarly some multiples of 5 are:

$25 = 5(5)$

$40 = 8(5)$

$60 = 12(5)$

Some numbers are multiples both of 3 and 5:

$30 = 10(3)$ and $6(5)$

Some numbers can be made by adding a multiple of 3 to a multiple of 5:

$17 = 12 + 5 = 4(3) + 1(5)$

but some other numbers, such as 4, cannot be made at all.

Investigate 3s and 5s.

Activity 6

To express a number in prime factor form, first divide by 2 until you get an odd number then try to divide by 3 and then 5, and so on.

The divisors are always prime numbers.

Example:

$$\begin{array}{r} 2\overline{)36} \\ 2\overline{)18} \\ 3\overline{)\ 9} \\ 3\overline{)\ 3} \\ 1 \end{array}$$

So 36 written as a product of its prime factors is $2 \times 2 \times 3 \times 3$ which can be written as $2^2 \times 3^2$.

Copy and complete the following table up to $n = 36$, writing the prime factors in index form:

Number n	Prime factors in index form
2	2^1
3	3^1
4	2^2
5	5^1
6	$2^1 \times 3^1$
⋮	⋮
36	$2^2 \times 3^2$

The complete list of factors for 36 is:

1, 2, 3, 4, 6, 9, 12, 18, 36

This can be written as $F(36) = 9$ because there are nine factors.

But 36 is also equal to $2^2 \times 3^2$.

Adding 1 to each of the powers gives $(2 + 1)$ and $(2 + 1)$.

Multiplying these numbers gives: $(2 + 1) \times (2 + 1) = 3 \times 3 = 9$ which is the same value as $F(36)$.

(a) Investigate prime factors and the number of factors for numbers up to 36 .

You might want to add a column to the table you have just completed:

Number n	Prime factors in index form	Number of factors
2	2^1	2
3	3^1	2
4	2^2	3
5	5^1	
6	$2^1 \times 3^1$	
⋮	⋮	
36	$2^2 \times 3^2$	

Check to see if there is a link between the indices or powers of the prime factors and the number of factors.

(b) What do you notice about numbers with only two factors?

(c) Which numbers have an odd number of factors?

The jail problem

- A high security jail has 100 prisoners.
- There is only one prisoner in each cell.
- The cells are numbered 1, 2, 3, ..., 99, 100.
- Each cell has a guard.
- The guards share the duties but one night, as they go off duty, they make a mistake:

Guard 1 goes along the cells and unlocks every one.

Guard 2 then locks the cells which are multiples of 2 i.e. cells 2, 4, 6, ...

Guard 3 then visits all the cells that are multiples of 3, i.e. 3, 6, 9, ... and unlocks any doors that are locked, and locks any door that is unlocked.

All the guards repeat this process as they go off duty, one by one.

After the last guard has left, which prisoners can escape? Explain why.

Number patterns

Look at this number pattern:

$49 \rightarrow 9^2 - 4^2 = 81 - 16 = 65$

$65 \rightarrow 6^2 - 5^2 = 36 - 25 = 11$

$11 \rightarrow 1^2 - 1^2 = 1 - 1 = 0$

Investigate two-digit numbers. What happens if you add the squares of the digits rather than finding the difference between them?

Gnomons

Here is part of a multiplication square.

The reversed L-shape is known as a gnomon. The numbers in each gnomon in the table add up to a cube number.

For example, 4^3 is equal to $4 + 8 + 12 + 16 + 12 + 8 + 4 = 64$

$1^3 \rightarrow$	1	2	3	4	5
$2^3 \rightarrow$	2	4	6	8	10
$3^3 \rightarrow$	3	6	9	12	15
$4^3 \rightarrow$	4	8	12	16	20
$5^3 \rightarrow$	5	10	15	20	25

(a) Now work out the values for 3^3 and 5^3.

Note:

$4 + 8 + 12 + 16 + 12 + 8 + 4$ can be written as

$4 \times 1 + 4 \times 2 + 4 \times 3 + 4 \times 4 + 4 \times 3 + 4 \times 2 + 4 \times 1$

and as 4 is clearly a common factor, this expression could be written as

$4(1 + 2 + 3 + 4 + 3 + 2 + 1)$.

(b) Try to write down the equivalent expressions for 3^3 and 5^3.

(c) Check that this method works for 10^3.
Also note that :
$1^3 = 1^2 = 1$
$1^3 + 2^3 = (1 + 2)^2 = 9$
$1^3 + 2^3 + 3^3 = (1 + 2 + 3)^2 = 36$

(d) Work out as simply as possibly
$1^3 + 2^3 + 3^3 + 4^3 + ... + 10^3$

Finally, some work on roots:

Whole numbers such as 1, 4, 9, 16, 25, … are **square numbers**, because, for example, 16 = 4 × 4 or '4 squared'.

The square root of 16, written as $\sqrt{16}$ = 4. Similarly $\sqrt{25}$ = 5 and $\sqrt{9}$ = 3.

You should know the square roots of the square numbers from 1 to 100.

The numbers 1, 8, 27, 64, … are called **cube numbers**, because they can be written as the cube of a whole number, for example 27 = 3 × 3 × 3 or '3 cubed', written as 3^3.

Because 64 = 4 × 4 × 4 or 4^3 the cube root of 64, $\sqrt[3]{64}$ = 4 and the cube root of 8, $\sqrt[3]{8}$ = 2.

Activity 10

Square and cube roots

Write down the values of:

1. (a) $\sqrt{81}$ (b) $\sqrt[3]{125}$ (c) $\sqrt{289}$ (d) $\sqrt[3]{216}$
2. (a) $\sqrt{(3^2 + 4^2)}$ (b) $\sqrt{(169 - 144)}$ (c) $\sqrt{(5^2 + 12^2)}$

Key points

- Prime numbers have no factors other than themselves and 1.

  Examples of prime numbers are 2, 3, 5, 7, 11, 13, 17, 19, 23.

- Factors are numbers that will divide into other numbers, so 3 is a factor of 9, but not a factor of 8.

- A multiple of a number will be in that number's 'times table', so 9 is a multiple of 3, 35 is a multiple of 5 and also of 7.

Revision exercise 9a

1. Write down the answers to these calculations.
   (a) 7^2
   (b) 11^2
   (c) $5^2 + 6^2$
   (d) $1^2 + 2^2 + 3^2$
   (e) $\sqrt{144}$
   (f) $\sqrt{169}$
   (g) $\sqrt{(5^2 - 4^2)}$
   (h) $\sqrt[3]{27}$
2. Express the following as products of prime factors.
   (a) 24   (b) 30   (c) 45   (d) 60   (e) 64

3. Investigate the last digits in the squares of the whole numbers from 1 to 20.
   Using these results, decide which of these numbers are definitely not square numbers.
   17 178   17 170   17 179   17 172   17 177

1. Four squares

$4^2 - 3^2 + 2^2 - 1^2 = 10$
$5^2 - 4^2 + 3^2 - 2^2 = 14$
$6^2 - 5^2 + 4^2 - 3^2 = 18$

Can you continue this pattern?

Does it always work?

What is $14\,263^2 - 14\,262^2 + 14\,261^2 - 14\,260^2$?

2. Co-primes

Two numbers are co-prime if the only integer which goes into, or is a factor of, or is a divisor of both of them is 1.

For example:

- 3 and 7 are co-prime because the only factor they have in common is 1
- 4 and 6 are not co-prime because they share a common factor of 2
- 14 and 21 are not co-prime because they share a common factor of 7
- 5 and 23 are co-prime because they have no common factor except 1.

1. Try to find four pairs of numbers that are co-prime and four pairs of numbers that are not co-prime.

2. Write down all the positive integers less than 10:
   1 2 3 4 5 6 7 8 9

   Of these,
   2, 4, 6 and 8 share a common factor of 2 with 10,
   and 5 divides into 10 so is also a factor.

   Therefore there are four numbers:
   1   3   7   9

   that are less than 10 and also co-prime with 10.

3. Copy this table and complete it for the integers up to 24.

Integer	Integers less than and co-prime with it	Number of these integers
2	1	1
3	1, 2	2
4	1, 3	2
5	1, 2, 3, 4	4
6	1, 5	2
7	1, 2, 3, 4, 5, 6	6
⋮		⋮
24	1, 5, 7, 11, 13, 17, 19, 23	8

What do you notice about the numbers in the right-hand column?

Denote the number of integers which are less than n and co-prime with n by $\Psi(n)$.

So $\Psi(10) = 4$

- Does $\Psi(3) \times \Psi(4) = \Psi(12)$?
- Does $\Psi(2) \times \Psi(6) = \Psi(12)$?
- Investigate $\Psi(m) \times \Psi(n) = \Psi(mn)$.

3. Grids

Here are two similar tasks.

The diagram shows a 3 × 5 rectangular grid.

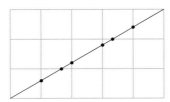

The diagonal of the grid passes through six grid lines.

Each point where the line crosses a grid line is marked with a dot.

- Investigate rectangular grids. Try to find a relationship between the size of the grid and the number of points where the diagonal crosses the grid lines.

The line also passes through seven squares, as shown.

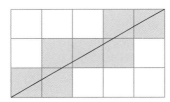

- Investigate rectangular grids. Try to find a relationship between the size of the grid and the number of squares through which the diagonal passes.

4. Perfect numbers

What numbers divide into 6 exactly?

The factors of 6 are 1, 2, 3, 6

and $1 + 2 + 3 = 6$

6 is called a **perfect number** because it is **equal to** the sum of its factors, not including itself.

- What is the next perfect number?
- See how many more perfect numbers you can find.
- Can you see a pattern or a rule?

An **abundant number** is **less than** the sum of its factors, excluding itself. For example:

the factors of 24 are 1, 2, 3, 4, 6, 8, 12, 24

and $1 + 2 + 3 + 4 + 6 + 8 + 12$

$= 36$

and $24 < 36$

so 24 is an abundant number.

A **deficient number** is **more than** the sum of its factors, excluding itself. For example:

the factors of 22 are 1, 2, 11, 22

and $1 + 2 + 11 = 14$

but $22 > 14$

so 22 is a deficient number.

- Investigate abundant and deficient numbers.

You will need to know:

- the terms perimeter, area, volume, radius, diameter and circumference
- the units for these quantities e.g. cm, cm^2 and cm^3 or m, m^2 and m^3
- how to find the area of a rectangle from the lengths of its sides
- how to find the volume of a cuboid by counting cubes
- how to draw triangles, knowing three sides or two sides and the included angle.

Area of a triangle

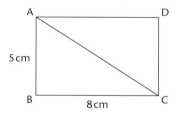

The area of the rectangle ABCD is $8 \times 5 = 40\,cm^2$.

The area of the triangle ABC is half the area of the rectangle ABC and so is equal to $20\,cm^2$.

So the area of the right-angled triangle $= \frac{1}{2} \times 8 \times 5 = 20\,cm^2$.

Triangle PQR is not right-angled but has been split into two right-angled triangles.

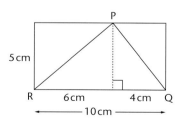

Each right-angled triangle is half a rectangle.

The area of the left-hand triangle is $\frac{1}{2} \times 6 \times 5 = 15\,cm^2$.

The area of the right-hand triangle is $\frac{1}{2} \times 4 \times 5 = 10\,cm^2$.

Therefore, the total area of the triangle is $10 + 15 = 25\,cm^2$.

Notice that the area of triangle PQR is half the area of the large rectangle.

So the area of the triangle is $\frac{1}{2} \times 10 \times 5 = 25\,\text{cm}^2$.

This shows that whether the triangle is right-angled or not the area can be found by the formula:

> area of triangle $= \frac{1}{2} \times$ base $\times$ perpendicular height
> or $A = \frac{1}{2} \times b \times h$

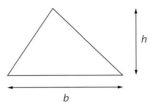

Try to explain why the formula still works for an obtuse-angled triangle.

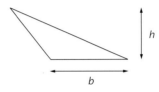

Examiner's tip

Always use the perpendicular height of the triangle, **never** the slant height. So in the triangle on the right, area $= \frac{1}{2} \times 6 \times 3 = 9\,\text{cm}^2$

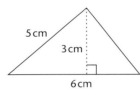

Remember that the units of area are always square units such as square centimetres or square metres, written cm^2 or m^2.

When using the formula, you can use any of the sides of the triangle as the base provided you use the perpendicular height that goes with it.

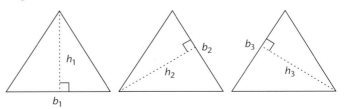

Area $A = \frac{1}{2} \times b_1 \times h_1$ Area $A = \frac{1}{2} \times b_2 \times h_2$ Area $A = \frac{1}{2} \times b_3 \times h_3$

1. Find the areas of these triangles.

(a)
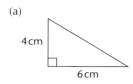
4 cm
6 cm

(b)
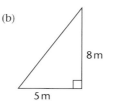
8 m
5 m

(c)
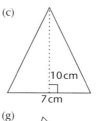
10 cm
7 cm

(d)
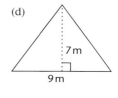
7 m
9 m

(e)
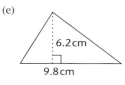
6.2 cm
9.8 cm

(f)
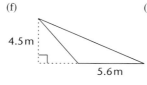
4.5 m
5.6 m

(g)
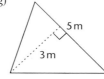
5 m
3 m

(h)
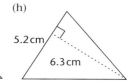
5.2 cm
6.3 cm

(i)
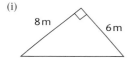
8 m
6 m

2. The vertices of a triangle are at A(2, 1), B(5, 1) and C(5, 7). Draw triangle ABC on squared paper and find its area.

3. The vertices of a triangle are at P(2, 2), Q(7, 2) and R(4, 6). Draw triangle PQR on squared paper and find its area.

4. The vertices of a triangle are at U(2, 6), V(9, 6) and W(5, 1). Draw triangle UVW on squared paper and find its area.

5. Using a ruler and compasses, draw an equilateral triangle with sides 5 cm. Measure its perpendicular height and find its area.

6. In triangle ABC, AB = 6 cm, BC = 8 cm and angle ABC = 40°. Draw the triangle accurately and find its area.

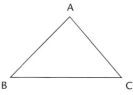

A

B   C

1. Find the areas of these triangles.

(a)

10 m
8 m

(b)
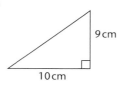
9 cm
10 cm

(c)
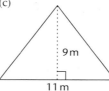
9 m
11 m

(d)
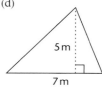
5 m
7 m

Chapter 10 Area of a triangle

Exercise 10.1b continued

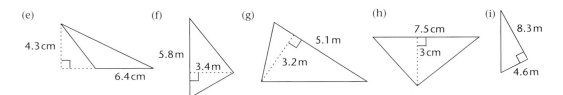

(e) 4.3 cm 6.4 cm

(f) 5.8 m 3.4 m

(g) 5.1 m 3.2 m

(h) 7.5 cm 3 cm

(i) 8.3 m 4.6 m

2. The vertices of a triangle are at A(2, 1), B(2, 7) and C(5, 3). Draw triangle ABC on squared paper and find its area.

3. The vertices of a triangle are at P(−2, 2), Q(3, 2) and R(5, 6). Draw triangle PQR on squared paper and find its area.

4. The vertices of a triangle are at U(−2, −1), V(−2, 4) and W(3, 6). Draw triangle UVW on squared paper and find its area.

5. In triangle ABC, AB = 7 cm, BC = 10 cm and CA = 6 cm. Using ruler and compasses, draw the triangle accurately and find its area.

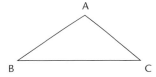

6. A triangle has an area of 12 cm² and a base of 4 cm. Find the perpendicular height associated with this base.

7. In triangle ABC, AB = 6 cm, BC = 8 cm and AC = 10 cm. Angle ABC = 90°.
   (a) Find the area of the triangle.
   (b) Find the perpendicular height BD.

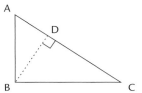

Circumference of a circle

Activity 1

Using a tape measure, or a piece of string and a ruler, measure the circumference and diameter of eight circular objects. Make sure you use the same units for both measurements.

Copy this table and complete it. For the fourth column use your calculator to work out the value of:

circumference C ÷ diameter d

Name of object	Circumference	Diameter	C ÷ d

The numbers in the fourth column of the table should all be about 3.

The approximate relationship $C \approx 3 \times d$ has been known for thousands of years. In fact, very accurate calculations have shown that instead of 3 the multiplier should be approximately 3.142. Even more accurate calculations have found this number to hundreds of decimal places. Because the number is a never-ending decimal, it is denoted by the Greek letter π.

Your calculator has the number which π represents stored in its memory.

Find the π button on your calculator and write down all the digits it shows. On some calculators you may have to press $=$ after pressing the π button.

The relationship between circumference and diameter can now be written as:

$C = \pi \times d$

> Remember, when two letters are written next to each other in algebra, it means multiply. The formula can be written shortly as $C = \pi d$.

If you know the radius of the circle instead of the diameter, use the fact that the diameter is double the radius.

$d = 2r$

The formula then becomes:

$C = \pi \times 2r$

Since multiplication can be done in any order, this can be written as $C = 2 \times \pi \times r$ or, using the shorter version:

$C = 2\pi r$

Examiner's tip

Can you think of a reason why $C \div d$ should be constant? It is because all circles are similar (the same shape), so the ratio of corresponding lengths will be constant. Try it with the sides of a rectangle 2 units $\times$ 1 unit, and similar rectangles with sides of 4 units $\times$ 2 units, 6 units $\times$ 3 units, and so on.

 Example 1   A circle has a diameter of 5 cm. Find its circumference.

Circumference $= \pi d = \pi \times 5 = 15.707\,96... = 15.7$ cm (to 1 d.p.)

 Example 2   A circle has a radius of 8 cm. Find its circumference.

Circumference $= 2\pi r = 2 \times \pi \times 8 = 50.265.... = 50.3$ cm (to 1 d.p.)

 Example 3   A circle has a circumference of 20 m. Find its diameter.

Circumference $= \pi d$ so $20 = \pi \times d$.

Solving the equation gives: $d = 20 \div \pi = 6.366... = 6.37$ m (to 2 d.p.)

Exercise 10.2a

1. Calculate the circumference of the circles with these diameters, giving your answers correct to one decimal place.
   (a) 12 cm   (b) 9 cm   (c) 20 m   (d) 16.3 cm   (e) 15.2 m
2. Find the circumferences of circles with these radii, giving your answers correct to one decimal place.
   (a) 5 cm   (b) 7 cm   (c) 16 m   (d) 18.1 m   (e) 5.3 m
3. The centre circle on a football pitch has a radius of 9.15 metres. Calculate the circumference of the circle.
4. The diagram shows a wastepaper bin in the shape of a cylinder. Anthea is going to decorate it with braid around the top rim. Calculate the length of braid she needs.

30 cm

5. The circumference of a circle is 50 cm. Calculate the diameter of the circle.

Exercise 10.2b

1. Calculate the circumferences of circles with these diameters, giving your answers correct to one decimal place.
   (a) 25 m   (b) 0.3 cm   (c) 17 m
   (d) 5.07 m   (e) 6.5 cm
2. Find the circumferences of circles with these radii, giving your answers correct to one decimal place.
   (a) 28 cm   (b) 3.2 cm   (c) 60 m
   (d) 1.9 m   (e) 73 cm
3. The radius of the earth at the equator is 6378 km. Calculate the circumference of the earth at the equator.
4. Bijan and Claire have a ride on the roundabout at a fair.
   Bijan rides on a motorcycle which is 2.5 metres from the centre of the roundabout. Claire rides on a horse which is 3 metres from the centre.

On each ride the roundabout goes round ten times.
How much further does Claire travel than Bijan?

5. A circular racetrack is 300 metres in circumference. Calculate the diameter of the racetrack.

Area of a circle

The diagram shows a circle, radius r, divided into 36 sectors of centre angle 10°.

The next diagram shows these sectors cut out and rearranged.

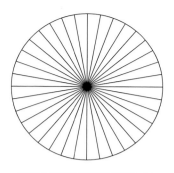

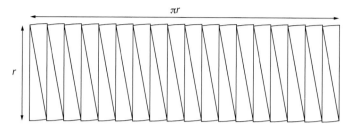

This shape is now very close to a rectangle.

The length of this rectangle will be approximately half the circumference. That is $2\pi r \div 2 = \pi r$.

The width of the rectangle is r.

The area of this rectangle = length × width = $\pi r \times r = \pi r^2$.

If you take more sectors, with smaller angles, the shape is even closer to a rectangle.

The formula for the area of a circle is therefore:

area of circle $A = \pi r^2$

Examiner's tip

One of the most common errors is to mix up diameter and radius. Every time you do a calculation make sure you have used the right one.

Examiner's tip

Make sure you can use the π button and the square button on your calculator so you do not have to write down long decimals before the final answer.

 Example 4 The radius of a circle is 6 cm. Calculate the area of the circle.

$A = \pi r^2 = \pi \times 6^2 = \pi \times 36 = 113.1 \, \text{cm}^2$ (to 1 d.p.)

 Example 5 The radius of a circle is 4.3 m. Calculate the area of the circle.

$A = \pi r^2 = \pi \times 4.3^2 = 58.1 \, \text{m}^2$ (to 1 d.p.)

 Example 6 The diameter of a circle is 18.4 cm. Calculate the area of the circle.

$r = 18.4 \div 2 = 9.2 \, \text{cm}$

$A = \pi r^2 = \pi \times 9.2^2 = 266 \, \text{m}^2$ (to the nearest m²)

 Example 7 The radius of a circle is 5 m. Calculate the area of the circle, leaving your answer as a multiple of π.

$A = \pi r^2 = \pi \times 5^2 = \pi \times 25 = 25\pi \, \text{m}^2$

Chapter 10 Area of a circle

Exercise 10.3a

In all of these questions, make sure you state the units of your answer.

1. Find the areas of circles with these radii.
   (a) 4 cm   (b) 16 m   (c) 11.3 m   (d) 13.6 m   (e) 8.9 cm
2. The radius of a circular fish pond is 1.5 m. Find the area of the surface of the water.
3. The diameter of a circular table is 0.8 m. Find the area of the table.
4. To make a table mat, a circle of radius 12 cm is cut from a square of side 24 cm as shown in the diagram on the right. Calculate the area of material that is wasted.
5. The radius of the circular face of a church clock is 1.2 m. Calculate the area of the clock face.
6. Use your calculator to find the area of a circle with radius 6.8 cm. Without using your calculator do an approximate calculation to check your answer.

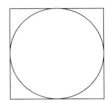

Exercise 10.3b

In all of these questions, make sure you state the units of your answer.

1. Find the areas of these circles.

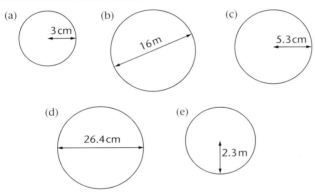

2. A circular mouse mat has a radius of 9 cm. Find the area of the mouse mat.
3. According to the *Guinness Book of Records*, the largest pizza ever made was 37.4 m in diameter. What was the area of the pizza?
4. A square has a side of 3.5 cm and a circle has a radius of 2 cm. Which has the bigger area? You must show your calculations.

5. Charlie is making a circular lawn with a radius of 15 m. The packets of grass seed say 'sufficient to cover 50 m²'. How many packets will she need?
6. The diagram shows a child's plastic ring. The diameter of the large circle is 30 cm. The diameter of the small circle is 20 cm. Calculate the area of the ring (shaded).

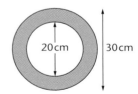

Examiner's tip

In the last two sections all the questions in the exercise have been about circumference, or all of them have been about area, so there has been no problem about which formula to use. In an examination you will have to choose the formula.

If you have trouble remembering which of the circle formulae gives circumference and which gives area, remember r^2 must give the units cm² or m² and so πr^2 must be the area formula.

Volume of a cuboid

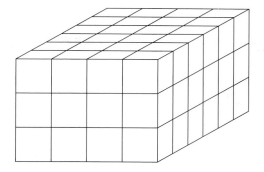

The diagram shows a cuboid made up of centimetre cubes.

On the top layer there are six rows of four cubes, so there are $6 \times 4 = 24$ centimetre cubes in the top layer.

There are three layers. So in all there are $3 \times 24 = 72$ cubes.

This means that the volume of the cuboid
= 72 cubic centimetres or 72 cm³.

The calculation for this was $6 \times 4 \times 3 = 72$, so the formula for the volume of a cuboid is:

Volume of a cuboid = V = length × width × height
or $V = lwh$

 Example   Calculate the volume of a cuboid with length 8.5 cm, width 6.4 cm and height 3.6 cm.

Volume = $8.5 \times 6.4 \times 3.6 = 195.84$ cm³

 Example   For the cuboid in the diagram, calculate:
(a) the volume   (b) the total surface area.

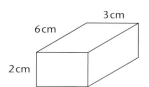

(a) Volume = $6 \times 3 \times 2 = 36$ cm³

(b) Area of top = $6 \times 3 = 18$ cm²
Area of side = $6 \times 2 = 12$ cm²
Area of end = $2 \times 3 = 6$ cm²
Total surface area of top + bottom + 2 sides + 2 ends
= $2 \times 18 + 2 \times 12 + 2 \times 6$
= $36 + 24 + 12$
= 72 cm²

Example 10

A concrete path is laid which is 20 m long, 1.5 m wide and 10 cm thick. Calculate the volume of concrete used.

Thickness $= 10 \div 100 = 0.1$ m

Volume $= 20 \times 1.5 \times 0.1 = 3$ m³

Exercise 10.4a

In all of these questions, make sure you state the units of your answer.

1. A classroom is in the shape of a cuboid, it has length 6.8 m, width 5.3 m and height 2.8 m. Calculate the volume of the room.

2. Measure the length, width and thickness of this text book. What is the volume of the text book?

3. For the cuboid shown, calculate:
   (a) the volume
   (b) the total surface area.

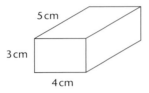

4. A rectangular fish pond is 2.5 m long and 1.4 m wide. The depth of the water is 40 cm. Calculate the volume of water in cm³. How many litres is this? (1 litre = 1000 cm³)

5. The diagram shows the net of a shoe box. Calculate the volume of the shoe box.

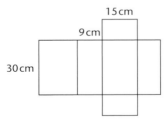

6. The kitchen cupboards in the picture have a total length of 2.6 m. The depth (front to back) of the cupboards is 0.6 m and the height is 0.85 m.
   (a) Calculate the volume of the cupboards.

   (b) A worktop rests on top of the cupboards. It has the same length and depth as the cupboards and is 4 cm thick. Calculate the volume of the worktop.

7. The volume of a cuboid is 72 m³. If the length is 6 m and the width is 4 m, what is the height?

8. A pane of glass is 120 cm wide and 80 cm high. If the thickness of the pane is 3 mm, what is the volume of glass?

In all of these questions, make sure you state the units of your answer.

1. Find the volume of a cube of side 20 cm.

2. A midi stereo system is 35 cm wide, 27 cm in depth (front to back) and 39 cm high. Assuming the system to be a cuboid, find its volume.

3. A freezer is 0.6 m wide, 0.6 m deep and 1.4 m high.
   (a) Assuming the freezer to be a cuboid, find its volume.
   (b) About 35% of the volume can be used for storing food. Calculate the volume of food that can be stored.

4. Measure the length, width and height of your bedroom. (If it is not approximately a cuboid choose another room in your home.) Calculate the volume of your bedroom.

5. Alan is laying concrete to make a garden path 18 metres long and 0.8 metres wide. If the concrete is to be 10 cm deep, how many cubic metres of concrete should he order?

6. For the cuboid shown, calculate:
   (a) the volume of the cuboid
   (b) the total surface area of the cuboid.

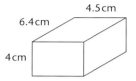

7. (a) A fish tank is 80 cm long, 45 cm wide and 40 cm high. Ashraf filled it with water to a depth of 35 cm. Find the volume of water he used.
   (b) Ashraf used a 2-litre jug to fill the tank with water. How many times did he fill the jug? (1 litre = 1000 cm^3)

8. (i)

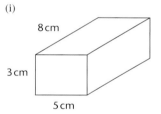

   (ii)

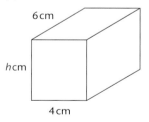

The two cuboids have the same volume.
(a) Find h.
(b) Which cuboid has the smaller surface area? Show your calculations.

Key points

- The area of a triangle
  $= \frac{1}{2} \times$ base $\times$ height $= \frac{1}{2} bh$.
- The circumference of a circle $= \pi d = 2\pi r$.
- The area of a circle $= \pi r^2$.
- The volume of a cuboid
  $=$ length $\times$ width $\times$ height $= lwh$.

- The units of area are square units such as square metres or square centimetres, written as m² and cm².
- The units of volume are cubic units such as cubic metres or cubic centimetres, written as m³ and cm³.

Revision exercise 10a

In all of these questions make sure you state the units of your answer.

1. A triangle has vertices at A(6, 2), B(6, 7) and C(3, 5). Draw triangle ABC on squared paper and calculate its area.

2. A pentagon has vertices at A(2, 1), B(6, 1), C(6, 3), D(5, 6) and E(2, 3). Draw the pentagon on squared paper and calculate the area of:
   (a) the rectangle ABCE
   (b) the triangle EDC
   (c) the pentagon ABCDE.

3. The picture shows a public footpath sign. By splitting the shape into a rectangle and triangle, find the area of the sign.

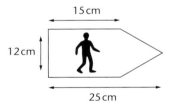

4. Calculate the area of a circle with a radius of 19.5 cm. Without using your calculator, make an approximate calculation to check your answer.

5. The minute hand of a church clock has a length of 1.2 m. Calculate the distance the end furthest away from the centre moves in 1 hour.

6. The diagram shows a shape made up of a rectangle and a semicircle.

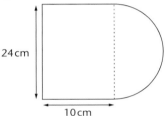

Calculate:
   (a) the total area of this shape
   (b) the perimeter of the shape.

7. The diagram shows a baked bean tin. The radius of the tin is 4 cm and the height is 11 cm. The label goes all the way round the tin, with a 2 cm overlap. The label is the same height as the tin. Calculate the length of the label and hence its area.

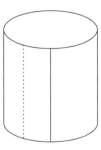

8. Lids for the baked bean tins in question 7 are stamped out of sheets of metal 40 cm by 32 cm, as shown in the diagram. Calculate the area of metal wasted.

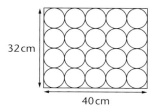

32 cm

40 cm

9. A netball has a circumference of 70 cm.
   (a) Calculate the diameter of the ball.
   A netball ring has a diameter of 38 cm.
   (b) How much space is there either side of the ball when it goes through the centre of the ring?

10. A room measures 4.2 m long by 3.2 m wide by 2.6 m high.
    (a) Calculate the volume of the room.
    (b) Calculate the area of the four walls and the ceiling.
    Jane is going to paint the walls and ceiling with emulsion paint.
    1 litre of paint covers 13 m². The total area of the windows and door is 5 m².
    (c) How much paint will she need?

Probability of an outcome not happening

When a fair die is thrown, this is called an **event**. In this example, there are six equally likely scores or **outcomes**: 1, 2, 3, 4, 5 and 6.

So the probability of throwing a six is $\frac{1}{6}$.

The probability of *not* throwing a 6 is the same as the probability of throwing 1 *or* 2 *or* 3 *or* 4 *or* 5, which is $\frac{5}{6}$.

Now $\frac{5}{6} = 1 - \frac{1}{6}$, so this gives the method of finding the probability that an outcome does not occur.

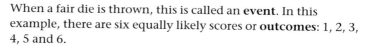

If the probability of an outcome happening is p, the probability of the outcome **not** happening $= 1 - p$.

Instead of writing 'the probability that an outcome happens', you can use the shorter P(outcome).

So in the above example: P(getting a six) $= \frac{1}{6}$

P(not getting a six) $= \frac{5}{6}$

Example 1

The probability that Nayim's school bus is late is $\frac{1}{8}$. What is the probability that Nayim's bus is not late?

Probability that Nayim's bus is not late
$= 1 -$ (probability that Nayim's bus is late)
$= 1 - \frac{1}{8}$
$= \frac{7}{8}$

Never write probabilities as '1 in 4', '1 to 4', '1 out of 4' or '1 : 4'. You will lose marks if you do.
Probabilities should be written as fractions or decimals, for example $\frac{1}{4}$ or 0.25. Fractions are usually preferable as they are exact. If, however, the probabilities in the question are given as decimals, then it is usually better to use decimals.
Percentages are acceptable but, since they usually involve extra work, there is little point in using them.

Example 2

The probability that it will rain tomorrow is 0.3. What is the probability that it will not rain tomorrow?

Probability that it will not rain tomorrow
$= 1 -$ (probability that it will rain tomorrow)
$= 1 - 0.3$
$= 0.7$

Exercise 11.1a

1. The probability of United winning their next game is 0.8. What is the probability of United not winning their next game?

2. The probability that I will have cereal for breakfast is $\frac{2}{7}$. What is the probability that I will not have cereal for breakfast?

3. The picture shows a fair spinner.

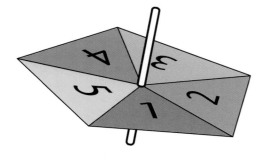

   (a) What is the probability of getting an even number with one spin?
   (b) What is the probability of getting an odd number with one spin?

4. There are 13 diamonds in a pack of 52 playing cards. I choose a card at random.
   (a) What is the probability that I choose a diamond?
   (b) What is the probability that I do not choose a diamond?

5. In a multiple-choice test paper there are five possible answers to each question, only one of which is right. If Obaid does not know an answer to a question, he guesses. Obaid guessed the answer to question 15.
   (a) What is the probability that Obaid got question 15 right?
   (b) What is the probability that Obaid got question 15 wrong?

6. The probability that the first ball drawn in the lottery draw is white is $\frac{9}{49}$. What is the probability that it is not white?

7. The probability that Peter goes out on a Friday night is 0.995. What is the probability that Peter stays in on a Friday night?

Exercise II.1b

1. The probability that Elma will pass her next Maths examination is 0.85. What is the probability that she will fail the examination?
2. Based on past experience, the probability that Kevin's school bus will be late tomorrow is 0.23. What is the probability that it is on time or early?
3. The probability that Jane will do the washing up tonight is $\frac{3}{8}$. What is the probability that she will not do the washing up?
4. The probability that Barry will get a motorcycle for his sixteenth birthday is 0.001. What is the probability that he will not get a motorcycle?
5. The probability that Zebedee will go to bed on time is $\frac{1}{25}$. What is the probability that he will not go to bed on time?
6. The probability that Liam will watch City on Saturday is 0.98. What is the probability that he will not watch City?
7. The probability that Umair will choose to study Geography at GCSE is $\frac{2}{7}$. What is the probability that he will not choose Geography?

Mutually exclusive outcomes

Mutually exclusive outcomes are outcomes which cannot happen together. For example if you toss a coin once then the outcomes 'the coin comes down heads' and 'the coin comes down tails' are mutually exclusive, since it is impossible for the coin to come down both heads and tails.

If outcomes are mutually exclusive and cover all the possibilities then the probabilities of those events must add up to 1.

So in the above example:

P(heads) + P(tails) = 1

This would be true even if the coin was not fair but was biased towards heads.

If P(heads) = 0.6 then P(tails) = 0.4 and P(heads) + P(tails) = 1 since you cannot throw both a head and a tail on the same coin on the same toss, and no other outcome is possible.

Example 3

The probability that City win their next game is 0.5. The probability that they lose the game is 0.2. What is the probability that the game will be drawn?

The outcomes 'win', 'lose' and 'draw' are mutually exclusive, as it is impossible for any two or all of them to happen together.

Also no outcomes, other than 'win', 'lose' and 'draw', are possible.

Therefore P(win) + P(lose) + P(draw) = 1

P(draw) = 1 − P(win) − P(lose) = 1 − 0.5 − 0.2 = 0.3

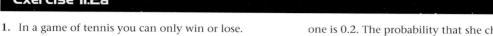

Exercise 11.2a

1. In a game of tennis you can only win or lose. A draw is not possible. Qasim plays Robin at tennis. The probability that Qasim wins is 0.7. What is the probability that Robin wins?

2. A bag contains red, white and blue balls. I pick one ball out of the bag at random. The probability that I pick a red one is $\frac{1}{12}$. The probability that I pick a white one is $\frac{7}{12}$. What is the probability that I pick a blue one?

3. For my next holiday I will go to Spain or France or the USA. The probability that I will go to Spain is $\frac{7}{20}$. The probability that I will go to France is $\frac{11}{20}$. What is the probability that I will go to the USA?

4. The probability that the school team will win their next match is 0.4. The probability that they will lose the match is 0.5. What is the probability that they will draw the match?

5. Sally has only grey, navy and black skirts in her wardrobe. She is choosing a skirt to wear to the cinema. The probability that she chooses a grey one is 0.2. The probability that she chooses a navy one is 0.15. What is the probability that she chooses a black one?

6. Max travels to school by bus, car or cycle or he walks. The probability that he travels by bus is $\frac{1}{11}$. The probability that he travels by car is $\frac{3}{11}$. The probability that he cycles is $\frac{2}{11}$. What is the probability that he walks?

7. Tariq hears on the weather forecast that the probability that it will rain tomorrow is 0.6. He says that this means that the probability that the sun will shine is 0.4. Give two reasons why he may not be correct.

Exercise 11.2b

1. A set of traffic lights may be on red, red and amber, amber or green. The probability that they are on red is 0.5. The probability that they are on red and amber is 0.05 and the probability that they are on amber is 0.05. What is the probability that the lights are on green?

2. Alex is choosing his GCSE options. In one pool, he can choose History or Geography or Business Studies. The probability that he chooses History is 0.3. The probability that he chooses Geography is 0.45. What is the probability that he chooses Business Studies?

3. The coach is selecting the netball team. She has three goal attacks to choose from, Raisa, Kimberley and Melanie. The probability that she chooses Raisa is $\frac{3}{8}$. The probability that she chooses Kimberley is $\frac{5}{8}$. What is the probability that she chooses Melanie?

4. There are four breakfast cereals in the cupboard: muesli, cornflakes, weetycrisps and frosties. Kim has decided to have cereal for breakfast. The probability that she has muesli is 0.05. The probability that she has weetycrisps is 0.4. The probability that she has frosties is 0.2. What is the probability that she will have cornflakes?

5. Greg is choosing his main course for his school dinner. There are three choices: burger and chips, tuna salad and tagliatelli. The probability that he chooses tuna salad is $\frac{1}{12}$. The probability that he chooses tagliatelli is $\frac{5}{12}$. What is the probability that he chooses burger and chips?

6. There are red, yellow and blue beads in a bag. The probability of choosing a red one is $\frac{1}{3}$. The probability of choosing a yellow one is $\frac{1}{4}$. What is the probability of choosing a blue one?

7. Outcomes A, B and C are mutually exclusive and one of them must happen. If $P(A) = 0.47$ and $P(B) = 0.31$, what is $P(C)$?

8. The probability that Rovers will win their next game is 0.4. Geri says this means that the probability that they will lose is 0.6. Why is she almost certainly wrong?

Relative frequency and probability

It is not always possible to find probabilities from looking at equally likely outcomes. You may have to work out the probability of throwing a six with a die which may be biased, the probability of a young driver having an accident, the probability that a person will visit a certain supermarket.

For this type of event you need to set up some sort of experiment, carry out a survey or look at past results.

Take the example of throwing a six with a die which may be biased. For a fair die (unbiased), the probability of getting a six $= \frac{1}{6} = 0.166... = 0.17$ approximately.

If you were to throw it ten times and get four sixes would this be evidence of bias?

The proportion of sixes is $\frac{4}{10} = 0.4$ which is very different from 0.17 but, in a small sample of trials, there may be runs of non-typical results. So you would not conclude that it is biased.

What about ten times in 50 throws? Here the proportion is $\frac{10}{50} = 0.2$ which is still quite a bit different from 0.17 but again you probably have not thrown it often enough. So you still would not conclude that it was biased.

What about 108 times in 600 throws? You have thrown it a large number of times and the proportion of sixes is $\frac{108}{600} = 0.18$. This is too close to 0.17 to conclude it was biased, now that you have thrown it so many times.

What about 100 times in 500 throws? Now you have thrown it a large number of times **and** the proportion $= \frac{100}{500} = 0.2$ which is significantly different from 0.17, but not so much that you should conclude that the die is biased.

The important question is, how many trials are necessary to ensure a representative result?

There is no fixed answer to this, other than 'the more the better'.

As a general rule, any event being examined should occur at least 20 times, but even this is probably a bare minimum.

In the last case above a six occurred 100 times in 500 throws. The proportion of sixes is $\frac{100}{500}$.

This fraction is called the **relative frequency**.

$$\text{Relative frequency} = \frac{\text{number of times an outcome occurs}}{\text{total number of trials}}$$

This is a measure of the proportion of the trials in which the outcome occurs. It is not itself a measure of probability. If, however, the number of trials is large enough, relative frequency can be used as an estimate to probability.

Remember that, however many trials have taken place, relative frequency is still only an estimate. In many cases, however, it is the only method of estimating probability.

Example 4 Ian carries out a survey on the colours of the cars passing his school. His results are shown in this table.

Colour	Black	Red	Blue	White	Green	Other	Total
Number of cars	51	85	64	55	71	90	416

Use these figures to estimate the probability that the next car that passes will be: (a) red   (b) not red.

Since the number of trials is large, use relative frequency as an estimate to probability.

(a) Relative frequency of a red car $= \frac{85}{416}$
    Estimate of probability $= \frac{85}{416}$ or 0.204 (to 3 d.p.)

(b) $416 - 85 = 331$
    Relative frequency $= \frac{331}{416}$
    Estimate of probability $= \frac{331}{416}$ or 0.796 (to 3 d.p.)

Exercise 11.3a

1. Using the figures in Ian's survey above, estimate the probability that the next car will be:
    (a) blue   (b) black or white.
    Give your answers correct to three decimal places.
2. Kim Lee tossed a coin ten times and it came down heads eight times. Kim Lee said that the coin was biased towards heads. Explain why she may not be right.
3. Solomon has a spinner in the shape of a pentagon. The sides are labelled 1, 2, 3, 4 and 5. Solomon spun the spinner 500 times. The results are shown in the table.

Number on spinner	1	2	3	4	5
Number of times	102	103	98	96	101

(a) What is the relative frequency of scoring:
    (i) 2   (ii) 4?
(b) Do the results suggest that Solomon's spinner is fair? Explain your answer.

4. In an experiment, a drawing pin is thrown. It can land either point up or point down. It lands point up 87 times in 210 throws. Use these figures to estimate the probability that, the next time it is thrown, it will land: (a) point up   (b) point down. Give your answers correct to two decimal places.

5. Denise carried out a survey about crisps. She asked 400 people, in the town where she lived, which was their favourite flavour of crisps. The results are shown in this table.

Flavour	Ready salted	Salt and vinegar	Cheese and onion	Prawn cocktail	Other
Number of people	150	75	55	50	70

   (a) Explain why it is reasonable to use these figures to estimate the probability that the next person Denise asks chooses salt and vinegar.

   (b) Use the figures to estimate the probability that the next person Denise asks will choose: (i) salt and vinegar   (ii) ready salted. Give your answers as fractions in their simplest form.

6. An insurance company finds that 203 drivers out of 572 in the age range 17−20 have an accident in the first year after passing their driving test. Use these figures to estimate the probability that a driver aged 17−20 will have an accident in the first year after pasing their test.

7. A large bag contains thousands of red beads and white beads. Describe carefully how you would find the probability of picking out a red bead from the bag.

Exercise 11.3b

1. When Tom is standing at the bus stop he notices that five out of the 20 cars he sees passing are Fords. He says that therefore the probability that the next car that passes will be a Ford is $\frac{1}{4}$. Explain why he is wrong.

2. Freya carries out a survey to find out how students in her school travel to school. She asks a random selection of 200 students. The results are in this table.

Method of travel	Bus	Car	Train	Cycle	Walk
Number of students	34	33	23	45	65

   (a) Explain why it is reasonable to estimate the probabilities of students travelling by the various methods from this survey.

   (b) Use these figures to estimate the probability that a student selected at random from the school:
   (i) travels by bus   (ii) cycles.

3. Noel has two coins which he suspects may be biased.

   (a) He tosses the first 600 times and throws 312 heads. Is there evidence to suggest that this coin is biased? Give your reasons. If there is, estimate the probability that the next throw is a head.

   (b) He tosses the second coin 600 times and throws 420 heads. Is there evidence to suggest that this coin is biased? Give your reasons. If there is, estimate the probability that the next throw is a head.

4. The table shows the results of a survey on the type of detergent households used to do their washing.

Type of detergent	Liquid	Powder	Tablet
Number of households	120	233	85

Use these figures to estimate, correct to two decimal places, the probability that the next household surveyed will use:

(a) liquid   (b) tablet.

5. Stewart made a five-sided spinner. Unfortunately he did not make the pentagon regular. In order to find the probabilities of getting each of the numbers he spun the spinner 400 times. His results are shown in this table.

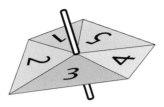

Number	1	2	3	4	5
Frequency	63	84	101	57	95

Use the figures in the table to estimate the probability of the spinner landing on:

(a) 1   (b) 3   (c) an even number.

6. A shopkeeper noticed from his till roll that, out of 430 customers that day, 82 had spent over £10. Use these figures to estimate the probability that his next customer will spend £10 or less.

7. Murphy's Law states that when you drop a piece of toast, it will land buttered side down nine times out of ten. Describe carefully an experiment you could carry out to test Murphy's Law.

Key points

- If the probability of an outcome happening is p, then the probability of the outcome not happening $= 1 - p$.
- If two outcomes A and B cannot occur together they are mutually exclusive.
- If two outcomes A and B are mutually exclusive and cover all possible outcomes, then $P(A) + P(B) = 1$. Similarly if events A, B and C are mutually exclusive and cover all possible outcomes then:
  $$P(A) + P(B) + P(C) = 1$$
  and so on.

- Relative frequency =

  $$\frac{\text{number of times an outcome occurs}}{\text{total number of trials}}$$

- If the number of trials is large enough, relative frequency can be used as an estimate to probability.

1. The probability that Waseem will get all his Maths homework correct is 0.65. What is the probability that he will not get it all correct?

2. The probability that James will play computer games tonight is $\frac{7}{9}$. What is the probability that he will not play computer games tonight?

3. In a game of badminton one player must win. A draw is not possible. Graeme plays Chris in three games of badminton. The probability that Graeme wins all three games is $\frac{1}{8}$. What is the probability that Chris wins at least one game?

4.

> **MENU**
> Main course
> Sausage & Chips
> Ham Salad
> Vegetable Lasagne

Melanie is choosing one main course from a choice of three on the canteen menu. The probability that she chooses sausage and chips is 0.1. The probability that she chooses ham salad is 0.6. What is the probability that she chooses vegetable lasagne?

5. There are green, black and white balls in a bag. One ball is selected at random. The outcomes are denoted by G, B and W.
P(G) = 0.43, P(B) = 0.28. Find P(W).

6. Mudassir travels to school in one of three ways. She walks, catches the bus or else her mother takes her by car. The probability that Mudassir walks is $\frac{7}{20}$. The probability that she goes by bus is $\frac{1}{4}$. What is the probability that Mudassir's mother takes her by car?

7. The three vegetables on the school dinner menu are peas, beans and cabbage. The probability that Steven chooses peas is 0.6. The probability that he chooses beans is 0.3. Chloe says that the probability that Steven chooses cabbage is therefore 0.1. Explain why she may not be right.

8. Over the last year Rebecca has been late for school 25 times in 190 days. Use these figures to estimate the probability that she will be on time the next school day. Give your answer correct to two decimal places.

9. The ages of the drivers of the last 250 cars to have crashes at an accident black spot are shown in the following table.

Age (years)	17−20	21−24	25−49	50−64	over 65
Number of crashes	40	35	105	45	25

(a) Use the table to estimate the probability that the next driver to have a crash at the black spot will be aged:
(i) 25−49 (ii) over 65.

(b) Explain why these figures do not necessarily mean that drivers over 65 years of age are the safest.

10. Rachel has a four-sided spinner with the sides numbered 1, 2, 3 and 4. She wants to test whether the spinner is a fair one. Describe carefully how you would advise her to do this.

You will need to know:

- about place value, for example 378.46 is 3 hundreds, 7 tens, 8 units, 4 tenths and 6 hundredths
- how to change fractions and percentages into decimals
- how to compare fractions
- about equivalent ratios, working with ratios and sharing in a given ratio
- how to add and subtract negative numbers.

Approximating numbers

There are two ways of approximating numbers, or rounding, when an exact answer or number is not required. Both are described below.

Method 1: rounding to the nearest whole number or to a given number of decimal places

Think of a number line.

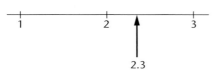

2.3

2.3 to the nearest whole number is 2.

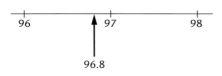

96.8

96.8 to the nearest whole number is 97.

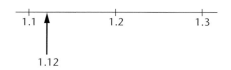

1.12 to the nearest tenth or one decimal place is 1.1.

Example ①

Write the following numbers correct to two decimal places (2 d.p.).

(a) 9.368   (b) 0.0438   (c) 84.655

(a) 9.368

> The first two decimal figures are 3 6, the third figure, the 8, is more than 5 so round up.

= 9.37 (to 2 d.p.)

(b) 0.0438

> The first two decimal figures are 0 4, the third figure is 3, which is less than 5, so round down.

= 0.04 (to 2 d.p.)

(c) 84.655

> The third decimal figure is 5 so the convention is to round up.

= 84.66 (to 2 d.p.)

Exercise 12.1a

1. Write these numbers first correct to two decimal places, then correct to one decimal place.
   (a) 5.481   (d) 0.5666   (g) 7.0064
   (b) 12.0782 (e) 9.017   (h) 0.0734
   (c) 0.214   (f) 78.044

2. Use your calculator to find the square root of 55. Write this value correct to:
   (a) one decimal place   (b) two decimal places.

3. Use your calculator to work out the value of $\sqrt{5^2 + 8^2}$.

   Write your answer correct to two decimal places.

4. Write these fractions as decimals correct to three decimal places.
   (a) $\frac{1}{3}$ (b) $\frac{2}{7}$ (c) $\frac{3}{11}$ (d) $\frac{4}{13}$

5. Work out the mean of the following numbers. Give your answer correct to one decimal place.
   4, 6, 8, 9, 11, 3, 2, 15

6. Calculate 13.8% of 67.7, giving your answer correct to two decimal places.

Exercise 12.1b

1. Write these numbers first correct to two decimal places, then correct to one decimal place.
   (a) 9.424   (d) 0.85   (g) 7.1111
   (b) 0.8413   (e) 7.093   (h) 8.081
   (c) 0.283   (f) 18.63

2. Use your calculator to find the square root of 75. Write this value correct to:
   (a) two decimal places   (b) one decimal place.

3. Use your calculator to work out the value of $\sqrt{4^2 + 5^2}$.
   Write your answer correct to two decimal places.

4. Write these fractions as decimals correct to three decimal places.
   (a) $\frac{1}{8}$ (b) $\frac{3}{7}$ (c) $\frac{4}{11}$ (d) $\frac{5}{13}$

5. Work out the mean of the following numbers. Give your answer correct to one decimal place.
   3.51, 5.21, 7.91, 8.31, 9.41, 11.71, 13.51

6. Calculate 14.2% of 93.4, giving your answer correct to two decimal places.

Rounding to the nearest 10, 100, ...

A number line may also help in rounding numbers to the nearest 10, 100 and so on.

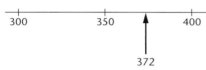

372 to the nearest 100 is 400.

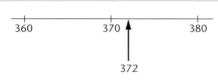

372 to the nearest 10 is 370.

Exercise 12.2a

1. Write these numbers correct to the nearest 10.
   (a) 456   (b) 254   (c) 123   (d) 998
   Write these numbers correct to the nearest 100.
   (e) 5678   (f) 9870   (g) 8801   (h) 151

   As with decimals the convention with a number in which the last digit is 5 is to round up so that, to the nearest 10, 35 would be 40 and 655 would be 660, and to the nearest 100, 550 would be 600 and 1350 would be 1400.

2. Round these numbers to the nearest 10 and then to the nearest 100.
   (a) 125   (b) 450   (c) 545   (d) 4555   (e) 1405

1. Round these numbers to the nearest 10.
   (a) 97   (b) 111   (c) 374   (d) 444
   Round these numbers to the nearest 1000.
   (e) 1234   (f) 8724   (g) 6789   (h) 9988
2. Round these numbers to the nearest 10 and then to the nearest 100.
   (a) 135   (b) 955   (c) 645   (d) 7555   (e) 1005

Method 2: rounding to a given number of significant figures

Sometimes you may be asked to give an answer to a given number of significant figures. Significant figures are counted from left to right, for example:

5 000 000 has one significant figure (1 s.f.). This is the 5.

5 300 000 has two significant figures (2 s.f.). These are the 5 and 3.

5 340 000 has three significant figures (3 s.f.). These are the 5, 3 and 4.

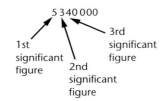

Rounding 5 340 000:
to one significant figure
   (rounding to the nearest million)   gives 5 000 000
to two significant figures   gives 5 300 000
to three significant figures   gives 5 340 000

> Note that zeros are kept to show the size of the number.

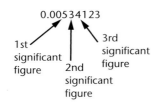

For a number less than 1, for example rounding 0.005 341 23:
to one significant figure   gives 0.005
to two significant figures   gives 0.0053
to three significant figures   gives 0.005 34

> Note that zeros are kept to show the size of the number.

Figures are often rounded to one or more significant figures in newspapers:
- '30 000 see Derby win' when the actual attendance could have been 29 877
- 'Cambridge man wins £3 million' when the actual figure could have been £3 132 677.

Exercise 12.3a

1. Write each of these to the number of significant figures (s.f.) shown.
   (a) 67 890 (to 3 s.f.)
   (e) 0.006 78 (to 2 s.f.)
   (b) 54.123 (to 3 s.f.)
   (f) 1.456 (to 2 s.f.)
   (c) 1789 (to 2 s.f.)
   (g) 0.0894 (to 1 s.f.)
   (d) 1 564 389 (to 5 s.f.)
   (h) 45.278 (to 3 s.f.)

2. Work these out.
   (a) 25% of 16 844 (to 2 s.f.)
   (b) 80% of 888 (to 3 s.f.)

3. A room is 3.65 m long, 2.44 m wide and 2.2 m high. Calculate the total wall area, correct to two significant figures.

4. The amounts of money taken in a charity shop, over a six-day period, are listed here.
   £245.77, £452.88, £189.52, £212.79, £181.83, £233.56
   (a) Calculate the average daily takings, correct to three significant figures.
   The shop is open for 51 weeks in the year.
   (b) Use your answer to part (a) to estimate the yearly takings. Give this answer correct to two significant figures.

5. Round each of these numbers to two significant figures.
   (a) 76 560
   (c) 0.8099
   (e) 0.707
   (b) 681
   (d) 3.086

6. Work these out on your calculator. Round your answers to two decimal places.
   (a) $19 \div 16$
   (c) 1.59^2
   (e) 0.35×1.35
   (b) $\sqrt{30}$
   (d) 1.3^3

7. Estimate the answers to these calculations by rounding the numbers to one significant figure. Compare your estimates with the calculations worked out on a calculator and with the answers rounded to two significant figures.
   (a) $5.89 \times 0.186\,75$
   (b) $19.258 \div 3.889$
   (c) 36.87×15.87
   (d) $9.7687 \div 0.0512$
   (e) $2.14 \times 5.8754 \times 0.9876$

Exercise 12.3b

1. Write each of these to the number of significant figures (s.f.) shown.
   (a) 12 340 (to 3 s.f.)
   (e) 0.007 98 (to 2 s.f.)
   (b) 76.456 (to 3 s.f.)
   (f) 1567 (to 2 s.f.)
   (c) 1654 (to 2 s.f.)
   (g) 0.0923 (to 1 s.f.)
   (d) 1 456 789 (to 5 s.f.)
   (h) 54.827 (to 3 s.f.)

2. Work these out.
   (a) 30% of 17 824 (to 2 s.f.)
   (b) 70% of 999 (to 3 s.f.)

3. A water tank is 4.45 m long, 3.24 m wide and 1.4 m high. Calculate the volume correct to two significant figures.

4. Find the average of these numbers.
   134.64, 157.92, 194.33, 254.45, 188.88
   Give this answer correct to two significant figures.

5. Work these out on your calculator.
   Round your answers to two decimal places.
   (a) $24 \div 17$
   (d) 1.5^3
   (b) $\sqrt{40}$
   (e) 0.42×1.42
   (c) 1.78^2

6. Estimate the answers to these calculations by rounding the numbers to one significant figure. Compare your estimates with the calculations worked out on a calculator, and with the answers rounded to two significant figures.
   (a) $4.89 \times 0.196\,54$
   (b) $24.342 \div 4.874$
   (c) 34.62×16.34
   (d) $7.9685 \div 0.0432$
   (e) $8.16 \times 5.974 \times 0.9325$

Ordering numbers

Ordering numbers means arranging them in size order. It is essential to work with numbers that are all of the same type, for example to compare fractions and decimals, change the fractions into decimals, using a calculator if necessary.

 Example Put these numbers in size order, smallest first.

$\frac{3}{5}$, 0.55, 0.7

$\frac{3}{5} = 0.6$ so the order is 0.55, 0.6, 0.7

To compare fractions with different denominators, either change them into decimals (the simplest approach), or change them into equivalent fractions.

 Example Write these fractions in size order, smallest first.

$\frac{2}{5}$, $\frac{3}{7}$, $\frac{4}{9}$

$\frac{2}{5} = 0.4$ $\frac{3}{7} = 0.428\,57$ $\frac{4}{9} = 0.444\,44$

or $\frac{2}{5} = \frac{126}{315}$ $\frac{3}{7} = \frac{135}{315}$ $\frac{4}{9} = \frac{140}{315}$

Make sure you understand why 315 is in the denominator of each fraction.

They were given in size order.

Exercise 12.4a

1. Make as many fractions as you can, choosing pairs of numbers from those listed.
   3   4   5   6
   Put your fractions in order, smallest first.
2. Write these decimals in order, starting with the smallest.
   0.000 280, 0.0098, 0.0126, 0.0042, 0.5, 0.0014
3. Write these numbers in order, largest first.
   15 065, 15 605, 156 005, 1 560 005, 15 565
4. Write these numbers in size order, from largest to smallest.
   0.6, 0.006 004, 0.0624, 0.62

Exercise 12.4b

1. Make as many fractions as you can, choosing pairs of numbers from those listed.
   6   7   8   9
   Put your fractions in order, smallest first.
2. Write these decimals in order, starting with the smallest.
   0.000 342, 0.0064, 0.0136, 0.004 14, 0.6, 0.0015
3. Write these numbers in order, largest first.
   14 104, 14 140, 141 401, 1 410 401
4. Write these numbers in order, from largest to smallest.
   0.4, 0.424, 0.042 424, 0.442, 0.0044

Ratio

Look back to Chapter 1, where ratios were introduced, and remind yourself of the work you did.

Ratios can be used to compare quantities.

The ratio of noughts to crosses in this box is 3 to 1 which can be written as 3 : 1,

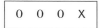

but the ratio of crosses to noughts is 1 : 3.

> The order in which a ratio is given is important.

If a ratio is written as n : 1 it means that the first quantity is n times the second quantity, so if a colour is made by mixing yellow and red paint in the ratio 5 : 1, there is five times as much yellow paint as red paint.

Maps often have their scale expressed in the form 1 : n, for example 1 : 50 000 means that 1 cm on the map represents 50 000 cm on the ground.

Exercise 12.5a

1. Copy these statements and complete them.
   (a) 1 : 4 is equivalent to 6 : □
   (b) 2 : 5 is equivalent to 12 : □
   (c) 4 : 5 is equivalent to □ : 40
   (d) 4 : 5 : 6 is equivalent to □ : 15 : □
2. Write these ratios in their simplest form.
   (a) 40p to £4   (b) 100 kg to 25 kg
   (c) 2 to 2.5   (d) 0.5 to 9
3. Write these ratios in the form 1 : n.
   (a) 4 : 20 (b) 9 : 27 (c) 3.5 : 14   (d) 0.1 : 30
4. A map is drawn with a scale of 1 : 30 000.
   Two towns are 7.6 cm apart on the map. How far apart are they on the ground?
5. The scale of a map is 1 : 100 000.
   The length of a road is 2.8 km. How long will the road be on the map?
6. The teacher–student ratio in a school is given as 1 : n.

   (a) If the ratio for a school is 1 : 17.6 about how many staff would you expect in a school with 700 students?
   (b) A school has 84 staff. About how many students would you expect the school to have?
7. Write the following ratios in the form 1 : n.
   (a) 2 cm to 1 m   (b) 1 litre to 5000 ml
   (c) 4 minutes to 4 hours
8. A prize is divided so that one person gets three times as much as the other.
   (a) Write this as a ratio.
   (b) If the prize money were £480 how much should each person get?
9. The Fibonacci series is 1 1 2 3 5 8 …
   (a) Continue the series for another five terms.
   (b) Use your calculator to investigate the ratio of each term to the previous one. What do you notice?

1. Copy these statements and complete them.
   (a) 1 : 5 is equivalent to 5 : □
   (b) 3 : 4 is equivalent to 12 : □
   (c) 3 : 5 is equivalent to □ : 80
   (d) 5 : 6 : 7 is equivalent to □ : 18: □
2. Write these ratios in their simplest form.
   (a) 30p to £3   (b) 100 g to 10 kg   (c) 3 to 3.6   (d) 0.5 to 5
3. Write these ratios in the form 1 : n.
   (a) 5 : 35   (b) 16 : 48   (c) 5.2 : 26   (d) 0.2 : 30
4. A map is drawn with a scale of 1 : 250 000.
   Two towns are 8.5 cm apart on the map. How far apart are they on the ground?
5. The scale of a map is 1 : 100 000.
   The length of a road is 3.2 km. How long will the road be on the map?
6. On a map, the distance between two towns is 12 cm. How far apart are they on the ground if the scale is 1 : 300 000?
7. Write the following ratios in the form 1 : □.
   (a) 20 g to 5 kg   (b) 1 m to 15 000 cm   (c) 7 days to 4 weeks
8. A prize is divided so that one person gets five times as much as the other.
   (a) Write this as a ratio.
   (b) If the prize money were £480 how much should each person get?

Negative numbers

Multiplying and dividing

When multiplying and dividing the following rules apply.

$(+) \times (+) = (+)$ $(+) \div (+) = (+)$
$(-) \times (-) = (+)$ $(-) \div (-) = (+)$
$(-) \times (+) = (-)$ $(-) \div (+) = (-)$
$(+) \times (-) = (-)$ $(+) \div (-) = (-)$

Example 4

$-3 \times -4 = +12$

$-7 \times +9 = -63$

$-24 \div -6 = 4$

Exercise 12.6a

1. (a) $-9 \times -6 =$ (b) $+12 \times -6 =$ (c) $-3 \times +14 =$
   (d) $-8 \times -50 =$ (e) $-7 \times -60 =$ (f) $+12 \times +9 =$
   (g) $+45 \div -15 =$ (h) $-88 \div +4 =$ (i) $+144 \div -9 =$

2. Copy this number square and complete it.

$\times$			4	
10	80	-60		-30
-7		42		21
	24	-18		-9
	-40	30	-20	15

3. Replace the □ with the numbers -6, -3 and 2 to make these statements true.
   (a) $\square \div \square \times \square = 1$ (b) $(\square \div \square) - \square = 0$
   (c) $(\square - \square) \times \square = -30$

4. If $x = -3$ find the value of:
   (a) $3x$ (b) $4 \div x$ (c) $x \div 4$ (d) $3x^2$
   (e) x^3 (f) $4x + 3$ (g) $5x^2 - 6$ (h) $(-4x)^2$

Exercise 12.6b

1. (a) $-5 \times -7 =$ (b) $+5 \times -6 =$ (c) $-6 \times +7 =$
   (d) $-9 \times -3 =$ (e) $-6 \times -8 =$ (f) $+8 \times +9 =$
   (g) $+16 \div -4 =$ (h) $-24 \div +3 =$ (i) $+56 \div -8 =$

2. Copy this number square and complete it.

$\times$		-7		
	-9	21		-15
6	18		-54	30
8		-56		40
-9	-27		81	

3. Replace the □ with the numbers -8, -4 and -2 to make these statements true.
   (a) $\square \div \square \times \square = -1$ (b) $(\square \div \square) + \square = 0$
   (c) $(\square - \square) \times \square = 16$

4. If $x = -4$ find the value of:
   (a) $5x$ (b) $8 \div x$ (c) $x \div -2$ (d) $2x^2$ (e) x^3 (f) $4x + 4$
   (g) $3x^2 - 8$ (h) $(-3x)^2$

Indices

Indices (or powers) are a form of mathematical shorthand:

$3 \times 3 \times 3 \times 3$ is written as 3^4 and
$2 \times 2 \times 2 \times 2 \times 2 \times 2 \times 2 \times 2$ is written as 2^8.

Multiplying numbers in index form

$$3^4 \times 3^8 = (3 \times 3 \times 3 \times 3) \times (3 \times 3 \times 3 \times 3 \times 3 \times 3 \times 3 \times 3)$$
$$= 3 \times 3 \times 3 \times 3 \times 3 \times 3 \times 3 \times 3 \times 3 \times 3 \times 3 \times 3$$
$$= 3^{12}$$

The indices or powers are added i.e. $3^4 \times 3^8 = 3^{4+8} = 3^{12}$

The rule is $n^a \times n^b = n^{a+b}$

Powers of numbers can also be raised to powers.

If the numbers are in brackets then powers are multiplied.

Example 5

$(3^4)^2 = (3^4) \times (3^4) = 3^8$

and $3^{4 \times 2} = 3^8$

The rule is $(n^a)^b = n^{a \times b}$

Dividing numbers in index form

$$2^6 \div 2^4 = (2 \times 2 \times 2 \times 2 \times 2 \times 2) \div (2 \times 2 \times 2 \times 2)$$
$$= 2 \times 2$$
$$= 2^2$$

so $2^6 \div 2^4 = 2^{(6-4)}$
$$= 2^2$$

The rule is $n^a \div n^b = n^{a-b}$

If the subtraction gives a negative answer, for example $2^4 \div 2^6$ then the rule gives the result 2^{-2}.

But $2^4 \div 2^6 = \dfrac{2 \times 2 \times 2 \times 2}{2 \times 2 \times 2 \times 2 \times 2 \times 2}$

$$= \dfrac{1}{2^2}$$

so $2^{-2} = \dfrac{1}{2^2}$

The rule is $n^{-a} = \dfrac{1}{n^a}$

One important fact can be shown from the following example.

Example 6

Work out $4^3 \div 4^3$.

Using the rule established above:

$4^3 \div 4^3 = 4^{3-3} = 4^0$

but $4^3 \div 4^3 = (4 \times 4 \times 4) \div (4 \times 4 \times 4) = 1$

therefore $4^0 = 1$

The rule is $n^0 = 1$

Finally look at the following example.

Example 7

Write these in index form.

(a) $5 \times 5 \times 6 \times 6 \times 6 \times 6$

(b) $3x^3 \times 4x^5$

(c) $25x^4 \div 5x^2$

(d) $4^3 \div 4^6$

(a) $5 \times 5 \times 6 \times 6 \times 6 \times 6 = 5^2 \times 6^4$

(b) $3x^3 \times 4x^5 = 12x^8$

(c) $25x^4 \div 5x^2 = 5x^2$

(d) $4^3 \div 4^6 = 4^{3-6} = 4^{-3}$

Exercise 12.7a

1. Write these in a simpler form, using indices.
   (a) $3 \times 3 \times 3 \times 3 \times 3$
   (b) $7 \times 7 \times 7$
   (c) $8 \times 8 \times 8 \times 8 \times 8$
   (d) $3 \times 3 \times 5 \times 5 \times 5$
   (e) $2 \times 2 \times 2 \times 3 \times 3 \times 4 \times 4 \times 4 \times 4 \times 4$

2. Write these in a simpler form, using indices.
   (a) $5^2 \times 5^3$ (b) $6^2 \times 6^7$ (c) $10^3 \times 10^4$
   (d) $3^6 \times 3^5$ (e) $8^3 \times 8^2$ (f) $5^{-2} \times 5^{-4}$

3. Work these out, giving your answers in index form.
   (a) $\dfrac{3^4}{3^5 \times 3^2}$ (b) $\dfrac{(2^3)^4}{2^5}$ (c) $\dfrac{5^2 \times 5^3}{5^4 \times 5^5}$

Exercise 12.7b

1. Write these in a simpler form, using indices.
   (a) $4 \times 4 \times 4 \times 4 \times 4$
   (b) $8 \times 8 \times 8$
   (c) $2 \times 2 \times 2 \times 2 \times 2$
   (d) $5 \times 4 \times 4 \times 4 \times 5$
   (e) $7 \times 7 \times 7 \times 8 \times 8 \times 9 \times 9 \times 9$

2. Write these in a simpler form, using indices.
   (a) $4^2 \times 4^3$ (b) $9^2 \times 9^7$ (c) $10^5 \times 10^2$
   (d) $3^5 \times 3^2$ (e) $8^4 \times 8^2$ (f) $6^2 \times 6^6$
   (g) $7^5 \div 7^3$ (h) $9^7 \div 9^9$ (i) $6^3 \div 6^2$
   (j) $3^6 \div 3^3$ (k) $4^4 \div 4^4$

3. Work these out, giving your answers in index form.
   (a) $\dfrac{4^6}{4^5 \times 4^4}$ (b) $\dfrac{(2^4)^2}{2^4}$ (c) $\dfrac{6^2 \times 6}{6^5 \times 6^4}$

Standard form

Standard form is a very important use of powers and indices. It is a way of writing numbers as a number from 1 to 10 multiplied by a power of 10.

Example 8

$$4\,000\,000 = 4 \times 10^6$$

$$0.03 = 3 \times 10^{-2}$$

Example 9

The speed of light is 300 000 000 m/s

$= 3 \times 100\,000\,000$ ← This line can be omitted.

$= 3 \times 10^8 \text{m/s}$

A virus is 0.000 000 000 56 cm in diameter

$= \dfrac{5.6}{10\,000\,000\,000}$ ← This line can be omitted.

$= 5.6 \times 10^{-10} \text{cm}$

Exercise 12.8a

1. Write these numbers in standard form.
   (a) 5000   (b) 50   (c) 70 000
   (d) 46   (e) 0.02   (f) 546 000
   (g) 0.000 45   (h) 16 million

2. These numbers are in standard form. Write them out in full.
   (a) 5×10^2 (b) 4×10^5 (c) 6×10^{-3}
   (d) 4.5×10^3 (e) 8.4×10^{-3} (f) 2.87×10^{-5}
   (g) 9.7×10^3 (h) 5.55×10^{-5}

Exercise 12.8b

1. Write these numbers in standard form.
   (a) 6000   (b) 80   (c) 30 000
   (d) 67   (e) 0.08   (f) 897 000
   (g) 0.000 54   (h) 18 million

2. These numbers are in standard form. Write them out in full.
   (a) 7×10^2 (b) 8×10^5 (c) 3×10^{-3}
   (d) 2.5×10^3 (e) 6.7×10^{-3} (f) 3.82×10^{-5}
   (g) 5.7×10^3 (h) 4.65×10^{-5}

Working with numbers in standard form

Always take care when working with numbers written in standard form. With multiplication and division you can deal with the numbers in the normal way.

Example 10

Work out $(3 \times 10^2) \times (4 \times 10^3)$.

$$(3 \times 10^2) \times (4 \times 10^3) = 3 \times 4 \times 10^2 \times 10^3$$
$$= 12 \times 10^{2+3}$$
$$= 12 \times 10^5$$
$$= 1.2 \times 10^6$$

Addition and subtraction are not so straightforward.

Example 11

Work out $(3 \times 10^2) + (4 \times 10^3)$.

It is safer to write out the numbers in the brackets first, add then change the answer back into standard form.

$$(3 \times 10^2) + (4 \times 10^3) = 300 + 4000$$
$$= 4300 = 4.3 \times 10^3$$

Exercise 12.9a

Work out these calculations. Give your answers in standard form.
(a) $(5 \times 10^3) + (7 \times 10^4)$
(b) $(7 \times 10^6) - (3 \times 10^3)$
(c) $(3 \times 10^3) + (3 \times 10^2)$
(d) $(6 \times 10^3) - (5 \times 10^2)$

Exercise 12.9b

Work out these calculations. Give your answers in standard form.
(a) $(4 \times 10^3) + (5 \times 10^4)$
(b) $(8 \times 10^6) - (4 \times 10^3)$
(c) $(2 \times 10^3) + (4 \times 10^2)$
(d) $(4 \times 10^3) - (3 \times 10^2)$

Powers and roots

The set of whole numbers 1, 4, 9, 16, 25, 36, … are **squares** of the counting numbers, for example:

$$1 = 1^2, 4 = 2^2, 9 = 3^2$$

Because $16 = 4^2$ the **square root** of 16 is 4, written as $\sqrt{16} = 4$. Similarly $\sqrt{36} = 6$ and $\sqrt{81} = 9$.

The numbers 1, 8, 27, 64, … are **cube** numbers because each of them can be written as the cube of a whole number, for example:

$$8 = 2^3, 27 = 3^3$$

Because $27 = 3^3$ the cube root of 27 is 3, written as $\sqrt[3]{27} = 3$.

Exercise 12.10a

1. Write down the square of each number.
   (a) 7   (b) 12   (c) 25   (d) 40   (e) $(6^2 - 5^2)$
2. Write down the square root of each number.
   (a) 49   (b) 121   (c) 169   (d) 289   (e) $(3^2 + 4^2)$
3. Write down the cube of each number.
   (a) 4   (b) 5   (c) 6   (d) 10   (e) 2^3
4. Write down the cube root of each number.
   (a) 343   (b) 729   (c) 1331   (d) 1 000 000
5. Work these out, giving your answers correct to two decimal places.
   (a) $\sqrt{56}$ (b) $\sqrt{27}$ (c) $\sqrt{60}$ (d) $\sqrt{280}$ (e) $\sqrt{678}$

1. Write down the square of each number.
   (a) 8   (b) 11   (c) 35   (d) 50   (e) $(7^2 - 5^2)$
2. Write down the square root of each number.
   (a) 81   (b) 144   (c) 196   (d) 361   (e) $(2^2 + 6^2)$
3. Write down the cube of each number.
   (a) 2   (b) 3   (c) 1.5   (d) 7   (e) 2^2
4. Write down the cube root of each number.
   (a) 216   (b) 512   (c) 1728   (d) 1000
5. Work these out, giving your answers correct to two decimal places.
   (a) $\sqrt{70}$ (b) $\sqrt{39}$ (c) $\sqrt{90}$ (d) $\sqrt{380}$ (e) $\sqrt{456}$

Reciprocal

Earlier in this chapter you saw this rule:

$$n^{-a} = \frac{1}{n^a}$$

When the power (the value of a) is 1, the rule becomes

$$n^{-1} = \frac{1}{n}$$

and $\frac{1}{n}$ is called the **reciprocal** of n.

Similarly, n is the reciprocal of $\frac{1}{n}$.

Examiner's tip

If you find the reciprocal of, or divide by, a very small number, the result is very large. (Try $1 \div 0.000\,000\,01$ on your calculator.) If you divide by 0 you will not get a number, but your calculator will probably show E for error. Mathematicians call the result of dividing by zero **infinity**.

Example Write down the reciprocal of (a) 4 (b) 25.

(a) The reciprocal of 4 is $\frac{1}{4}$.

(b) The reciprocal of 25 $= \frac{1}{25}$.

Exercise 12.11a

1. Write down the reciprocal of each of these numbers.
   (a) 3   (b) 6   (c) 49   (d) 100   (e) 640
2. Write down the numbers of which these are the reciprocals.
   (a) $\frac{1}{16}$ (b) $\frac{1}{9}$ (c) $\frac{1}{52}$ (d) $\frac{1}{67}$ (e) $\frac{1}{1000}$

Exercise 12.11b

1. Write down the reciprocal of each of these numbers.
   (a) 4   (b) 9   (c) 65   (d) 10   (e) 4.5
2. Write down the numbers of which these are the reciprocals.
   (a) $\frac{1}{3}$ (b) $\frac{1}{10}$ (c) $\frac{1}{25}$ (d) $\frac{1}{71}$ (e) $\frac{1}{100}$

Key points

- 23 to the nearest 10 is 20.
- 379 to the nearest 100 is 400.
- 4.569, correct to two decimal places, is 4.57.
- 343, correct to two significant figures, is 340.
- 4 : 16 is the same ratio as 1 : 4.

- Multiplying or dividing a negative number by a negative number gives a positive answer.
- In standard form, $432 = 4.32 \times 10^2$.
- $2^3 = 8$ and $^3\sqrt{8} = 2$.
- Reciprocals are written as, for example, the reciprocal of $9 = \frac{1}{9}$.

Revision exercise 12a

1. Write the following numbers correct to two decimal places.
   (a) 7.897   (b) 13.1234   (c) 0.243
   (d) 0.6772

2. Work out the value of $\sqrt{6^2 + 5^3}$.
   Write your answer correct to two decimal places.

3. Write these fractions as decimals correct to three decimal places.
   (a) $\frac{3}{7}$ (b) $\frac{2}{11}$ (c) $\frac{2}{13}$ (d) $\frac{7}{13}$

4. Calculate 17.5% of £167.75, giving your answer correct to two decimal places.

5. Write these numbers to the nearest 10.
   (a) 127   (b) 543   (c) 995   (d) 1239
   Write these numbers to the nearest 100.
   (e) 7898   (f) 9820   (g) 8850   (h) 51

6. Write each of these correct to the stated number of significant figures (s.f.).
   (a) 6789 (to 3 s.f.)   (b) 57.123 (to 3 s.f.)
   (c) 1897 (to 2 s.f.)   (d) 1 576 398 (to 5 s.f.)
   (e) 0.005 88 (to 2 s.f.)   (f) 1.756 (to 2 s.f.)
   (g) 0.0812 (to 1 s.f.)   (h) 40.278 (to 3 s.f.)

7. Write these decimals in order, starting with the smallest.
   0.000 980, 0.0098, 0.0926, 0.9042, 0.9, 0.914

8. Write these ratios in the form $1 : n$.
   (a) 6 : 24   (b) 36 : 72   (c) 2.5 : 7.5
   (d) 0.1 : 3

9. A map is drawn with a scale of 1 : 25 000. Two towns are 8.6 cm apart on the map. How far apart are they on the ground?

10. Write the following ratios in the form $1 : n$.
    (a) 10 m to 2 km   (b) 10 g to 5000 kg
    (c) 40p to £4

11. Without using a calculator, work these out.
    (a) $-8 \times -9 =$ (b) $+25 \times -6 =$
    (c) $-6 \times +3 =$ (d) $-12 \times -3 =$
    (e) $-6 \times -5 =$ (f) $+8 \times +12 =$
    (g) $+36 \div -4 =$ (h) $-24 \div +8 =$
    (i) $+56 \div -7 =$

12. If $x = -5$ find the value of each of these expressions.
    (a) $4x$ (b) $10 \div x$ (c) $x \div 2$ (d) $3x^2$
    (e) x^3 (f) $4x + 3$ (g) $5x^2 - 6$ (h) $(-4x)^2$

13. Write these numbers in a simpler form.
    (a) $7^2 \times 7^3$ (b) $6^3 \times 6^6$ (c) $10^9 \times 10^3$
    (d) $3^4 \times 3^8$ (e) $8^9 \div 8^3$ (f) $6^7 \div 6^9$
    (g) $4^2 \div 4^3$ (h) $9^6 \div 9^2$

14. Write these numbers in standard form.
    (a) 7600   (b) 89.9   (c) 60 000
    (d) 466   (e) 0.056   (f) 564 600
    (g) 0.0055   (h) 24 million

15. These numbers are in standard form. Write them out in full.
    (a) 6×10^3 (b) 5×10^2
    (c) 7×10^{-3} (d) 4.5×10^2
    (e) 8.4×10^{-2} (f) 2.87×10^{-3}
    (g) 4.7×10^3 (h) 5.55×10^{-2}

16. Work out these calculations. Give your answers in standard form.
    (a) $(6 \times 10^3) + (8 \times 10^3)$
    (b) $(7 \times 10^4) - (2 \times 10^3)$

> ## You will need to know:
>
> - that the sum of the angles of a triangle is 180°
> - in an isosceles triangle the angles opposite the equal sides are themselves equal
> - the properties of alternate and corresponding angles.

Exterior angle of a triangle

If one of the sides of a triangle is extended, to form an angle outside the triangle, this angle is called the exterior angle.

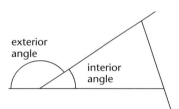

Activity 1

Draw a large triangle with sides 10 cm, 12 cm and 14 cm. Extend the length of the base.

Cut off two of the corners, as shown by the shaded angles.

Try to fit them together, next to the remaining angle at the base.

What do you notice?

They should fit over the exterior angle.

Repeat for a different sized triangle with a different exterior angle.

Does the same thing happen?

You have demonstrated that:

> The exterior angle of a triangle is equal to the sum of the interior opposite angles.

Try to write a proof of the result you have just found, using some or all of the statements given below. Give reasons if you can.

Note that there is more than one proof.

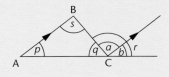

$a + b = r$	because angles on a straight line add up to 180°
$p = b$	because they are alternate angles
$p + s = r$	because they are vertically opposite angles
$s = a$	because they are corresponding angles
$p + s = a + b$	because angles in a triangle add up to 180°

Exercise 13.1a

1. Calculate the sizes of all the angles marked with letters.

(a)
(b)

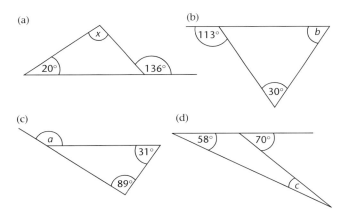

(c)
(d)

2. Write down as many different ways as you can to find the value of e once you have found the value of d.

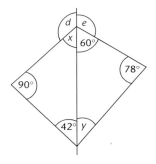

3. Calculate the sizes of all the angles marked with letters.

(a)   (b)

(c)

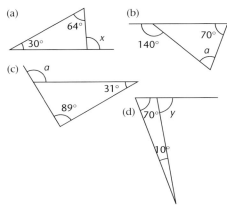

(d)

4. Calculate the sizes of all the angles marked with letters.

(a)   (b)

(c)   (d)

(e)

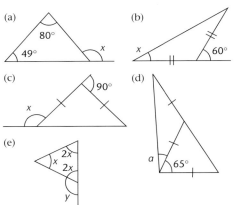

Exercise 13.1b

1. Calculate the sizes of all the angles marked with letters.

   (a)

   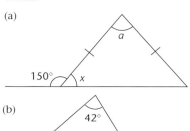

   (b)

   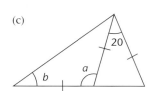

   (c)

   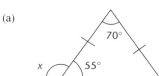

2. Calculate the sizes of all the angles marked with letters.

   (a)

   70°
   x   55°

   (b)

   42°
   x   110°

   (c)

   a
   b   100° 80°

3. Calculate the sizes of all the angles marked with letters.

   (a)

   p
   37°   144°

   (b)

   60°   m

4. Calculate the sizes of all the angles marked with letters.

   (a)

   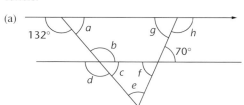

   (b)

   x
   122°   y

   (c)

   y   x
   85°
   62°   z

Examiner's tip

Remember all the angle properties you have met. There are several ways of calculating some of these angles.

Angle in a semicircle

- Draw a circle of radius 6 cm.
  On the circumference, mark three points that can be joined up to form a right-angled triangle.
  Repeat for circles with different radii.
  What do you notice? Write down your idea.

- Now test your idea by drawing a right-angled triangle with sides 6 cm, 8 cm and 10 cm.
  Use a protractor or angle measurer to draw the angles as accurately as you can.
  Draw a circle that passes through the three vertices of the triangle.
  Was your idea correct?
  Check by drawing a different right-angled triangle and then drawing a circle round it.

- An alternative approach is to use a set-square or a piece of card with a right-angled corner, and two nails.
  You could use two points marked on the paper if you prefer to.
  Place the set-square between the two nails so that the edges of the set-square which form the right angle touch the nails.

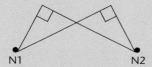

- Mark a point to represent the position of the tip of the right angle, when the edges of the set-square are firmly against the nails.
  Repeat for different positions of the set-square.
  (a) What shape are the marks forming?
  (b) What can you say about the line joining the two nails (or points)?

Notes
(a) You should find that if you repeat the activity enough times the pencil marks at the right-angled corner will trace a circle.
(b) The line joining the two nails (or points) is a diameter.
(d)–(e) If the angle is acute, the points form the major part of a circle.

> The angle in a semicircle is 90°.

(c) What would happen if the angle at the corner were acute?
(d) What would happen if the angle at the corner were obtuse?
(e) In each case, state what shape would be formed.

Now you can prove it. Study the diagram, then work through the statements given below.

Copy statements 1–4 and complete them, filling in the reasons.
1. The lines OA, OP and OB are equal because …
2. The angles marked a are equal to each other because …
3. The angles marked b are equal to each other because …
4. In triangle APB, $a + a + b + b = 180°$ because …

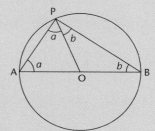

In other words $2(a + b) = 180°$
so angle APB $= a + b = 90°$

- Now use a cylindrical tin, or something similar, to draw a circle.
  Use the fact that the angle in a semicircle is 90° to think of a method for finding the centre of the circle which you have drawn.

Explain your method.

If the process is continued below the imaginary line joining N1 and N2 the shape formed makes two parts of two equal circles which join but do not make a complete circle. If the angle is obtuse then a similar result is obtained but from the smaller part of a circle.

> You have just demonstrated an important fact!

Exercise 13.2a

In question 1–3, points A and B are at opposite ends of a diameter of the circle with its centre at O.

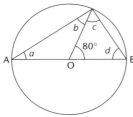

1. (a) Without calculating their values, write down a fact that is true about a and b, and also about c and d.
   (b) Now write down their values, explaining how you worked them out.
   (c) Write down another angle fact that you could use to check that your answer to part (b) is correct.

2. Calculate the value of a.

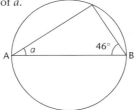

3. (a) Calculate the sizes of the angles marked with letters in this diagram.

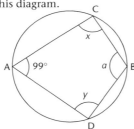

   (b) Why can't the straight line joining C and D be a diameter?

4. APQBRS is a regular hexagon. Prove that AB is a diameter.

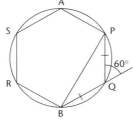

Exercise 13.2b

1. Calculate the size of the angles marked with letters.

   (a)   (b)

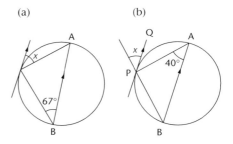

2. If AB is a diameter, what can you say about angles C and D?

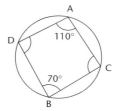

3. These three diagrams show semicircles. Find the sizes of all the angles marked with letters.

   (a)   (b)

   (c)

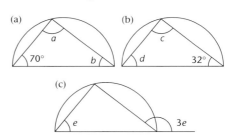

4. Find the sizes of all the angles marked with letters in this diagram, if O is the centre of the circle.

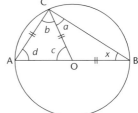

Angles in polygons

Activity 4

- What is the maximum number of internal right angles any particular polygon can have?

  Investigate for polygons with up to eight sides.

- Is there a relationship between the number of sides and the maximum number of right angles?

- Is there a relationship between the number of sides a polygon has and the total of its interior angles?

  Interior and exterior angles of polygons were explained in Chapter 6.

- Think of some 'What if … ?' questions. Investigate them.

In Chapter 6, it was stated that for any polygon,
interior angle + exterior angle = 180°.

This fact can be used to work out the number of sides of a regular polygon, if its interior angle is known.

Example 1 Find the number of sides of a regular polygon with an interior angle of 144°.

If the interior angle = 144° the exterior angle = 36°.

The sum of the exterior angles = 360°.

Therefore the number of sides = 360° ÷ 36° = 10.

Exercise 13.3a

1. Work out the number of sides of the regular polygons with these interior angles.
   (a) 150°   (b) 135°   (c) 140°   (d) 60°
   (e) 156°

2. Is it possible to make a tessellation which has six identical regular polygons fitting round a point?

   (a) If so, what size must the interior angle be?
   (b) How many sides will the polygon have?

3. Calculate the interior angle of a regular polygon with:
   (a) 100 sides   (b) 1000 sides.

1. The interior angle of a regular polygon is 160°.
   (a) How many sides does it have?
   (b) Will it tessellate?
2. The interior angle of a regular polygon is $13x$ and the exterior angle is $2x$.
   (a) Calculate x.
   (b) Find the number of sides of the polygon.
3. (a) Copy this table for regular polygons, and complete it.

Name of shape	Number of sides	Exterior angle	Interior angle
equilateral triangle	3		
square	4		
pentagon	5		
hexagon	6		
heptagon	7		
octagon	8		
nonagon	9		
decagon	10		
dodecagon	12		

   (b) Draw a graph plotting the interior angle against the number of sides. What do you notice?
   (c) Is it possible to have a regular polygon with $3\frac{1}{2}$ sides?
   (d) Follow these steps and try to draw it. You will need about half a page for the drawing.

   (i) From the graph of interior angle plotted against number of sides, when there are $3\frac{1}{2}$ sides the angle will be about 77°.
   You could check this by calculation.
   (ii) Follow the sketch below as you read the instructions.
   Draw a line 10 cm long across the page (line 1). At one end, measure and draw the second line (line 2), also 10 cm long, at an angle of 77° to the first line. Continue drawing lines 3, 4, 5, 6 and 7 in the same way. Line 7 should bring you back to the start.

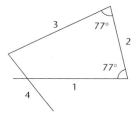

   (e) As an improper fraction, $3\frac{1}{2}$ is $\frac{7}{2}$. Can you see a link with your drawing and the figures 7 and 2?
   (f) Investigate other polygons with fractional numbers of sides. Try a polygon with $2\frac{1}{2}$ sides.

Angles in circles

(a) Draw a circle, radius 6 cm. Mark the centre O.
Mark three points, A, B and C, as shown. Join AO, AC, BO, BC.
Measure angle AOB and angle ACB. What do you notice?
Try with other points on the circle.
Suggest two general results.

(b) Now draw a straight line to touch the circle at A. This is a tangent.
Measure the angle between the tangent and the radius at A.
Try this at other points.
Suggest a general result.

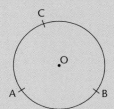

(c) Draw another circle.
Mark four points round the circumference, A, B, C and D, join them in order.
Measure angles ABC and CDA. Add them.
Measure angles BCD and DAB. Add them.
Try this for another circle and quadrilateral.
Suggest a general result.

General results

- The angle subtended by an arc (or chord) of a circle is twice the angle subtended by the same arc at the circumference.
- Angles subtended at the circumference by the same arc (or chord) are equal.
- The angle between a tangent and the radius at the point of contact is 90°.
- The opposite angles of a cyclic quadrilateral add up to 180°.

Definitions:
- 'subtended by' means made, or based on

- a cyclic quadrilateral has all its vertices on the circumference of a circle

Exercise 13.4a

Work out the sizes of all the angles marked with letters. In each case, O is the centre of the circle.

1.
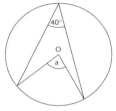
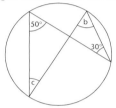
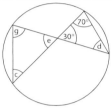

2.
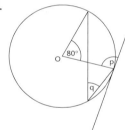
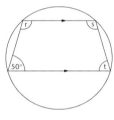
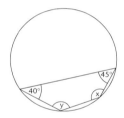

Exercise 13.4b

Work out the sizes of all the angles marked with letters. In each case, O is the centre of the circle.

1.
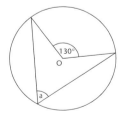
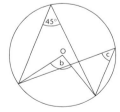
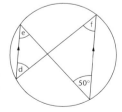

2.
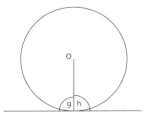
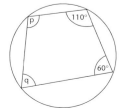

Key points

- The exterior angle of a triangle is equal to the sum of the interior opposite angles.
- The angle in a semicircle is 90°.
- The exterior angle of an *n*-sided regular polygon = 360 ÷ *n*.
- Interior angle = 180° – exterior angle.
- The angle at the centre is twice the angle at the circumference.

- Angles on the same arc (or chord) are equal.
- The angle between the radius and the tangent at the point of contact is a right angle.
- Opposite angles of a cyclic quadrilateral add up to 180°.

Revision exercise 13a

1. Find the size of the angles marked with letters.

(a)   (b)   (c)

2. In each diagram, O is the centre of the circle. Find the size of the angles marked *x*.

(a) 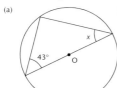 (b) 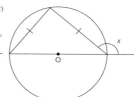 (c)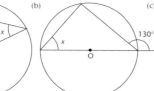

3. Calculate the size of the interior angle of a regular polygon with 20 sides.
4. Calculate the number of sides of a regular polygon with an interior angle of 168°.
5. Find the sizes of the angles marked with letters. In each case, O is the centre of the circle.

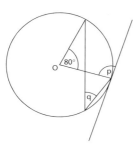 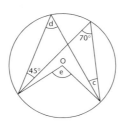 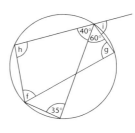

You will need to know:

- how to find the mode, median and mean of non-grouped data
- how to find the range of a set of ungrouped data
- how to construct frequency polygons
- how to draw bar charts for discrete data.

Grouped discrete data

When working with large amounts of data, it is often easier to see the pattern of the data if they are grouped. For example, this is a list of goals scored in 20 matches.

1   1   3   2   0   0   1   4   0   2
2   0   6   3   4   1   1   3   2   1

Instead, the information could be set out like this.

Number of goals	Frequency
0	4
1	6
2	4
3	3
4	2
5	0
6	1

Mode

From the table, it is easy to identify the number of goals with the greatest frequency. This is the mode of the data. Here the mode is 1 goal.

Mean

The table can also be used to calculate the mean.

There are: four matches with 0 goals $4 \times 0 = 0$ goals
           six matches with 1 goal $6 \times 1 = 6$ goals
           four matches with 2 goals $4 \times 2 = 8$ goals
and so on.

To find the total number of goals scored altogether, multiply each number of goals by its frequency and then add the results.

Then dividing by the total number of matches (20) gives the mean.

The working for this is shown in this table.

Number of goals	Frequency	Number of goals × frequency
0	4	0
1	6	6
2	4	8
3	3	9
4	2	8
5	0	0
6	1	6
Totals	20	37

Mean = 37 ÷ 20 = 1.85 goals.

In this example you have been given the original data. Check the answer by adding up the list of goals scored and dividing them by 20!

Examiner's tip

When given a table of grouped data, add an extra column if necessary to help you work out the values multiplied by their frequencies.

Example Work out the mean, mode and range for the number of children in the houses in Berry Road, listed in this table.

Number of children (c)	Frequency (number of houses)	c × frequency
0	6	0
1	4	4
2	5	10
3	7	21
4	1	4
5	2	10
Totals	25	49

Mean = 49 ÷ 25 = 1.96 children
Mode = 3 children
Range = 5 – 0 = 5 children

Exercise 14.1a

1. Ben has counted the number of sweets in ten packets. Here are the results.

    12 11 10 10 12 13 12 11
    12 11

    (a) Make a frequency table for Ben's results.
    (b) What is the mode of his results?
    (c) Use the frequency table to calculate the mean number of sweets.
    (d) Use the original list to calculate the mean, to check your results.

2. Here are some results for the number of crisps in a bag.

Number of crisps	25	26	27	28	29
Frequency	4	9	16	7	4

    (a) How many bags of crisps were counted?
    (b) What was the total number of crisps in these bags?
    (c) What was the mean number of crisps in these bags?

3. (a) What is the mode for the data in question 2?
    (b) What is the range for these data?

4. Find the mean value of x in these data.

x	7	8	9	10	11	12
Frequency	6	0	12	23	8	16

5. Here are the numbers of letters a postman delivered to the houses in Selly Road one morning.

Number of letters	0	1	2	3	4	5	6
Number of houses	4	5	7	2	6	0	2

    (a) How many houses are there in Selly Road?
    (b) What was the mode of the number of letters delivered there?
    (c) What was the mean number of letters delivered there? Give your answer correct to one decimal place.

Exercise 14.1b

1. Jenny counted the numbers of peas in ten pods. Here are the results.

    5 6 4 5 6 5 4 4 3 5

    (a) Make a frequency table for Jenny's results.
    (b) What is the mode of her results?
    (c) Use your frequency table to calculate the mean number of peas.
    (d) Use the original list to calculate the mean, to check your results.

2. Here are some results for the number of matches in a box.

Number of matches	43	44	45	46	47
Frequency	6	8	17	15	4

    (a) How many boxes of matches were counted?
    (b) What was the total number of matches in these boxes?

    (c) What was the mean number of matches in these boxes?

3. (a) What is the mode for the data in question 2?
    (b) What is the range for these data?

4. Find the mean value of x in these data.

x	5	6	7	8	9	10
Frequency	13	9	0	13	24	3

5. In a game, scores from 1 to 10 are possible. Dipta had 60 goes and obtained these scores.

Score	1	2	3	4	5	6	7	8	9	10
Frequency	3	4	8	7	11	9	1	4	5	8

    (a) What was Dipta's modal score?
    (b) Calculate Dipta's mean score.

Representing continuous data

When data involve measurements they are always grouped, even if they don't look like it. For instance, a length L given as 18 cm to the nearest centimetre means $17.5 \leqslant L < 18.5$. Any length between these values will count as 18 cm. Often, however, the groups are larger, to make handling the data easier. For example, when recording the heights, h cm, of 100 students in year 11, groups such as $180 \leqslant h < 185$ may be used.

Bar graph

Continuous data may be represented on a bar graph, using proper scales on both axes. Where the groups are not of the same width, the graph is called a **histogram**, and the **area** of each bar represents the frequency. In this chapter, however, only bars of equal width are considered, and the height of the bars represents the frequency, as with bar charts you have studied before.

Example

Draw a bar graph to represent this information about the heights of students in year 11 at Sandish School.

Height (h cm)	Frequency
$155 \leqslant h < 160$	2
$160 \leqslant h < 165$	6
$165 \leqslant h < 170$	18
$170 \leqslant h < 175$	25
$175 \leqslant h < 180$	9
$180 \leqslant h < 185$	4
$185 \leqslant h < 190$	1

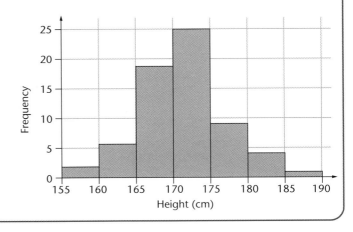

Examiner's tip

Check that you have labelled both scales carefully and that the boundaries of your bars match the boundaries of the groups

Frequency polygons

A frequency polygon may also be used to represent the data. However, in this case, only one point is used to represent each group. The midpoint value of each group is chosen, as it is an average value for the group.

Examiner's tip

To work out the midpoint of a group, add together its boundary values and divide by 2.

Example 3

Show the heights of the year 11 students in Sandish School in a frequency polygon.

The midpoint of the $155 \leqslant h < 160$ group is
$$\frac{155 + 160}{2} = 157.5.$$

So the points are plotted at h-values of 157.5, 162.5, 167.5 and so on.

Examiner's tip

It can be helpful to add a column to the frequency table, like this.

Height (h cm)	Frequency	Midpoint
$155 \leqslant h < 160$	2	157.5
$160 \leqslant h < 165$	6	162.5
$165 \leqslant h < 170$	18	167.5
$170 \leqslant h < 175$	25	172.5
$175 \leqslant h < 180$	9	177.5
$180 \leqslant h < 185$	4	182.5
$185 \leqslant h < 190$	1	187.5

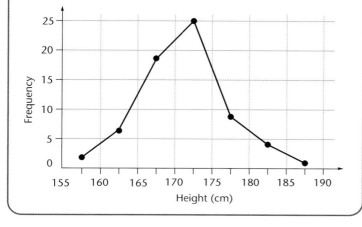

Calculating with grouped continuous data

Finding the mean

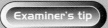
When working out the mean of grouped data in a table, you do not know the exact value for each item of data, so again the midpoint value is chosen to represent each group, and this is used to calculate an estimate of the mean. The midpoint is multiplied by the frequency of the group, as in calculating the mean of grouped discrete data.

Calculate an estimate of the mean height of the students in year 11 at Sandish School.

Height (h cm)	Frequency	Midpoint	Midpoint × frequency
$155 \leqslant h < 160$	2	157.5	315
$160 \leqslant h < 165$	6	162.5	975
$165 \leqslant h < 170$	18	167.5	3015
$170 \leqslant h < 175$	25	172.5	4312.5
$175 \leqslant h < 180$	9	177.5	1597.5
$180 \leqslant h < 185$	4	182.5	730
$185 \leqslant h < 190$	1	187.5	187.5
Totals	65		11 132.5

Mean = 11 132.5 ÷ 65 = 171.3 cm, correct to one decimal place.

Mode and range of grouped continuous data

The **modal class** may be found when data are given as a table, frequency polygon or bar graph. It is the class with the highest frequency.

The **range** cannot be stated accurately from grouped data. For instance, the height of the tallest student in the example might be 189.7 cm or 185.0 cm. As with the mean, the midpoints of the groups are used to estimate the range.

Estimating the median of a grouped continuous distribution is covered in Chapter 18.

> **Example** ⑤   For the heights of the year 11 students in Sandish School,
>
> (a) state the modal class
>
> (b) estimate the range.
>
> (a) The modal class is the one with the largest frequency, which is
>   170 cm ⩽ height < 175 cm or 170–175 cm.
>
> (b) An estimate of the range is found by finding the difference between the
>   midpoint values of the top and bottom groups in the table. This gives
>   range = 187.5 – 157.5 = 30 cm.

Exercise 14.2a

1. State the boundaries of these intervals.
   (a) 18 cm, to the nearest cm
   (b) 35 m, to the nearest m
   (c) masses to the nearest gram: 5–9, 10–14
   (d) times to the nearest second: 2–3, 4–5
2. State the midpoints of these intervals.
   (a) 10 cm < length ⩽ 20 cm
   (b) 2.0 m ⩽ length ⩽ 2.5 m
   (c) 80 kg ⩽ mass < 85 kg
   (d) masses to the nearest kg: 81–85, 86–90
   (e) times to the nearest second: 31–40, 41–50
3. (a) Calculate an estimate of the mean of these times.

Time (seconds)	0–2	2–4	4–6	6–8	8–10
Frequency	4	6	3	2	7

   (b) Draw a bar graph of this distribution.
4. (a) Calculate an estimate of the mean of these heights.

Height (cm)	50–60	60–70	70–80	80–90	90–100
Frequency	15	23	38	17	7

   (b) Draw a frequency polygon to represent this distribution.
5. Calculate an estimate of the mean of these lengths.

Length (m)	1.0–1.2	1.2–1.4	1.4–1.6	1.6–1.8	1.8–2.0
Frequency	2	7	13	5	3

6. Draw a bar graph to show this information.

Length (y cm)	Frequency
$10 \leqslant y < 20$	2
$20 \leqslant y < 30$	6
$30 \leqslant y < 40$	9
$40 \leqslant y < 50$	5
$50 \leqslant y < 60$	3

Chapter 14   Calculating with grouped continuous data

Exercise 14.2a continued

7. For the data in question 6:
   (a) state the modal class
   (b) calculate an estimate of the mean.

8. Draw a frequency polygon to show these data.

Mass of tomato (t g)	Frequency
$35 < t \leqslant 40$	7
$40 < t \leqslant 45$	13
$45 < t \leqslant 50$	20
$50 < t \leqslant 55$	16
$55 < t \leqslant 60$	4

9. For the data in question 8:
   (a) estimate the range
   (b) calculate an estimate of the mean.

10. The bar graph shows the masses of a sample of 50 eggs.

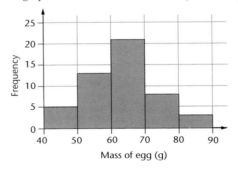

   (a) Make a frequency table for this information.
   (b) Calculate an estimate of the mean mass of these eggs.

Exercise 14.2b

1. State the boundaries of these intervals.
   (a) 20 cm, to the nearest cm
   (b) 41 m, to the nearest m
   (c) masses to the nearest gram: 15–16, 17–18
   (d) times to the nearest second: 11–15, 16–20

2. State the midpoints of these intervals.
   (a) 15 cm < length $\leqslant$ 20 cm
   (b) 12.0 cm $\leqslant$ length < 12.5 cm
   (c) 100 kg $\leqslant$ mass < 105 kg
   (d) masses to the nearest kg: 100–104, 105–109
   (e) times to the nearest second: 24–26, 27–29

3. (a) Calculate an estimate of the mean of these times.

Time (seconds)	0–20	20–40	40–60	60–80	80–100
Frequency	4	9	13	8	6

   (b) Draw a bar graph of this distribution.

4. (a) Calculate an estimate of the mean of these heights.

Height (m)	0–2	2–4	4–6	6–8	8–10
Frequency	12	26	34	23	5

   (b) Draw a frequency polygon to represent this distribution.

5. Calculate an estimate of the mean of these lengths.

Length (cm)	3.0–3.2	3.2–3.4	3.4–3.6	3.6–3.8	3.8–4.0
Frequency	3	8	11	5	3

6. Draw a bar graph to show this information.

Mass (w kg)	Frequency
$30 \leqslant w < 40$	5
$40 \leqslant w < 50$	8
$50 \leqslant w < 60$	2
$60 \leqslant w < 70$	4
$70 \leqslant w < 80$	1

7. For the data in question 6:
   (a) state the modal class
   (b) calculate an estimate of the mean.

8. Draw a frequency polygon to show these data.

Length (x cm)	Frequency
$0 < x \leqslant 5$	8
$5 < x \leqslant 10$	6
$10 < x \leqslant 15$	2
$15 < x \leqslant 20$	5
$20 < x \leqslant 25$	1

9. For the data in question 8:
   (a) estimate the range
   (b) calculate an estimate of the mean.

10. The bar graph shows the heights of students in year 9.

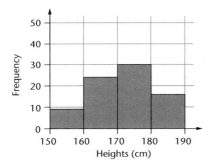

   (a) Make a frequency table for this information.
   (b) Calculate an estimate of the mean height.

Mean, median or mode?

Each of these terms may be called an average. Sometimes, you need to decide which of them is the best to use in a given situation.

The mean is what most people think of as the average, but it may be 'fairer' to use the mode or median.

Look at these annual wages, for example.

Annual wage (£)	Number of employees
10 000–15 000	2
15 000–20 000	18
20 000–25 000	12
25 000–30 000	4
30 000–35 000	0
35 000–40 000	2

The modal group is £15 000–20 000.

Calculating the mean gives:

Midpoint	Frequency	Midpoint × frequency
12 500	2	25 000
17 500	18	315 000
22 500	12	270 000
27 500	4	110 000
32 500	0	0
37 500	2	75 000
Totals	38	795 000

Mean = 795 000 ÷ 38 = 20 921.0526 = £20 900 to the nearest £100.

Which gives a better idea of the average here, the mean or the mode? It depends on the purpose for which you want to use the average. If you were arguing for a pay rise you would probably use the mode. If you were the management, you would be more likely to use the mean.

If a distribution is quite symmetrical, there is not much difference between the mean, median and mode. If the distribution is weighted to one side (or **skewed**), then it matters much more which is chosen. In this case, the mean may not give the most representative average. Be prepared to give a reason why you have chosen to use a particular average.

Key points

- To find the mean of grouped data, multiply each value by its frequency, add the results, and divide by the sum of the frequencies. For continuous data, use the midpoints of each group as the value for the group.
- For bar graphs of continuous data, label both axes with a scale. Make sure the edges of each bar are at the boundaries of their group.

- To use a frequency polygon to represent continuous data, plot the midpoints of each group.
- To estimate the range of continuous data, subtract the midpoints of the end groups.
- The modal class for continuous data is the group with the largest frequency.

Revision exercise 14a

1. The numbers of matches won by a school's sports teams during one term are shown in this table.

Number of matches won	0	1	2	3	4	5	6	7	8
Number of teams	2	1	0	2	4	6	3	0	2

Calculate the mean number of matches won by the teams.

2. A Biology class counted the number of daisies in 10-cm squares of grass on the field. Here are their results.

Number of daisies	0	1	2	3	4	5	6
Frequency	2	2	5	8	9	3	1

(a) How many 10-cm squares were counted?
(b) What was the total number of daisies found in these squares?
(c) What was the mode?
(d) What was the median?

3. The bar chart shows the results of a survey about the number of videos watched during one week.

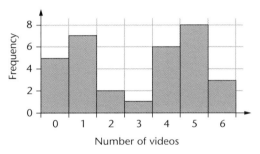

(a) State the mode.
(b) State the range.
(c) Draw up a frequency table to show these results.
(d) Calculate the mean number of videos watched.

4. Sanjit asked some people how many chocolates they had eaten at a party. These are the results.

Number of chocolates	2	3	4	5	6	7	8	9	10
Frequency	1	10	5	4	2	1	0	0	2

(a) State the mode.
(b) How many people took part in the survey?
(c) Find the median.
(d) Calculate the mean.
(e) Why does the mean not necessarily give the 'best' average here?

5. Harry picked and measured some runner beans. These were their lengths.

Length (L cm)	Frequency
$10 < L \leqslant 15$	3
$15 < L \leqslant 20$	7
$20 < L \leqslant 25$	11
$25 < L \leqslant 30$	8
$30 < L \leqslant 35$	1

(a) Draw a bar graph to show this information.
(b) Estimate the range of the length of these runner beans.

6. Calculate an estimate of the mean length of the runner beans in question 5.

7. Lisa timed her little brother when he was playing with his new toys over Christmas.

Time (t minutes)	Frequency
$0 < t \leqslant 10$	2
$10 < t \leqslant 20$	5
$20 < t \leqslant 30$	7
$30 < t \leqslant 40$	10
$40 < t \leqslant 50$	4

(a) Draw a frequency polygon for these data.
(b) Calculate an estimate of the mean of these times.

8. A class held a hopping race and recorded the distances they travelled before putting the other foot down. Here are the results.

Length (x m)	Frequency
$0 < x \leqslant 4$	1
$4 < x \leqslant 8$	4
$8 < x \leqslant 12$	8
$12 < x \leqslant 16$	5
$16 < x \leqslant 20$	2

For the above data:
(a) state the modal class
(b) estimate the range
(c) calculate an estimate of the mean.

9. Kim and Petra asked their class, 'How much exercise have you had this week?' These were the results.

Time of exercise (h hours)	Number of people
$0 \leqslant h < 1$	3
$1 \leqslant h < 2$	8
$2 \leqslant h < 5$	12
$5 \leqslant h < 10$	5
$h \geqslant 10$	0

(a) How many people were in the survey?
(b) What are the midpoints of the classes
(i) $2 \leqslant h < 5$ and (ii) $5 \leqslant h < 10$?
(c) Calculate an estimate of the mean time of exercise.

10. Some batteries were tested to see how long they lasted. Here are the results.

Time (hours)	0–2	2–4	4–6	6–8	8–10
Frequency	4	5	12	16	3

(a) Calculate an estimate of the mean time these batteries lasted.
(b) In what circumstances would the mode be a suitable average to use here?

You will need to know:

- the terms 'object' and 'image' as they apply to transformations
- how to recognise and draw the reflection of a simple shape in a mirror line
- how to recognise and draw an enlargement of a shape using a centre and scale factor of enlargement
- how to rotate a simple shape about its centre or the origin through 90°, 180° or 270°
- how to recognise and draw simple translations
- the lines $y = x$ and $y = -x$ (or $y + x = 0$).

Drawing reflections

To draw a **reflection** you need an object and a mirror line.

Reflect the L-shape in the given mirror line.

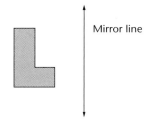

Method 1

Remember to keep the ruler at right angles to the mirror line.

For each vertex or corner of the L-shape, use a ruler to measure the perpendicular distance from the mirror line. Then measure the same distance on the other side of the mirror line to find the corresponding image point.

Method 2

Method 2 is often the easier one when the mirror line is sloping, as it is difficult to keep the ruler at right angles to the mirror line.

Using tracing paper, trace the object point and the mirror line. Turn the tracing paper over and line up the tracing of the mirror line with the original mirror line, but with the object on the other side. Using a pin or compass point, prick through the corners of the object onto the paper below. Remove the tracing paper and join up the pinpricks to draw the image.

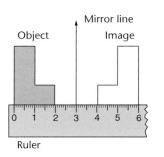

If the mirror line is easy to work with, for example the *x*-axis on a graph, it may be possible to plot the reflection by counting squares. It is still advisable to use tracing paper to check the reflection.

In mathematics all mirror lines are regarded as 'double-sided'. This means that any shape that crosses the mirror line will have part of its reflection on each side of the line.

Recognising reflections

It should be easy to recognise when a transformation is a reflection. If there is any doubt, check that the tracing paper needs to be **turned over** before it will fit on the image. Finding the mirror line can be more difficult.

Example 2

Describe the transformation that maps shape ABC onto shape PQR.

It should be fairly obvious that the transformation is a reflection but this could be checked using tracing paper.

To find the mirror line, put a ruler between two corresponding points (B and Q) and mark a point halfway between them, at (3, 3).

Repeat this for two other corresponding points (C and R). The midpoint is (4, 4).

Join the two midpoints to find the mirror line. The mirror line is $y = x$.

The transformation is a reflection in the line $y = x$.

Again, the result can be checked using tracing paper.

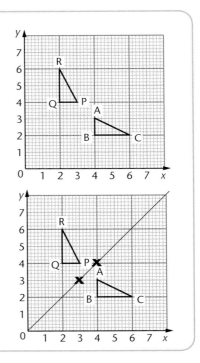

Rotations

A **rotation** involves turning the object about a point, called the **centre of rotation**.

Drawing rotations

Examiner's tip

Tracing paper is always stated as optional extra material in examinations. When doing transformation questions, **always** ask for it.

 Example 3
Rotate triangle ABC through 90° clockwise about C.

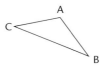

Measure an angle of 90° clockwise from the line AC.

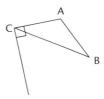

Trace the shape ABC. Put a pencil or pin at C to hold the tracing to the diagram at that point. Rotate the tracing paper until AC coincides with the new line you have drawn. Use another pin or the point of your compasses to prick through the other corners (A and B).

Join up the new points to form the image.

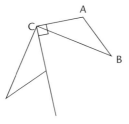

If the centre of rotation is not on the object then the method is slightly more difficult.

Rotate the triangle ABC through 90°
clockwise about the point O.

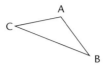

Join O to a point on the object (C).
Measure an angle of 90° clockwise from OC
and draw a line.

Trace the triangle ABC and the line OC.
Rotate the tracing about O until the line
OC coincides with the new line you have
drawn. Use a pin or the point of your
compasses to prick through the corners
(A, B and C). Join up the pin holes to form
the image.

For other angles of rotation (e.g. 120° clockwise), the first angle is
measured as the stated angle (e.g. 120°) instead of 90° but
otherwise the method is the same.

Examiner's tip

Always remember to state
whether the rotation is
clockwise or anticlockwise.

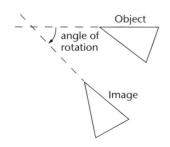

If the centre of rotation is easy to work with, for example the origin,
then you may be able to draw the rotation by counting squares but it
is always best to check using tracing paper.

Recognising rotations

It is usually easy to recognise when a transformation is a
rotation, as it should be possible to place a tracing of the object
over the image without turning the tracing paper over.

To find the angle of rotation, find a pair of sides that correspond
in the object and the image. Measure the angle between them.
You may need to extend both of these sides to do this.

If the centre of rotation is not on the object, its position may not
be obvious. The easiest method to use is trial and error, either by
counting squares or using tracing paper. In a later chapter, you
will learn a method which will find the centre directly, without
trial (or error!).

Example 5 Describe fully the transformation that maps flag A onto flag B.

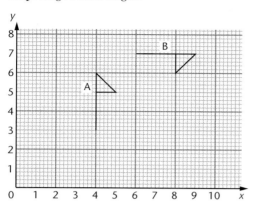

It is clear that the transformation is a rotation and that the angle is 90° clockwise. You may need to make a few trials, using tracing paper and a compass point centred on different points, to find that the centre of rotation is (7, 4).

If you did not spot it, try it now with tracing paper.

Exercise 15.1a

Label the diagrams you draw in this exercise carefully and keep them, as you will need them in a later exercise.

1. Draw a triangle with vertices at (1, 0), (1, −2) and (2, −2). Label it A. Draw the reflection of triangle A in the line $y = 1$. Label it B.

2. On the same grid as for question 1, reflect triangle B in the line $y = x$. Label the new triangle C.

3. On a new grid draw a triangle with vertices at (2, 5), (3, 5) and (1, 3). Label it D. Draw the reflection of triangle in the line $x = \frac{1}{2}$. Label it E.

4. On the same grid as for question 3, reflect triangle E in the line $y = -x$. Label the new triangle F.

5. Using graph paper, copy this diagram. Rotate the flag G through 90° clockwise about the point (1, 2). Label the new flag H.

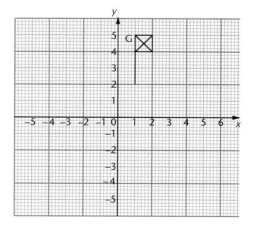

6. On the same grid as for question 5, rotate the flag H through 180° about the point (2, −1). Label the new flag I.

7. Draw a triangle with vertices at (0,1), (0, 4) and (2, 3). Label it J. Rotate triangle J through 90° anticlockwise about the point (2, 3). Label the new triangle K.

8. On the same grid as for question 7, rotate triangle K through 90° clockwise about the point (2, −1). Label the new triangle L.

9. Study the diagram below.

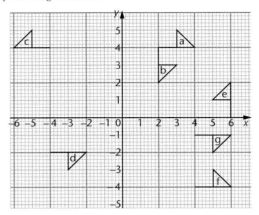

Describe fully the single transformation that maps:

(a) flag a onto flag b

(b) flag a onto flag c

(c) flag b onto flag d

(d) flag b onto flag e

(e) flag e onto flag f

(f) flag f onto flag g.

For questions 1–6, either use the worksheet or copy the diagrams carefully, making them larger if you wish.

1. Reflect the trapezium in the given mirror line.

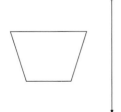

2. Reflect the triangle in the given mirror line.

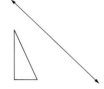

3. Reflect the triangle in the given mirror line.

4. Rotate the triangle through 180° about the point C.

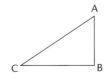

5. Rotate the triangle through 90° clockwise about the point O.

6. Rotate the triangle through 120° clockwise about the point O.

7. Study the diagram below.

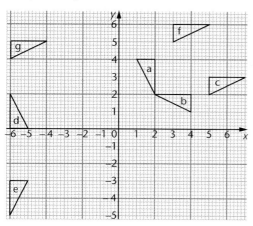

Describe fully the single transformation that maps:

(a) triangle a onto triangle b
(b) triangle a onto triangle c
(c) triangle a onto triangle d
(d) triangle d onto triangle e
(e) triangle e onto triangle f
(f) triangle d onto triangle g.

Translations

In a **translation**, every point of an object moves the same distance in the same direction. The object and the image look identical with no turning or reflection. It looks just as if the object has moved to a different position.

Drawing translations

To draw a translation, all you need to know is how far across the page and how far up the page to move the object.

Example 6   Translate the shape on the diagram 5 cm to the right and 3 cm down.

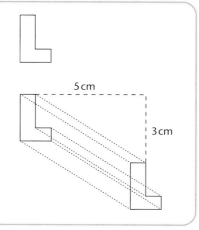

The dotted lines show that every point in the object has 'moved' 5 cm to the right and 3 cm down to the corresponding point in the image.

Column vectors

When working with translations, it is usual to work on a grid so it is not necessary to measure the movements.

On a grid, the movements can be described as a movement in the x-direction and a movement in the y-direction. They are written in the form of a **column vector**, for example $\begin{pmatrix} 5 \\ -3 \end{pmatrix}$.

In a column vector:
the top number represents the x-movement
the bottom number represents the y-movement.

The directions are the same as for coordinates.

If the top number is positive, move to the right.

If the top number is negative, move to the left.

If the bottom number is positive, move up.

If the bottom number is negative, move down.

Example 7

Translate the shape on the grid through the vector $\begin{pmatrix} -3 \\ 4 \end{pmatrix}$.

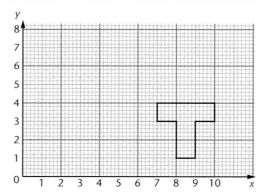

The movement represented by the vector $\begin{pmatrix} -3 \\ 4 \end{pmatrix}$ is 3 units to the left and 4 units up, so every point in the object 'moves' this amount to form the corresponding point in the image.

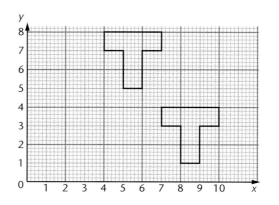

Recognising translations

It should be easy to recognise when a transformation is a translation, as the object and image look identical with no turning or reflecting. Having stated that the transformation is a translation, you need to find the column vector.

Identify a point on the object and its corresponding point on the image. Count:

- how many units left or right and
- how many units up or down

that point has moved. Write these movements as a column vector.

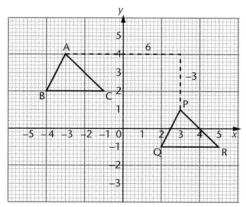

Example 8   Describe fully the transformation that maps triangle ABC onto triangle PQR.

Point A translates on to point P. This is a movement of 6 to the right and 3 down. The transformation is a translation through the vector $\begin{pmatrix} 6 \\ -3 \end{pmatrix}$.

Examiner's tip

Try not to mix up the words 'transformation' and 'translation'. Transformation is the general name for all the changes made to shapes. Translation is the particular transformation that has been described here.

Enlargements

An **enlargement** produces an image that is exactly similar in shape to the object, but is larger or smaller.

Drawing enlargements

Enlarge the triangle ABC with scale factor $2\frac{1}{2}$ and centre of enlargement O, to form triangle PQR.

Draw lines from O to A, O to B and O to C and extend them.

Measure the lengths OA, OB and OC. These are 2.0 cm, 1.5 cm and 2.9 cm respectively.

Multiply these lengths by 2.5 to give OP = 5.0 cm, OQ = 3.7 cm and OR = 7.2 cm.

Measure these distances along the extended lines OA, OB and OC, and mark P, Q and R.

Join P, Q and R to form the triangle.

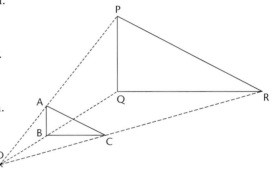

Enlarge the shape DEF with scale factor $\frac{1}{2}$ and centre of enlargement O.

The steps are exactly the same as for Example 9 except that, instead of being multiplied by 2.5, the distances are multiplied by 0.5.

Check that OU = $\frac{1}{2}$OD, OV = $\frac{1}{2}$OE and OW = $\frac{1}{2}$OF.

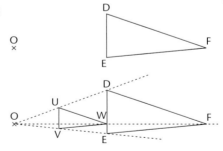

Notice that the length of each side in triangle UVW is half the length of the corresponding side in triangle DEF. In Mathematics this is still called an enlargement even though the image is smaller than the object. It just means that the scale factor is less than 1.

Examiner's tip

When you have drawn your enlargement check that the ratio of the sides of the image to the corresponding sides in the object is equal to the scale factor (in the case of Example 9, $2\frac{1}{2}$). If it is not, you have probably measured some or all of your distances from the points of the object and not from O.

Recognising enlargements

Example II

Describe fully the transformation that maps triangle DEF onto triangle STU.

The shapes are similar, so clearly the transformation is an enlargement.

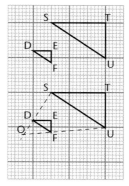

Since the lengths of the sides of triangle STU are 3 times the lengths of the corresponding sides of triangle DEF, the scale factor is 3. All that remains to be found is the centre of enlargement.

Join SD and extend it. Join UF and extend it to cross the extended line SD. The point where the lines cross, O, is the centre of enlargement.

The transformation is an enlargement, scale factor 3, centre of enlargement O.

If you were working on a grid, you would describe the centre of enlargement by stating the coordinates of the point.

Exercise 15.2a

Label the diagrams you draw in this exercise carefully and keep them, as you will need them in a later exercise.

1. Draw a triangle with vertices at (1, 2), (1, 4) and (2, 4). Label it A.
   Draw the translation of triangle A through the vector $\binom{5}{2}$. Label it B.

2. On the same grid as for question 1, translate triangle B through the vector $\binom{2}{-4}$. Label the new triangle C.

3. On a new grid, draw a triangle with vertices at (0, 2), (1, 4) and (3, 2). Label it D.
   Draw the translation of triangle D through the vector $\binom{-4}{2}$. Label it E.

4. On the same grid as for question 3, translate triangle E through the vector $\binom{8}{0}$. Label the new triangle F.

Draw a set of axes. Label the x-axis from 0 to 13 and the y-axis from 0 to 15. Use it to answer questions 5 and 6.

5. Draw a triangle with vertices at (1, 2), (2, 4) and (1, 3). Label it G.
   Draw the enlargement of triangle G with scale factor 2 and centre the origin. Label it H.

6. On the same grid as question 5, draw the enlargement of triangle H with scale factor 3 and centre of enlargement (0, 5). Label it I.

7. Copy the diagram. Enlarge the flag J with scale factor $1\frac{1}{2}$ and centre of enlargement (1, 2). Label the new flag K.

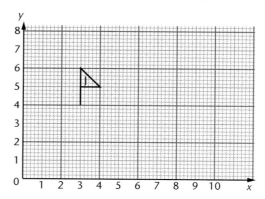

8. On the same grid you drew for question 7 enlarge the flag K with scale factor 2 and centre of enlargement (2, 8). Label the new flag L.

9. Study the diagram below.

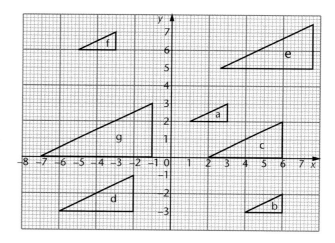

Describe fully the single transformation that maps:
(a) triangle a onto triangle b
(b) triangle a onto triangle c
(c) triangle c onto triangle d
(d) triangle a onto triangle e
(e) triangle a onto triangle f
(f) triangle g onto triangle a.

For questions 1–6, either use the worksheet or copy the diagrams carefully, making them larger if you wish.

1. Translate flag A through the vector $\begin{pmatrix} -2 \\ -5 \end{pmatrix}$. Label the new flag B.

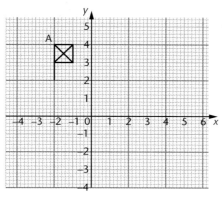

2. On the same grid you drew for question 1, translate flag B through the vector $\begin{pmatrix} 7 \\ 4 \end{pmatrix}$. Label the new flag C.

3. On the same grid as for question 1, translate flag C through the vector $\begin{pmatrix} -5 \\ 1 \end{pmatrix}$. Label the new flag D. What do you notice?

   Try to explain the result.

4. Enlarge the shape with centre O and scale factor 3.

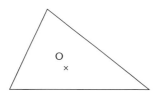

5. Enlarge the shape with centre O and scale factor $\frac{1}{3}$.

6. Enlarge the shape with centre O and scale factor $1\frac{1}{2}$.

Exercise 15.2b continued

7. Study the diagram below.

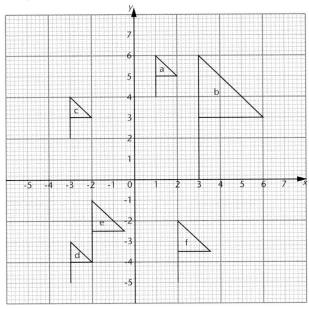

Describe fully the transformation that maps:

(a) flag a onto flag b   (b) flag a onto flag c   (c) flag c onto flag d

(d) flag d onto flag e   (e) flag e onto flag f   (f) flag b onto flag d.

Combining transformations

Sometimes when one transformation is followed by another, the result is equivalent to a single transformation. For example, in the following diagram, triangle A has been translated through the vector $\binom{2}{5}$ onto triangle B.

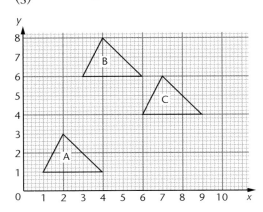

Triangle B has then been translated through the vector $\begin{pmatrix} 3 \\ -2 \end{pmatrix}$ onto triangle C.

Notice that triangle A could have been translated directly onto triangle C through the vector $\begin{pmatrix} 5 \\ 3 \end{pmatrix}$.

So the first transformation followed by the second transformation is equivalent to the single transformation, translation through the vector $\begin{pmatrix} 5 \\ 3 \end{pmatrix}$.

Examiner's tip

Make sure you do the transformations in the right order, as it usually makes a difference.

If a question asks for a single transformation, do not give a combination of two transformations as this does not answer the question and will usually score no marks.

Example 12

Find the single transformation that is equivalent to a reflection in the line $x = 1$, followed by a reflection in the line $y = -2$.

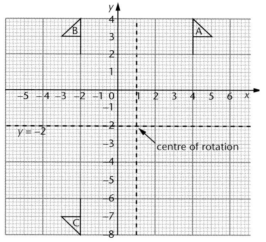

In the diagram, reflecting the object flag A in the line $x = 1$ gives flag B.

Reflecting flag B in the line $y = -2$ gives flag C.

The transformation that maps A directly onto C is a rotation through 180°.

The centre of rotation is $(1, -2)$, which is where the mirror lines cross.

> Use tracing paper to check this.

The transformation is a rotation through 180° about the centre of rotation $(1, -2)$.

> A rotation of 180° is the only rotation for which you do not need to state the direction, as 180° clockwise is the same as 180° anticlockwise.

In this exercise you will need some of the diagrams you drew in Exercises 15.1a and 15.2a.

1. Look back at the diagram you drew for questions 1 and 2 of Exercise 15.2a. Describe fully the single transformation that is equivalent to a translation through the vector $\begin{pmatrix} 5 \\ 2 \end{pmatrix}$ (A onto B) followed by a translation through the vector $\begin{pmatrix} 2 \\ -4 \end{pmatrix}$ (B onto C).

2. Look back at the diagram you drew for questions 3 and 4 of Exercise 15.2a. Describe fully the single transformation that is equivalent to a translation through the vector $\begin{pmatrix} -4 \\ 2 \end{pmatrix}$ (D onto E) followed by a translation through the vector $\begin{pmatrix} 8 \\ 0 \end{pmatrix}$ (E onto F).

3. Look at your answers to the last two questions. Try to make a general statement about the result of translating through the vector $\begin{pmatrix} a \\ b \end{pmatrix}$ followed by translation through the $\begin{pmatrix} c \\ d \end{pmatrix}$ vector.

4. Look back at the diagram you drew for questions 5 and 6 of Exercise 15.2a. Describe fully the single transformation that is equivalent to an enlargement scale factor 2, centre the origin (G onto H) followed by an enlargement scale factor 3, centre (0, 5) (H onto I).

5. Look back at the diagram you drew for questions 7 and 8 of Exercise 15.2a. Describe fully the single transformation that is equivalent to an enlargement, scale factor $1\frac{1}{2}$, centre the point (1, 2) (J onto K) followed by an enlargement, scale factor 2, centre (2, 8) (K onto L).

6. Look again at the answers to the last two questions. Try to make a general statement about the result of enlarging with scale factor p followed by enlarging with scale factor q.

7. Look back at the diagram you drew for questions 1 and 2 of Exercise 15.1a. Describe fully the single transformation that is equivalent to a reflection in the line $y = 1$ (A onto B) followed by a reflection in the line $y = x$ (B onto C).

8. Look back at the diagram you drew for questions 3 and 4 of Exercise 15.1a. Describe fully the single transformation that is equivalent to a reflection in the line $x = \frac{1}{2}$ (D onto E) followed by a reflection in the line $y = -x$ (E onto F).

9. Look again at the answers to the last two questions. Try to make a general statement about the result of reflection in a mirror line followed by reflection in an intersecting mirror line.

10. Look back at the diagram you drew for questions 5 and 6 of Exercise 15.1a. Describe fully the single transformation that is equivalent to a rotation through 90° clockwise about the point (1, 2) (G onto H) followed by a rotation through 180° about the point (2, −1) (H onto I).

11. Look back at the diagram you drew for questions 7 and 8 of Exercise 15.1a. Describe fully the single transformation that is equivalent to a rotation through 90° anticlockwise about the point (2, 3) (J onto K) followed by a rotation through 90° clockwise about the point (2, −1) (K onto L).

In this exercise, carry out the transformations on a simple object shape of your choice.

1. Describe fully the single transformation that is equivalent to a reflection in the x-axis followed by reflection in the y-axis.

2. Describe fully the single transformation that is equivalent to a reflection in the line $x = 1$ followed by a reflection in the line $x = 5$.

3. Describe fully the single transformation that is equivalent to a reflection in the line $y = 2$ followed by a reflection in the line $y = 6$.

4. Look again at your answers to the last two questions. Try to make a general statement about the result of reflecting in a mirror line followed by a reflection in a parallel mirror line.

5. Describe fully the single transformation that is equivalent to an enlargement, scale factor 2 and centre the origin, followed by a translation through the vector $\begin{pmatrix} 3 \\ 2 \end{pmatrix}$.

6. Describe fully the single transformation that is equivalent to a rotation through 90° clockwise about the origin, followed by a translation through the vector $\begin{pmatrix} 4 \\ 0 \end{pmatrix}$.

7. Describe fully the single transformation that is equivalent to a reflection in the x-axis followed by a rotation through 90° anticlockwise about the origin.

8. Describe fully the single transformation that is equivalent to a reflection in the y-axis followed by a rotation through 90° anticlockwise about the origin.

9. Describe fully the single transformation that is equivalent to a reflection in the line $y = x$ followed by a reflection in the line $y = -x$.

10. Describe fully the single transformation that is equivalent to a rotation through 90° clockwise about the point (2, 1) followed by a rotation through 90° anticlockwise about the point (3, 4).

Key points

- Use tracing paper to carry out or check rotations and reflections.
- When describing transformations, always give the name of the transformation first and then the extra information required.

Name of transformation	Extra information
Reflection	Mirror line
Rotation	Angle, direction, centre of rotation
Translation	Column vector
Enlargement	Scale factor, centre of enlargement

- When asked to describe a single transformation, do not give a combination of transformations.

Revision exercise 15a

1. Draw a triangle with vertices at (1, 4), (1, 6) and (2, 6). Label it A. Reflect triangle A in the line $y = x$. Label it B.

2. On the same grid you drew for question 1, rotate triangle B through 90° anticlockwise about the point (5, 5). Label the new triangle C.

3. Look again at the diagrams you drew for the last two questions. Describe fully the single transformation that is equivalent to reflection in the line $y = x$ followed by a rotation through 90° anticlockwise about the point (5, 5).

4. On a new grid, draw a triangle with vertices at (4, 1) (6, 1) and (4, 2). Label it D. Translate triangle D through the vector $\begin{pmatrix} 2 \\ 3 \end{pmatrix}$. Label it E.

5. On the same grid you drew for question 4, enlarge triangle E with scale factor 2 and centre of enlargement (5, 7). Label the new triangle F.

6. Look again at the diagrams you drew for the last two questions. Describe fully the single transformation that is equivalent to translation through the vector $\begin{pmatrix} 2 \\ 3 \end{pmatrix}$ followed by enlargement with scale factor 2 and centre of enlargement (5, 7).

7. Study the diagram below.

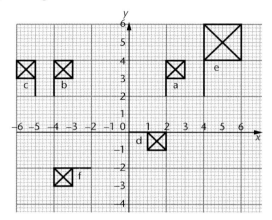

Describe fully the single transformation that maps:

(a) flag a onto flag b

(b) flag b onto flag c

(c) flag a onto flag d

(d) flag d onto flag b

(e) flag e onto flag a

(f) flag a onto flag f

You will need to know how to:

- multiply and divide negative numbers
- find a percentage of a quantity
- express a number as a percentage of another number
- calculate with fractions, percentages and decimals
- work out squares and square roots
- do straightforward calculations using a calculator.

Fractions

In a fraction $\frac{a}{b}$, a is called the **numerator** and b is the **denominator**.

Equivalent fractions

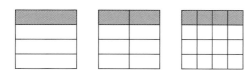

These squares can be divided into equal parts in different ways.

The fraction represented by the shaded parts can be thought of as $\frac{1}{4}$ or $\frac{2}{8}$ or $\frac{4}{16}$.

These three fractions are equal in value and are **equivalent**.

You can find equivalent fractions by multiplying or dividing the numerator *and* the denominator by the same number.

Example ❶ Fill in the missing numbers in these equivalent fractions.

(a) $\dfrac{1}{2} = \dfrac{2}{\square} = \dfrac{\square}{14}$ (b) $\dfrac{3}{8} = \dfrac{\square}{24} = \dfrac{15}{\square}$

(a) $\dfrac{1}{2} = \dfrac{2}{4} = \dfrac{7}{14}$ Multiply the first fraction by $\frac{2}{2}$, then multiply it by $\frac{7}{7}$.

(b) $\dfrac{3}{8} = \dfrac{9}{24} = \dfrac{15}{40}$ Multiply the first fraction by $\frac{3}{3}$, then multiply it by $\frac{5}{5}$.

You may be asked to write a fraction in its **lowest terms**.

This means finding the smallest possible denominator.

This is sometimes called **cancelling** the fractions.

 Write these fractions in their lowest terms.

(a) $\frac{6}{10}$ (b) $\frac{8}{12}$ (c) $\frac{15}{20}$

(a) $\frac{6}{10} = \frac{3}{5}$ Divide the numerator and denominator by 2.

(b) $\frac{8}{12} = \frac{2}{3}$ $\frac{8}{12}$ is also equivalent to $\frac{4}{6}$ but it can be simplified further to $\frac{2}{3}$.

(c) $\frac{15}{20} = \frac{3}{4}$ Divide top and bottom by 5.

Adding fractions

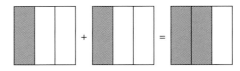

In this diagram, each square is divided into thirds.

The diagram shows $\frac{1}{3} + \frac{1}{3} = \frac{2}{3}$.

Counting squares or columns, 1 column + 1 column = 2 columns. If the denominator is the same in the fractions, just add the numerators.

In this diagram there are 12 small squares in each rectangle.

The diagram shows $\frac{1}{3} + \frac{1}{4}$.

Counting squares, 4 squares + 3 squares = 7 squares or $\frac{7}{12}$

To add two fractions, change them to the same type, so they have the same denominator.

$\frac{1}{3} = \frac{4}{12} \quad \frac{1}{4} = \frac{3}{12} \quad \frac{4}{12} + \frac{3}{12} = \frac{7}{12}$

Example 3

Add the fractions.

(a) $\frac{2}{5} + \frac{1}{5}$ (b) $\frac{1}{5} + \frac{3}{10}$ (c) $\frac{1}{6} + \frac{3}{4}$

(a) $\frac{2}{5} + \frac{1}{5} = \frac{3}{5}$ They have the same denominator so just add the numerators.

(b) $\frac{1}{5} + \frac{3}{10}$ They need to be changed so that they both have the same denominator.

  $= \frac{2}{10} + \frac{3}{10}$ Change $\frac{1}{5}$ to $\frac{2}{10}$, so they both have 10 as the denominator.

  $= \frac{5}{10} = \frac{1}{2}$ Write the answer in the lowest terms.

(c) $\frac{1}{6} + \frac{3}{4}$ This time they both need to be changed, to have the same denominator.

  $= \frac{2}{12} + \frac{9}{12} = \frac{11}{12}$ $\frac{1}{6} = \frac{2}{12} = \frac{3}{18} = ..., \frac{3}{4} = \frac{6}{8} = \frac{9}{12} = ...$

Subtracting fractions

In this diagram there are 20 small squares in each rectangle.

The diagram shows $\frac{1}{4} - \frac{1}{5}$.

5 squares $-$ 4 squares $= 1$ square $= \frac{1}{20}$

This can be written as $\frac{1}{4} - \frac{1}{5} = \frac{5}{20} - \frac{4}{20} = \frac{1}{20}$

Change both so that they have the same denominator, as with adding, but then subtract the numerators.

Examiner's tip

When adding or subtracting fractions, change to the same type (same denominator) and add or subtract the numerators. The most common error is to add the denominators and add the numerators.

Example 4

Work these out.

(a) $\frac{3}{5} - \frac{2}{5}$ (b) $\frac{3}{4} - \frac{2}{3}$ (c) $\frac{5}{6} + \frac{1}{4} - \frac{1}{3}$

(a) $\frac{3}{5} - \frac{2}{5} = \frac{1}{5}$ They have the same denominator so just subtract the numerators.

(b) $\frac{3}{4} - \frac{2}{3} = \frac{9}{12} - \frac{8}{12} = \frac{1}{12}$ 4 and 3 both divide into 12, so make 12 the denominator for both. Multiply $\frac{3}{4}$ by 3 top and bottom, and $\frac{2}{3}$ by 4 top and bottom.

(c) $\frac{5}{6} + \frac{1}{4} - \frac{1}{3}$ All the denominators divide into 12.

  $= \frac{10}{12} + \frac{3}{12} - \frac{4}{12}$

  $= \frac{9}{12} = \frac{3}{4}$

Chapter 16 Fractions

1. Fill in the blanks in these equivalent fractions.
   (a) $\frac{1}{7} = \frac{2}{\square} = \frac{\square}{35}$ (b) $\frac{4}{9} = \frac{16}{\square} = \frac{\square}{72}$

2. Write these fractions in their lowest terms.
   (a) $\frac{6}{8}$ (b) $\frac{12}{15}$ (c) $\frac{12}{24}$ (d) $\frac{12}{54}$

For the rest of the questions, give all answers in their lowest terms.

3. Add these fractions.
   (a) $\frac{2}{3} + \frac{1}{3}$ (b) $\frac{1}{3} + \frac{1}{2}$ (c) $\frac{3}{5} + \frac{1}{4}$ (d) $\frac{1}{6} + \frac{2}{3}$ (e) $\frac{2}{5} + \frac{3}{8}$ (f) $\frac{3}{4} + \frac{1}{6}$

4. Subtract these fractions.
   (a) $\frac{2}{7} - \frac{1}{7}$ (b) $\frac{5}{6} - \frac{1}{3}$ (c) $\frac{2}{3} - \frac{1}{4}$ (d) $\frac{11}{12} - \frac{2}{3}$ (e) $\frac{5}{8} - \frac{1}{3}$ (f) $\frac{7}{9} - \frac{5}{12}$

5. Work these out.
   (a) $\frac{4}{5} + \frac{7}{10} - \frac{3}{5}$ (b) $\frac{3}{5} + \frac{5}{6} - \frac{2}{3}$ (c) $\frac{2}{3} + \frac{3}{4} - \frac{1}{2}$ (d) $\frac{2}{5} + \frac{5}{8} - \frac{3}{4}$
   (e) $\frac{1}{5} + \frac{3}{10} - \frac{1}{2}$ (f) $\frac{3}{7} + \frac{5}{14} - \frac{1}{2}$

1. Fill in the blanks in these equivalent fractions.
   (a) $\frac{1}{6} = \frac{4}{\square} = \frac{\square}{12}$ (b) $\frac{2}{3} = \frac{\square}{6} = \frac{12}{\square} = \frac{\square}{24}$

2. Write these fractions in their lowest terms.
   (a) $\frac{8}{16}$ (b) $\frac{9}{15}$ (c) $\frac{10}{25}$ (d) $\frac{24}{30}$

For the rest of the questions, give all answers in their lowest terms.

3. Add these fractions.
   (a) $\frac{2}{7} + \frac{4}{7}$ (b) $\frac{1}{3} + \frac{1}{6}$ (c) $\frac{2}{3} + \frac{1}{4}$ (d) $\frac{1}{5} + \frac{3}{4}$ (e) $\frac{3}{8} + \frac{1}{5}$ (f) $\frac{3}{4} + \frac{2}{5}$

4. Subtract these fractions.
   (a) $\frac{3}{4} - \frac{1}{4}$ (b) $\frac{1}{2} - \frac{1}{3}$ (c) $\frac{3}{4} - \frac{3}{5}$ (d) $\frac{3}{4} - \frac{1}{6}$ (e) $\frac{3}{5} - \frac{1}{2}$ (f) $\frac{7}{8} - \frac{2}{3}$

5. Work these out.
   (a) $\frac{3}{5} + \frac{2}{5} - \frac{7}{10}$ (b) $\frac{1}{4} + \frac{3}{8} - \frac{1}{6}$ (c) $\frac{1}{6} + \frac{2}{3} - \frac{1}{4}$ (d) $\frac{5}{8} + \frac{3}{5} - \frac{3}{4}$
   (e) $\frac{3}{4} - \frac{5}{6} + \frac{2}{3}$ (f) $\frac{3}{20} - \frac{2}{5} + \frac{3}{4}$

Mixed numbers

Look at this calculation.

$$\frac{2}{3} + \frac{2}{3} = \frac{4}{3}$$

As you can see the result of this addition is a fraction which is 'top-heavy'. It is usual to write fractions like this as **mixed numbers**.

$$\frac{4}{3} = 1\frac{1}{3}$$

To change a top-heavy fraction to a mixed number, divide the denominator into the numerator and write the remainder as a fraction over the denominator.

Example **5** Change these fractions to mixed numbers.

(a) $\frac{7}{4}$ (b) $\frac{11}{5}$ (c) $\frac{24}{7}$

(a) $\frac{7}{4} = 1\frac{3}{4}$ $7 \div 4 = 1$ with 3 left over.

(b) $\frac{11}{5} = 2\frac{1}{5}$ $11 \div 5 = 2$ with 1 left over.

(c) $\frac{24}{7} = 3\frac{3}{7}$ $24 \div 7 = 3$ with 3 left over.

The most common error is to put the remainder over the numerator rather than the denominator.

Mixed numbers can be changed to top-heavy fractions. Just reverse the process.

Example **6** Change these mixed numbers to top-heavy fractions.

(a) $3\frac{1}{4}$ (b) $2\frac{3}{5}$ (c) $3\frac{5}{6}$

(a) $3\frac{1}{4} = 3 + \frac{1}{4} = \frac{12}{4} + \frac{1}{4} = \frac{13}{4}$

Change the whole number to quarters and then add. Another way to think of it is to multiply the whole number by the denominator and add on the numerator. $(3 \times 4 + 1 = 13)$

(b) $2\frac{3}{5} = \frac{13}{5}$ $2 \times 5 + 3 = 13$

(c) $3\frac{5}{6} = \frac{23}{6}$ $3 \times 6 + 5 = 23$

To add or subtract mixed numbers deal with the whole numbers first.

Example **7** Work these out.

(a) $1\frac{1}{6} + 2\frac{1}{3}$ (b) $2\frac{3}{4} + \frac{3}{5}$ (c) $3\frac{2}{3} - 1\frac{1}{6}$ (d) $4\frac{1}{5} - 1\frac{1}{2}$

(a) $1\frac{1}{6} + 2\frac{1}{3} = 3 + \frac{1}{6} + \frac{1}{3}$

$= 3 + \frac{1}{6} + \frac{2}{6}$

$= 3\frac{3}{6} = 3\frac{1}{2}$

Add the whole numbers, then deal with the fractions in the normal way.

(b) $2\frac{3}{4} + \frac{3}{5} = 2 + \frac{15}{20} + \frac{12}{20}$

$= 2 + \frac{27}{20}$

$= 2 + 1 + \frac{7}{20}$

$= 3\frac{7}{20}$

You end up with a top-heavy fraction which you have to change to a mixed number, and then add the whole numbers.

(c) $3\frac{2}{3} - 1\frac{1}{6} = 3 - 1 + \frac{2}{3} - \frac{1}{6}$

$= 2 + \frac{4}{6} - \frac{1}{6}$

$= 2\frac{3}{6} = 2\frac{1}{2}$

Subtract the numbers and then the fractions.

(d) $4\frac{1}{5} - 1\frac{1}{2} = 3 + \frac{2}{10} - \frac{5}{10}$

$= 2 + \frac{10}{10} + \frac{2}{10} - \frac{5}{10}$

$= 2\frac{7}{10}$

Working out $\frac{2}{10} - \frac{5}{10}$ gives a negative answer of $\frac{-3}{10}$. Change one of the whole numbers into $\frac{10}{10}$, then subtract.

Chapter 16 Mixed numbers

1. Change these top-heavy fractions to mixed numbers.
   (a) $\frac{7}{4}$ (b) $\frac{12}{5}$ (c) $\frac{17}{3}$ (d) $\frac{15}{4}$ (e) $\frac{25}{2}$

2. Change these mixed numbers to top-heavy fractions.
   (a) $1\frac{1}{2}$ (b) $2\frac{3}{5}$ (c) $5\frac{3}{8}$ (d) $2\frac{4}{7}$ (e) $9\frac{1}{4}$

3. Add. Write your answers as simply as possible.
   (a) $1\frac{1}{3} + 3\frac{1}{4}$ (b) $3\frac{1}{5} + \frac{7}{10}$ (c) $1\frac{3}{4} + 4\frac{2}{5}$ (d) $2\frac{5}{6} + 7\frac{4}{9}$
   (e) $\frac{2}{7} + \frac{1}{2} + \frac{5}{14}$ (f) $1\frac{1}{2} + \frac{3}{4} + 2\frac{3}{8}$

4. Subtract. Write your answers as simply as possible.
   (a) $2\frac{4}{5} - 1\frac{3}{5}$ (b) $5\frac{3}{8} - 2\frac{1}{4}$ (c) $3\frac{2}{3} - \frac{1}{2}$ (d) $3\frac{3}{5} - 1\frac{3}{4}$
   (e) $5\frac{1}{6} - 3\frac{2}{3}$ (f) $5\frac{1}{5} - \frac{2}{3}$

5. Work these out. Write the answers as simply as possible.
   (a) $\frac{1}{4} + \frac{2}{3} + \frac{1}{2}$ (b) $2\frac{1}{3} + 2\frac{1}{4} - 1\frac{5}{6}$ (c) $3\frac{3}{5} - \frac{1}{4} + \frac{1}{2}$
   (d) $2\frac{3}{8} - \frac{1}{2} + 3\frac{1}{4}$ (e) $4\frac{1}{5} + 1\frac{3}{10} - \frac{4}{5}$ (f) $2\frac{3}{7} - \frac{1}{2} - 1\frac{2}{7}$

6. Faisal cut two pieces of wood $3\frac{3}{8}$ inches and $5\frac{1}{4}$ inches long from a piece 10 inches long. How long was the piece that was left?

1. Change these top-heavy fractions to mixed numbers.
   (a) $\frac{9}{2}$ (b) $\frac{14}{3}$ (c) $\frac{17}{4}$ (d) $\frac{23}{6}$ (e) $\frac{35}{8}$

2. Change these mixed numbers to top-heavy fractions.
   (a) $5\frac{1}{2}$ (b) $2\frac{3}{10}$ (c) $2\frac{3}{7}$ (d) $4\frac{2}{3}$ (e) $4\frac{5}{6}$

3. Add. Write your answers as simply as possible.
   (a) $1\frac{1}{2} + 2\frac{1}{6}$ (b) $1\frac{4}{5} + 2\frac{1}{10}$ (c) $6\frac{1}{6} + 1\frac{4}{9}$ (d) $2\frac{4}{7} + 1\frac{2}{3}$
   (e) $\frac{4}{5} + 1\frac{3}{4} + 2\frac{1}{2}$ (f) $6\frac{1}{3} + 1\frac{4}{9} + 1\frac{2}{9}$

4. Subtract. Write your answers as simply as possible.
   (a) $2\frac{2}{3} - 1\frac{1}{6}$ (b) $3\frac{5}{8} - 1\frac{1}{4}$ (c) $2\frac{4}{5} - \frac{1}{2}$ (d) $4\frac{2}{5} - 1\frac{1}{4}$
   (e) $8\frac{1}{6} - 5\frac{3}{8}$ (f) $1\frac{1}{4} - \frac{5}{8}$

5. Work these out. Write the answers as simply as possible.
   (a) $2\frac{1}{3} + 3\frac{1}{2} - \frac{5}{6}$ (b) $1\frac{3}{4} - \frac{5}{6} + 2\frac{1}{2}$ (c) $3\frac{1}{6} - 2\frac{1}{8} + \frac{3}{4}$
   (d) $4\frac{1}{3} - \frac{4}{5} + \frac{2}{5}$ (e) $3\frac{3}{4} - 2\frac{1}{2} + 1\frac{5}{8}$ (f) $4\frac{1}{3} - 1\frac{5}{6} - 2\frac{1}{2}$

6. Caroline bought a piece of ribbon 24 inches long.
   She cut off two pieces, each $5\frac{5}{8}$ inches long. How long was the piece she had left?

Multiplying and dividing fractions

Multiplying fractions

You can think of a fraction as the number 1 multiplied by the numerator and divided by the denominator. For example,
$\frac{2}{3} = 1 \times 2 \div 3$ or $\frac{1}{1} \times \frac{2}{3} = \frac{1 \times 2}{1 \times 3}$.

To multiply fractions, multiply the numerators and multiply the denominators, then simplify if possible. If the fractions are mixed numbers change them to top-heavy fractions and then multiply.

Example 8

Work these out.

(a) $\frac{3}{5} \times \frac{1}{2}$ (b) $\frac{3}{8} \times \frac{4}{9}$ (c) $1\frac{3}{5} \times 3\frac{3}{4}$

(a) $\frac{3}{5} \times \frac{1}{2} = \frac{3}{10}$

> Multiply the numerators and multiply the denominators. The answer is already in its lowest terms.

(b) $\frac{3}{8} \times \frac{4}{9} = \frac{12}{72} = \frac{1}{6}$

> The answer simplifies to $\frac{1}{6}$ but you could divide by the common factors (cancel) first, for example $\frac{3}{8}^{1} \times \frac{4}{9}^{1} = \frac{1}{2} \times \frac{1}{3} = \frac{1}{6}$ as 3 divides into 3 and 9; 4 divides into 4 and 8.

(c) $1\frac{3}{5} \times 3\frac{3}{4} = \frac{8}{5} \times \frac{15}{4}$

> 5 divides into 5 and 15; 4 divides into 4 and 8.

$= \frac{2}{1} \times \frac{3}{1} = 6$

Examiner's tip

Note: $\frac{1}{2} \times \frac{1}{3} = \frac{1}{6}$
A common error is to multiply 1×1 and get 2.

Examiner's tip

When cancelling, cancel a term in the numerator with one in the denominator.

Remember that, when simplifying or cancelling, you can just cross out the numbers you are cancelling and write in the quotients, for example, $\frac{3}{8} \times \frac{4}{9} = \frac{3}{8}^{1} \times \frac{4}{9}^{1} = \frac{1}{2} \times \frac{1}{3}$.

Dividing fractions

Multiplying by $\frac{1}{2}$ is the same as dividing by 2, so dividing by $\frac{1}{2}$ is the same as multiplying by 2.

This can be extended, for example, $4 \div \frac{2}{3} = 4 \times \frac{3}{2} = \frac{12}{2} = 6$

When dividing fractions, turn the second fraction upside-down and then multiply.

Example 9

Work these out.

(a) $\frac{4}{5} \div \frac{3}{10}$ (b) $\frac{9}{10} \div \frac{3}{4}$ (c) $2\frac{1}{4} \div 1\frac{1}{2}$

(a) $\frac{4}{5} \div \frac{3}{10} = \frac{4}{5} \times \frac{10}{3}$

       $= \frac{4}{1} \times \frac{2}{3} = \frac{8}{3} = 2\frac{2}{3}$

> Turn the second fraction upside-down and multiply.
>
> Cancel 10 and 5 by 5, multiply and change to a mixed number.

(b) $\frac{9}{10} \div \frac{3}{4} = \frac{9}{10} \times \frac{4}{3}$

       $= \frac{3}{5} \times \frac{2}{1} = \frac{6}{5} = 1\frac{1}{5}$

> Cancel 9 and 3 by 3; 4 and 10 by 2.

(c) $2\frac{1}{4} \div 1\frac{1}{2} = \frac{9}{4} \div \frac{3}{2}$

> Change to top-heavy fractions.

       $= \frac{9}{4} \times \frac{2}{3} = \frac{3}{2} \times \frac{1}{1}$

> Cancel 9 and 3 by 3; 2 and 4 by 2.

       $= \frac{3}{2} = 1\frac{1}{2}$

Exercise 16.3a

Work these out.

1. $\frac{1}{4} \times \frac{2}{3}$
2. $\frac{2}{3} \times \frac{3}{5}$
3. $\frac{4}{9} \times \frac{1}{2}$
4. $\frac{2}{3} \times \frac{1}{3}$
5. $\frac{4}{5} \div \frac{3}{10}$
6. $\frac{1}{4} \div \frac{3}{8}$
7. $\frac{3}{4} \div \frac{5}{6}$
8. $4\frac{1}{2} \times 2\frac{1}{6}$
9. $1\frac{1}{2} \times 3\frac{2}{3}$
10. $2\frac{1}{3} \div 1\frac{1}{3}$
11. $2\frac{2}{5} \div 1\frac{1}{2}$
12. $1\frac{1}{4} \times 3\frac{1}{5} \div 1\frac{1}{2}$

Exercise 16.3b

Work these out.

1. $\frac{1}{2} \times \frac{5}{6}$
2. $\frac{3}{5} \times \frac{5}{6}$
3. $\frac{2}{3} \times \frac{5}{8}$
4. $\frac{3}{5} \times \frac{5}{12}$
5. $\frac{3}{8} \div \frac{1}{4}$
6. $\frac{2}{3} \div \frac{5}{6}$
7. $\frac{3}{4} \div \frac{3}{8}$
8. $3\frac{1}{3} \times 2\frac{2}{5}$
9. $2\frac{2}{5} \times \frac{3}{4}$
10. $3\frac{1}{8} \div 1\frac{1}{4}$
11. $2\frac{1}{4} \div 3\frac{1}{2}$
12. $3\frac{1}{3} \times 1\frac{1}{4} \div 2\frac{1}{2}$

Adding and subtracting negative numbers

Adding a negative number is the same as subtracting a positive number.

$$4 + (-2) = 4 - 2 = 2$$

Subtracting a negative number is the same as adding a positive number.

$$4 - (-2) = 4 + 2 = 6$$

When the numbers to be added are both negative, add them and put the negative sign in front.

$$-4 - 5 = -9$$

If they have different signs, subtract them and give the answer the same sign as the larger number.

$$-4 + 6 = +2$$
$$-5 + 2 = -3$$

When adding or subtracting negative and positive numbers, it is best to:

  total the positive numbers
  total the negative numbers separately
  then find the difference between the two totals,
  remembering to give the answer the correct sign.

You can already multiply and divide negative numbers. Now the steps can be combined.

Example 10

Work these out without using a calculator.

(a) $(-3 \times -4) + (-2 \times 3)$

(b) $-5 + 4 - 6 + 7 + 5 + 1 - 3 - 5 - 2$

(c) $\dfrac{5 \times -4 + 3 \times -2}{-6 + 4}$

(a) $(-3 \times -4) + (-2 \times 3) = (+12) + (-6) = 12 - 6 = +6 = 6$

(b) $-5 + 4 - 6 + 7 + 5 + 1 - 3 - 5 - 2 = +17 - 21 = -4$

(c) $\dfrac{5 \times -4 + 3 \times -2}{-6 + 4} = \dfrac{-20 + (-6)}{-2} = \dfrac{-26}{-2} = 13$

Example 11

Use a calculator to work these out, correct to three significant figures.

(a) $(-3.4)^2 - 2 \times 4.6$

(b) $4.7 \times 2.8 + (3 \times -17.1)$

(a) $(-3.4)^2 - 2 \times 4.6 = 11.56 - 9.2 = 2.36$

(b) $4.7 \times 2.8 + (3 \times -17.1) = 13.16 - 51.3$
$= -38.14 = -38.1$ (to 3 s.f.)

Both of these can be worked out directly on the calculator.

Exercise 16.4a

Work these out without using a calculator.

1. $(-4 \times -3) - (-2 \times +1)$
2. $(-7 \times -2) + (4 \times -2)$
3. $(-15 \div 2) - (4 \times -6)$
4. $-4 + 3 + 2 + 3 + 4 - 5 - 6 - 9 + 1$
5. $\dfrac{-2 + 12}{-5}$
6. $\dfrac{-4 \times -3}{-4 + 3}$

Use a calculator to work these out.

7. $-4.73 + 2.96 - 1.71 + 3.62$
8. $(-4.6 \times 7.2) + (3.1 \times -4.3)$
9. $\dfrac{-4.7 + 2.6}{-5.7}$
10. $\dfrac{7.92 \times 1.71}{-4.2 + 3.6}$

Exercise 16.4b

Work these out without using a calculator.

1. $(-2 \times +3) + (-3 \times +4)$
2. $(-1 \times -4) + (-7 \times -8)$
3. $(24 \div -3) - (-5 \times -4)$
4. $-6 - 2 - 3 + 5 - 7 + 4 - 2 + 8$
5. $\dfrac{-3 + 7}{-2}$
6. $\dfrac{-7 \times -12}{-8 + 4}$

Use a calculator to work these out.

7. $-14.7 + 6.92 - 1.41 - 2.83$
8. $(-1.2 \times -2.4) - (9.2 \times -3.6)$
9. $\dfrac{-4.72}{-1.4} \times \dfrac{8.61}{-7.21}$
10. $\dfrac{3.14 - 8.16}{-8.25 \times 3.18}$

Using a calculator

You already know how to use a calculator to do basic calculations.

In the last exercise you may have answered question 10 by working out the numerator and denominator separately and then dividing them, but you could do it all on your calculator, by using brackets or by various other means.

Calculators vary and you need to practise with yours to see what it can do.

Examiner's tip

Where it is possible to split the question up, it is useful to do so to check, but make sure you do not round the answer too early.

Unless the question states otherwise it is best to give your final answer to 3 s.f.

Example 12 Use a calculator to work these out.

(a) $\dfrac{14.73 + 2.96}{15.25 - 7.14}$ (b) $\sqrt{17.8^2 + 4.3^2}$

(a) $\dfrac{14.73 + 2.96}{15.25 - 7.14} = 2.1812 = 2.18$ (to 2 d.p.)

There are various ways to do this.
One is to key in:

$\boxed{1}\boxed{4}\boxed{\cdot}\boxed{7}\boxed{3}\boxed{+}\boxed{2}\boxed{\cdot}\boxed{9}\boxed{6}\boxed{=}$
$\boxed{\div}\boxed{(}\boxed{1}\boxed{5}\boxed{\cdot}\boxed{2}\boxed{5}\boxed{-}\boxed{7}\boxed{\cdot}\boxed{1}$
$\boxed{4}\boxed{)}\boxed{=}$

(b) $\sqrt{17.8^2 + 4.3^2} = 18.312 = 18.3$ (to 3 s.f.)

Key in:

$\boxed{(}\boxed{1}\boxed{7}\boxed{\cdot}\boxed{8}\boxed{x^2}\boxed{+}$
$\boxed{4}\boxed{\cdot}\boxed{3}\boxed{x^2}\boxed{)}\boxed{\sqrt{}}\boxed{=}$

These all need practice.

Some of the more common operations you will be asked to use are powers and roots (using the $\boxed{y^x}$ button and the $\boxed{\sqrt[x]{}}$ or $\boxed{y^{1/x}}$ button) and sine (sin), cosine (cos) and tangent (tan) of an angle.

Examiner's tip

When using sin, cos, tan, make sure the calculator is in degree mode.

Example 13 Use a calculator to evaluate these.

(a) 4.2^3 (b) $\sqrt[5]{15}$ (c) $\cos 73°$
(d) $\cos^{-1} 0.897$ (e) $(3.7 \times 10^{-5}) \div (8.3 \times 10^6)$

Remember that $\cos^{-1}$ is also called inv cos or arc cos.

(a) $4.2^3 = 74.088 = 74.1$ (to 3 s.f.)

This can be done by keying $4.2 \times 4.2 \times 4.2$, but using the y^x button is quicker.
Key in: $\boxed{4}\boxed{\cdot}\boxed{2}\boxed{y^x}\boxed{3}\boxed{=}$

(b) $\sqrt[5]{15} = 1.7187 = 1.72$ (to 3 s.f.)

Key in: $\boxed{5}\boxed{\text{2nd F}}\boxed{y^x}\boxed{1}\boxed{5}\boxed{=}$
This may vary, depending on your calculator. $\sqrt[5]{15}$ is the same as $15^{1/5}$.

(c) $\cos 73° = 0.2923 = 0.292$ (to 3 s.f.)

Key in: $\boxed{\cos}\boxed{7}\boxed{3}\boxed{=}$
On some calculators, you need to key in $\boxed{7}\boxed{3}\boxed{\cos}$

(d) $\cos^{-1} 0.897 = 26.23° = 26.2°$ (to 3 s.f.)

Key in: $\boxed{\text{2nd F}}\boxed{\cos}\boxed{\cdot}\boxed{8}\boxed{9}\boxed{7}\boxed{=}$
On some calculators, you need to key in $\boxed{\cdot}\boxed{8}\boxed{9}\boxed{7}\boxed{\text{2nd F}}\boxed{\cos}$

(e) $(3.7 \times 10^{-5}) \div (8.3 \times 10^6) = 4.46 \times 10^{-12}$ (to 3 s.f.)

Key in: $\boxed{3}\boxed{\cdot}\boxed{7}\boxed{\text{EXP}}\boxed{5}\boxed{+/-}\boxed{\div}\boxed{8}\boxed{\cdot}\boxed{3}\boxed{\text{EXP}}\boxed{6}\boxed{=}$
This may be different on your calculator.

Chapter 16 Using a calculator

Exercise 16.5a

Work these out.

1. (a) 3.8^4 (b) $31.8^{1/4}$

2. (a) $\sin 46.2°$ (b) $\tan 51.6°$ (c) $\cos 31.6°$
   (d) $\sin 12° - \cos 31°$

3. (a) $\cos^{-1} 0.832$ (b) inv tan 3.60   (c) $\sin^{-1} 0.910$
   (d) $\sin^{-1} \dfrac{16.3}{43.9}$

4. $43.7^3 + 17.1^2$

5. $\dfrac{3.4 \times \sin 47.1°}{\sin 19.2°}$

6. $4.7 \times 10^5 \times 7.9 \times 10^{-4}$

7. $(9.2 + 15.3)^2$

8. $\dfrac{6.2}{2.6} + \dfrac{5.4}{3.9}$

9. $\dfrac{2.6 + 4.25}{7.8 \times 3.6^2}$

10. $\dfrac{19.4 \times 6.3 - 2.61}{8.1 + 7.94}$

Exercise 16.5b

Work these out.

1. (a) 7.31^5 (b) $12.2^{1/5}$

2. (a) $\sin 14.6°$ (b) $\cos 71.3°$ (c) $\tan 15.9°$ (d) $\sin 247°$

3. (a) $\tan^{-1} 3.21$ (b) $\sin^{-1} 0.464$ (c) $\cos^{-1} 0.141$
   (d) $\sin^{-1} \dfrac{\sqrt{3}}{2}$

4. $2.7^2 + 3.6^2 - 2 \times 2.7 \times 3.6 \times 0.146$

5. $3.1 \times 4.2 \times \sin 41.2° \div 2$

6. $\sqrt{6.39^2 - 4.27^2}$

7. $(2.2 \times 10^{-2}) \div (5.3 \times 10^4)$

8. $3 \cos 14.2° - 5 \sin 16.3°$

9. $\left(2.4 \times 3.1 - \dfrac{6.8}{3.4}\right)^2$

10. $\frac{1}{2}(-5 + \sqrt{5^2 + 1200})$

Ratio and proportion

You have already done some work involving simple ratios. This is now extended.

Example 14

Adrian made a fruit drink with 9 parts orange juice, 4 parts lemon juice and 2 parts grapefruit juice. He made 2 litres of fruit drink. How much orange juice did he need?

	orange	:	lemon	:	grapefruit	total
Parts	9		4		2	15
Quantity	x					2000 ml

Using ratios, $x : 9 = 2000 : 15$

$x = \frac{2000}{15} \times 9 = 1200 \, ml$

Drawing up a table often helps to decide which parts you need to use.

This could have been worked out without a calculator, and you may be asked to do so.

Exercise 16.6a

Answer the first five questions without using a calculator.

1. Split £100 in the ratio 2 : 3 : 5.
2. The angles of a triangle are in the ratio 1 : 2 : 3. What are the sizes of the angles?
3. John and Qasim share £20 in the ratio 2 to 3. How much does John get?
4. Paint is mixed as 3 parts red to 5 parts white. How much of this paint can be made with 2 litres of white?
5. Susan and Chika invest £4000 and £6000 in a business venture and agree to share the profits in the ratio of their investment. They make a profit of £250 in the first year. How much does Chika receive?

For the rest of the questions you can use a calculator.

6. Maureen and Sheena's earnings are in the ratio 3 : 5. They earn £352 all together. How much does Sheena earn?
7. At Carterknowle Church Autumn Bazaar £875 was raised. It was agreed to share the profits between the church and the local charity for the homeless, in the ratio 5 to 1. How much did the charity receive? Give the answer to the nearest pound.

Examiner's tip

A common error, when using a calculator, is to round at too early a stage. In Example 14, if the answer is rounded after dividing by 15 to give 133 and this is then multiplied by 9, the answer comes out as 1197 instead of 1200.

8. To make her own breakfast cereal, Sally mixes bran, currants and wheatgerm in the ratio 8 to 3 to 1 by mass. How much bran, to the nearest 10 grams, does she use to make 500 grams of cereal?

9. David, Michael and Iain employed a gardener and agreed to pay him in the ratio of the time he spent on each garden. He spent 2 hours 20 minutes in David's garden, 3 hours 30 minutes in Michael's garden and 4 hours 10 minutes in Iain's garden. David paid £12.60. How much did the other two pay?

10. In a local election the votes were Labour 1200, Conservative 5312, Lib-Dems 878. Write this ratio in the form $1 : n : m$. Correct n and m to three significant figures.

Exercise 16.6b

Do not use a calculator for the first five questions.

1. Share £75 in the ratio 8 to 7.

2. A firm uses first and second class stamps in the ratio 9 to 1. During a week they used 250 stamps altogether. How many first class stamps did they use?

3. A metal alloy is made up of copper, iron and nickel in the ratio $3 : 4 : 2$. How much copper is there in 450 g of the alloy?

4. Vicki and Inderjit share the winnings from a raffle in the ratio 2 to 3. Vicki received £15. How much did they win altogether?

5. Old 2-stroke scooters used to mix petrol with oil in the ratio 25 to 1. How much oil had to be mixed with 5 litres of petrol?

Use a calculator for the rest of the questions.

6. In a school there are 875 pupils and 41 teachers. Write the pupil : teacher ratio in the form $n : 1$. Express n correct to three significant figures.

7. Shahida spends her pocket money on sweets, magazines and savings in the ratio $2 : 3 : 7$. She receives £15 a week. How much does she spend on sweets?

8. Doreen and Joan invested £5000 and £7500 respectively in a firm. They shared the profits in the ratio of their investment. Doreen received £320. How much did Joan receive?

9. Alec and Pat share a house. They agree to share the rent (to the nearest pound) in the ratio of the area of their bedrooms. The area of Alec's floor is 17 m² and the area of Pat's is 21 m². The rent is £320 a week. How much do they each pay?

10. In a questionnaire the three possible answers are 'Yes', 'No' and 'Don't know'. The answers from a group of 456 people are in the ratio 10 : 6 : 3. How many 'Don't knows' are there?

Repeated proportional changes

You already know how to increase or decrease an amount by a percentage or a fraction.

If an amount was increased by 20%, you work out 20% and add it on.

Alternatively, you could multiply by 1.2 (finding 120%).

This second method is vital if you need to deal with successive increases over a number of years.

Example 15
Selena invested £4000 and received 5% interest a year which was added on each year. How much had she in total after: (a) 1 year (b) 6 years?

A 5% increase means that the new amount is $100 + 5 = 105\%$ of the old amount each year.

105% is 1.05, so multiply by 1.05 each year.

(a) After one year the total amount is £4000 × 1.05 = £4200

(b) After 6 years the total amount is
$4000 \times 1.05 \times 1.05 \times 1.05 \times 1.05 \times 1.05 \times 1.05$
$= 4000 \times (1.05)^6$
$= £5360.38$ (to the nearest penny)

It is the same for fractions. If an amount is increased by $\frac{1}{10}$ each year, the new amount is $1 + \frac{1}{10} = \frac{11}{10}$ of the old amount each year.

Example 16

Andrew said he would increase his giving to charity by $\frac{1}{25}$ each year. He gave £120 at the start. How much did he give at the end of the fifth year?

Each year he gave $\frac{26}{25} \times$ what he gave in the previous year. $(1 + \frac{1}{25} = \frac{26}{25})$

At the end of the fifth year he gave £120 $\times (\frac{26}{25})^5$

Key in:

$= £145.998$
$= £146.00$ to the nearest penny.

If the change is a decrease, subtract from 100% or 1.

So 'reduce by three per cent' means 'find 97% or multiply by 0.97'
'reduce by one fifth ($\frac{1}{5}$)' means 'multiply by four fifths ($\frac{4}{5}$)'.

 Example 17

The distance that Patrick can walk in a day is reducing by $\frac{1}{15}$ each year. This year he can walk 12 miles in a day. How far will he be able to walk in five years' time?

The distance reduces by $\frac{1}{15}$ in 1 year, so multiply by $\frac{14}{15}$ for each year.

In five years he will be able to walk $12 \times (\frac{14}{15})^5 = 8.499 = 8.50$ miles.

Exercise 16.7a

1. What do you multiply a quantity by if it is increased by:
   (a) 6%   (b) 9%   (c) 17.5%   (d) 1.25%   (e) $\frac{1}{5}$ (f) $\frac{2}{9}$?

2. What do you multiply a quantity by if it is decreased by:
   (a) 6%   (b) 9%   (c) 17.5%   (d) 1.25%   (e) $\frac{1}{5}$ (f) $\frac{2}{9}$?

3. Peter invests £1000 and 5% of the balance is added to the amount each year. How much will be in the account after six years? Give the answer to the nearest pound.

4. A population of bacteria is estimated to increase by 12% every 24 hours. The population was 2000 at midnight on Friday. What was the population (to the nearest whole number) by midnight the following Wednesday?

5. The insurance premium for Della's car was £360. The firm reduced it by 12% for each year she had no claim. What was the cost after six years with no claims? Give the answer to the nearest pound.

6. Mr Costa was offered an 8% rise every year whilst he worked at the same firm. This year he earned £28 500. How much will he earn after four rises? Give the answer to the nearest pound.

7. At Premda department store they said they would reduce the price of goods still not sold by $\frac{1}{3}$ for each day of the sale.
   A coat was offered originally at £65. What was its price after three days, to the nearest penny?

8. An investment firm says it will add $\frac{1}{5}$ to your money each year. If you invested £3000, how much would it amount to after 10 years? Give the answer to the nearest pound.

9. Clement claimed that if you get a 7% increase each year you will double your money after 10 years. Is this true? Show figures to justify your answer.

10. Ambrose invested £3500 in a six-year bond that added 5% to the amount each year for the first three years and 7.5% each year for the next three years. What is the amount in the bond, to the nearest penny:
    (a) after three years   (b) after six years?

Exercise 16.7b

1. What do you multiply a quantity by if it is increased by:
   (a) 4%   (b) 18%   (c) 12.5%   (d) 5.6%   (e) $\frac{1}{6}$ (f) $\frac{3}{5}$?

2. What do you multiply a quantity by if it is decreased by:
   (a) 4%   (b) 18%   (c) 12.5%   (d) 5.6%   (e) $\frac{1}{6}$ (f) $\frac{3}{5}$?

3. Interest of 4% was added to an investment of £1500 each year for four years. How much was it then worth? Give the answer to the nearest pound.

4. Martyn had shares worth £8000. They increased in value by 7.5% each year. What was their value after 10 years? Give the answer to the nearest pound.

5. Cathy said she would withdraw $\frac{1}{5}$ of the money she had in the bank every time she made a withdrawal. She had £187.50 in the bank to start with. How much did she have after three withdrawals?

6. At Patnik shoe shop they offered to decrease the price of a pair of shoes by $\frac{1}{4}$ each day until they were sold. They were priced at £47 to start with. Jean bought them after they had been reduced four times. How much did she pay? Give the answer to the nearest penny.

7. Tony says his narrow boat is increasing in value by 6% a year. It was worth £25 000 in 1999. How much would it be worth, to the nearest hundred pounds, in 2005 (six years later) if he is correct?

8. Sheila joined a keep-fit club that claimed you would reduce your running time by 1% each week. She could run 500 metres in 12 minutes to start with. According to the club, how long would it take her after five weeks? Give the answer to the nearest second.

9. It is claimed that the number of rabbits in Freeshire is increasing by $\frac{1}{12}$ each year. It is estimated that there are 1700 rabbits now. How many will there be after four years if the statement is true? Give the answer to three significant figures.

10. Mordovia has high inflation. In 1999 it was 15% a month for the first six months and 12.5% for the next six months.
A car cost 78 000 scuds (their unit of currency) in January 1999. How much did it cost:
(a) after six months (b) in January 2000?
Give the answers to the nearest whole number.

Finding the value before a percentage change

If a quantity is increased by 5%, you multiply by 1.05 to get the new amount.

To find the original amount from the new amount just reverse the process and divide by 1.05.

Example 18

Irene paid £38.70 for a skirt in a sale. This was after it had been reduced by 10%.

What was the original price of the skirt?

New price = 0.9 × original price

Original price = new price ÷ 0.9
= £38.70 ÷ 0.9 = £43

Example 19

Berwyn received an increase of $\frac{1}{5}$ in his salary. After the increase he was earning £31 260 a year. What was his salary before the rise?

New salary = $\frac{6}{5}$ × old salary

Old salary = new salary ÷ $\frac{6}{5}$ = new salary × $\frac{5}{6}$

Old salary = £31 260 × $\frac{5}{6}$ = £26 050

Examiner's tip

Always start with 'new amount = something × old amount' and then rearrange as necessary.

Exercise 16.8a

In this exercise, some of these questions ask for the original amount and some ask for the new amount.

1. A price of £50 is increased by 7.5%. What is the new price?
2. A quantity is decreased by 3%. It is now 38.8. What was it to start with?
3. A coat was advertised at £79. In a sale the price was reduced by 5%. What was the new price?
4. Mr Diffom made a profit of £13 250 in the year 2000. This was an increase of 6% on his profit in 1999. What was his profit in 1999?
5. In a local election in 1997, Labour received 1375 votes. This was increased by 12% in 1998. How many people voted Labour in 1998?
6. Save-a-lot supermarket advertised jam at $\frac{1}{5}$ off. A jar cost £1.80 after the reduction. What did it cost originally?

7. Stephen was given a rise of 7%. His salary after the rise was £28 890. What was it before the rise?

8. Between 1978 and 1979 house prices increased by 12.5%. A house was valued at £27 000 in 1979. What was its value in 1978?

9. A holiday cost £564, including VAT at 17.5%. What was the cost without VAT?

10. At Jack's café all prices were increased by 5% (to the nearest penny).
    (a) A cup of tea cost 75p before the increase. What is the new price?
    (b) The new price of a cup of coffee is £1.30. What did it cost before the increase?

Exercise 16.8b

In this exercise, some of these questions ask for the original amount and some ask for the new amount.

1. A 'best score' of 70 is increased by $\frac{2}{5}$. What is it now?

2. After an increase of 12%, a quantity is 84 tons. What was it before the increase?

3. In a sale everything is reduced by 5%. A pair of shoes costs £47.50 in the sale. How much did they cost before the sale?

4. A newspaper increased its circulation by 3% and the new number sold was 58 195. What was it before the increase?

5. Santos sold his car for £8520. This was 40% less than he paid for it five years before. What did he pay for it?

6. A charity's income has been reduced by $2\frac{1}{2}\%$. Its income is now £8580. What was it before the reduction?

7. It was announced that the number of people unemployed had decreased by 3%. The number who were unemployed before the decrease was 2.56 million. How many are now unemployed? Give the answer to three significant figures.

8. The cost of a car, including VAT at 17.5%, is £12 925. What is the cost without VAT?

9. A car firm claims that for its latest model, the number of miles per gallon of fuel has increased by $\frac{1}{5}$. The new model travels 48 miles per gallon. How many miles per gallon did the old model travel?

10. At Percival's sale the price of everything is reduced by $7\frac{1}{2}\%$, rounded to the nearest penny.
    (a) A pair of boots cost £94.99 before the sale. What is the price in the sale?
    (b) Delia is charged £13.87 for a blouse in the sale. What was its original price?

Key points

- When adding or subtracting fractions, change both to the same denominator and add or subtract the numerators.
- If mixed numbers are involved, deal with the whole numbers separately.
- When multiplying fractions, multiply the numerators and multiply the denominators and cancel down.
- When dividing fractions, invert the second fraction and multiply.
- If mixed numbers are involved, change to top-heavy fractions before multiplying or dividing.
- When collecting two negative and/or positive numbers:
  - if the signs are the same add the numbers and the answer takes the same sign as both the numbers
  - if the signs are different, subtract the numbers and the answer takes the sign of the bigger number.
- When a quantity is increased by 5% a year for six years, multiply by 1.05^6.
- When a quantity is decreased by 3% a year for four years, multiply by 0.97^4.
- To find the quantity before a 2% increase, divide the new quantity by 1.02.
- To find a quantity before a 4% decrease, divide by 0.96.

Revision exercise 16a

1. Work these out.
   (a) $\frac{1}{2} + \frac{2}{3}$
   (b) $1\frac{2}{5} + 3\frac{1}{4}$
   (c) $\frac{5}{9} - \frac{1}{6}$
   (d) $3\frac{1}{4} - 2\frac{2}{3}$
   (e) $\frac{2}{3} + 4\frac{1}{2} - 2\frac{5}{6}$
   (f) $\frac{2}{3} \times \frac{3}{5}$
   (g) $\frac{3}{8} \div \frac{1}{4}$
   (h) $2\frac{1}{2} \times 3\frac{1}{5}$
   (i) $3\frac{1}{5} \div 2\frac{2}{3}$
   (j) $4\frac{1}{3} \times 1\frac{1}{4} \div 2\frac{1}{6}$

2. To make a frame John uses four pieces of wood: two are $4\frac{1}{4}$ inches long and two are $6\frac{2}{3}$ inches long. He cut them off a piece of wood 24 inches long. How much wood was left?

3. Work these out without using a calculator.
   (a) $(-2 \times -5) + (-6 \times 3)$
   (b) $4 + 6 - 8 - 7 + 1 + 3 - 5 - 6$
   (c) $\frac{-3 \times -2 + 7 \times -2}{-8 + 6}$

4. Use a calculator to work these out. Where answers are not exact, give them correct to three significant figures.
   (a) $-2.73 + 12.6 - 11.91 + 13.2$
   (b) $(-4.5 \times 8.3) + (6.1 \times -4.3)$
   (c) $\frac{-4.7 + 3.6}{-7.5}$
   (d) $\cos 14.2°$
   (e) $\sin^{-1} 0.365$
   (f) $\tan 71.2°$
   (g) 3^9
   (h) $12.3^3 - 2.6^3$
   (i) $\frac{3.2 \sin 12.3°}{\sin 28.2°}$
   (j) $\sqrt{5^2 + 8^2}$
   (k) $\frac{7.92 \times 1.71}{4.2 + 3.6}$
   (l) $(4.1 - 3 \times 2.6)^3$
   (m) $3 \cos 12° - 2 \sin 12°$
   (n) $\tan^{-1} \frac{4.3}{2.9}$
   (o) $(14.6 - 3.2^2)^{\frac{1}{4}}$

5. Kelly, Eileen and Susie share £1500 in the ratio $3:4:5$. How much does Kelly receive?

6. To make 12 scones Maureen uses 5 ounces of flour. How many scones can she make using 8 ounces of flour? Answer to the nearest whole number.

7. The voting in an election was in the ratio $8:4:3$ for Labour, Conservative and Others. If 6328 people voted Labour, how many voted altogether?

8. A bacteria culture is growing at 5% a day. There are 1450 bacteria on Tuesday. How many are there three days later?

9. A paper reported that the number of people taking their main holiday in Britain has reduced by 10% per year over the last five years. There were 560 people from a small town who took their main holiday in Britain five years ago. If the report is true, how many of them do so now? Give the answer to the nearest person.

10. Damien sold his bicycle for £286, at a loss of 45% on what he paid for it. How much did he pay?

11. A magazine sold 1020 copies this month. This is an increase of $\frac{1}{9}$ on the number sold last month. How many were sold last month?

12. At the Star theatre, the seating was changed and the number of seats in the stalls was increased by a third. There are now 312 seats in the stalls. How many were there before the increase?

You will need to know how to:

- write a simple formula using letters
- collect together simple algebraic terms
- find the nth term of a simple sequence
- expand single brackets
- solve linear equations
- add, subtract, multiply and divide fractions and negative numbers.

Most of these topics were covered in Chapter 3.

Substituting numbers in a formula

Numbers that can be substituted in a formula may be positive, negative, decimals or fractions.

Example 1

If $W = 4p - 5q^2$, find W when:

(a) $p = 2$, $q = -3$

(b) $p = 22.5$, $q = 3.4$

(c) $p = \frac{3}{4}$, $q = \frac{2}{5}$.

(a) $W = (4 \times 2) - (5 \times -3 \times -3)$
$= 8 - (5 \times 9) = 8 - 45$
$= -37$

(b) $W = (4 \times 22.5) - (5 \times 3.4 \times 3.4)$
$= 90 - 57.8$
$= 32.2$

In part (b) you would use a calculator.

(c) $W = 4 \times \frac{3}{4} - 5 \times \frac{2}{5} \times \frac{2}{5}$
$= 3 - \frac{4}{5}$
$= 2\frac{1}{5}$

Examiner's tip

Take special care when negative numbers are involved.

Examiner's tip

Remember $5b^2$ means $5 \times b \times b$ not $5 \times b \times 5 \times b$.

Examiner's tip

Work out each term separately and then collect together.

Example 2

The formula for the surface area of a cylinder is $S = 2\pi rh + 2\pi r^2$.

Find the surface area when $r = 5.7$ and $h = 4.6$.

Give the answer to three significant figures.

$$S = (2 \times \pi \times 5.7 \times 4.6) + (2 \times \pi \times 5.7^2)$$
$$= 164.7\ldots + 204.1\ldots = 368.8\ldots$$

$S = 369$ (to 3 s.f.)

Examiner's tip

Use the π key on your calculator. Write down the intermediate values but leave them in your calculator to avoid making errors through rounding too early.

Example 3

If $S = ut + \frac{1}{2}at^2$, find S when $u = 3$, $t = 4$, $a = -5$.

$$S = (3 \times 4) + (\tfrac{1}{2} \times -5 \times 4^2) = 12 - 40 = -28$$

Example 4

If $P = ab + 4b^2$, find P when $a = \frac{4}{5}$ and $b = \frac{3}{8}$, giving your answer as a fraction.

$$P = (\tfrac{4}{5} \times \tfrac{3}{8}) + (4 \times \tfrac{3}{8} \times \tfrac{3}{8}) = \tfrac{3}{10} + \tfrac{9}{16} = \tfrac{24}{80} + \tfrac{45}{80} = \tfrac{69}{80}$$

Exercise 17.1a

Work out each of the formulae in questions 1–7 for the values given. Do not use a calculator.

1. $V = ab - ac$ when $a = 3$, $b = -2$, $c = 5$
2. $P = 2rv + 3r^2$ when $r = 5$, $v = -2$
3. $T = 5s^2 - 2t^2$ when $s = -2$, $t = 3$
4. $M = 2a(3b + 4c)$ when $a = 5$, $b = 3$, $c = -2$
5. $R = \dfrac{2qv}{q + v}$ when $q = 3$, $v = -4$
6. $L = 2n + m$ when $n = \frac{2}{3}$, $m = \frac{5}{6}$
7. $D = a^2 - 2b^2$ when $a = \frac{4}{5}$, $b = \frac{2}{5}$
8. Use a calculator to find the value of $M = \dfrac{ab}{2a + b^2}$ (correct to three significant figures) when $a = 2.75$, $b = 3.12$.

9. The distance S metres fallen by a pebble is given by the formula $S = \frac{1}{2}gt^2$, where t is in seconds.

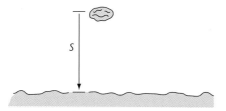

Find S when: (a) $g = 10$, $t = 12$ (b) $g = 9.8$, $t = 2.5$.
Use a calculator in part (b) only.

10. The surface area of a cuboid with sides x, y and z is given by the formula $A = 2xy + 2yz + 2xz$.

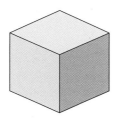

Find the surface area when $x = 5$, $y = 4.5$, $z = 3.5$.

Exercise 17.1b

Work out each of the formulae in questions 1–7 for the values given. Do not use a calculator.

1. $A = a^2 + b^2$ when $a = 5$, $b = -3$
2. $P = 2c^2 - 3cd$ when $c = 2$, $d = -5$
3. $B = p^2 - 3q^2$ when $p = -4$, $q = -2$
4. $T = (4a - 5b)^2$ when $a = -2$, $b = -1$
5. $Q = x(y^2 - z^2)$ when $x = -2$, $y = 7$, $z = -3$
6. $S = ab + 5b^2$ when $a = \frac{3}{4}$, $b = \frac{4}{5}$
7. $R = a + 2b$ when $a = 1\frac{5}{6}$, $b = \frac{2}{3}$
8. The elasticity of an elastic string is given by the formula $E = \frac{\lambda x^2}{2a}$.

   Find E (correct to three significant figures) when $\lambda = 3.4$, $x = 5.7$, $a = 2.5$.
9. The area of cross-section of a tree trunk is given by the formula $A = \frac{P^2}{4\pi}$ where P is the distance round the trunk.

Find the area of cross-section when $P = 56$ cm. Use $\pi = 3.14$ and give the answer correct to three significant figures.

10. The focal length of a lens is given by the formula $f = \frac{uv}{u + v}$. Find the focal length when $u = 6$, $v = -7$.

Collecting like terms and simplifying expressions

Remember that
$$a + a + a = 3a$$
$$4a + 3b + b - a = 3a + 4b$$
$$a \times b = ab$$
$$a \times a \times a = a^3$$

Complicated terms such as ab^2, a^2b and a^3 cannot be collected together unless they are exactly the same type.

Example 5

Simplify these expressions by collecting together the like terms.

(a) $2x^2 - 3xy + 2yx + 3y^2$ (b) $3a^2 + 4ab - 2a^2 - 3b^2 - 2ab$

(c) $3 + 5a - 2b + 2 + 8a - 7b$

(a) $2x^2 - 3xy + 2yx + 3y^2 = 2x^2 - xy + 3y^2$

> The two middle terms are like terms because xy is the same as yx.

(b) $3a^2 + 4ab - 2a^2 - 3b^2 - 2ab = a^2 + 2ab - 3b^2$

> Here the a^2 terms, the ab terms and the single b^2 term can be collected but they cannot then be combined.

(c) $3 + 5a - 2b + 2 + 8a - 7b = 5 + 13a - 9b$

> Here there are three different types which can be collected separately but not together.

Examiner's tip

Errors are often made by trying to go too far, for example $2a + 3b$ cannot be simplified any further.
A further common error is to work out $4a^2 - a^2$ as 4. The answer is $3a^2$.
Remember that ab is the same as ba.

Example 6

Simplify these.

(a) $3(4a - 5) + 2(3a + 2)$ (b) $2x(6x + 2y) - 5y(x - 3y)$

(a) $3(4a - 5) + 2(3a + 2) = 12a - 15 + 6a + 4$ Multiply out the brackets.

                       $= 18a - 11$ Collect like terms.

(b) $2x(6x + 2y) - 5y(x - 3y) = 12x^2 + 4xy - 5xy + 15y^2$ Take care with the signs in the second bracket.

                       $= 12x^2 - xy + 15y^2$ Collect like terms.

Simplify where possible.

1. $2a + 3b - 2b + 3a$
2. $4ab - 3ac + 2ab - ac$
3. $a^2 + 3b^2 - 2a^2 - b^2$
4. $2x^2 - 3xy - xy + y^2$
5. $4b^2 + 3a^2 - 2b^2 - 4a^2$

6. $5a^3 + 4a^2 + 3a$
7. $a^3 + 3a^2 + 2a^3 + 4a^2$
8. $9abc + 4cab - 5bca + 6cba$
9. $3x^3 - 2x^2 + 4x^2 - 3x^3$
10. $a + 3b - b + 2a - 3a - 2b$
11. $2(a - 2) + 3(a + 4)$

12. $3(3x + 7) + 5(2x - 6)$
13. $4(2b + 5) - 2(3b + 2)$
14. $3(2x - 3) - (3x - 8)$
15. $2x(x + y) + 3y(x - 4y)$

Simplify where possible.

1. $3a - 4b + 2a - 2b$
2. $9a^2 - 3ab + 5ab - 6b^2$
3. $4ab + 2bc - 3ba - bc$
4. $2p^2 - 3pq + 4pq - 5p^2$
5. $3ab + 2ac + ad$

6. $9ab - 2bc + 3bc - 7ab$
7. $a^3 + 3a^3 - 6a^3$
8. $3ab^2 - 4ba^2 + 7a^2b$
9. $8a^3 - 4a^2 + 5a^3 - 2a^2$
10. $abc + cab - 3abc + 2bac$
11. $2(c - 4) + 5(2c + 3)$

12. $3(x - y) + 4(2x - 3y)$
13. $2(2x + 4) - 3(x + 2)$
14. $5(x + 7) - 2(2x - 1)$
15. $a(a - 2b) - b(2a + 4b)$

Multiplying out two brackets

Expressions such as $a(3a - 2b)$ can be multiplied out to give
$a(3a - 2b) = (a \times 3a) - (a \times 2b) = 3a^2 - 2ab$.

This can be extended to working out expressions such as
$(2a + b)(3a + b)$.

Each term of the first bracket must be multiplied by each term of
the second bracket.

$(2a + b)(3a + b)$
$= 2a(3a + b) + b(3a + b)$ Expanding the first bracket.
$= 6a^2 + 2ab + 3ab + b^2$
$= 6a^2 + 5ab + b^2$ Notice that the middle two terms are **like terms** and so can be collected.

> **Examiner's tip**
>
> Most errors are made in multiplying out the second bracket when the sign in front is negative.

> **Examiner's tip**
>
> Apart from multiplying out the brackets, you may sometimes be asked to simplify, expand or remove the brackets, which all mean the same thing.

Example 7   Multiply out the brackets.

(a) $(2a + 3)(a - 1)$ (b) $(5a - 2b)(3a - b)$ (c) $(2a - b)(a + 2b)$

(a) $(2a + 3)(a - 1) = 2a(a - 1) + 3(a - 1) = 2a^2 - 2a + 3a - 3$

$= 2a^2 + a - 3$ Be careful with the signs.

(b) $(5a - 2b)(3a - 2b) = 5a(3a - b) - 2b(3a - b)$

$= 15a^2 - 5ab - 6ab + 2b^2$

$= 15a^2 - 11ab + 2b^2$ Note that it is $-2b$ times the bracket.

(c) $(2a - b)(a + 2b) = 2a(a + 2b) - b(a + 2b)$

$= 2a^2 + 4ab - ab - 2b^2$

$= 2a^2 + 3ab - 2b^2$ Note that it is $-b$ times the bracket.

In each of examples 6 and 7, the two brackets have resulted in three terms.

There are two other types of expansions of two brackets that you need to know.

Example 8

Expand the brackets.

(a) $(2a - 3b)^2$ (b) $(2a - b)(2a + b)$

(a) $(2a - 3b)^2 = (2a - 3b)(2a - 3b)$

$= 2a(2a - 3b) - 3b(2a - 3b) = 4a^2 - 6ab - 6ab + 9b^2$

$= 4a^2 - 12ab + 9b^2$ Note that $-6ab - 6ab = -12ab$

(b) $(2a - b)(2a + b) = 2a(2a + b) - b(2a + b) = 4a^2 + 2ab - 2ab - b^2$

$= 4a^2 - b^2$

Note that we only get two terms here because the middle terms cancel each other out. This type is known as the difference of two squares because: $(A - B)(A + B) = A^2 - B^2$

Examiner's tip

Take care with the negative signs.

Examiner's tip

The important thing in Example 8a is to make sure that you write the brackets separately and that you end up with three terms.

Examiner's tip

Some people can multiply out two brackets without writing down anything. However, you are more likely to make an error by missing steps and so it is worth showing every step in an examination.

Exercise 17.3a

Multiply out the brackets.

1. $(x + 2)(x + 3)$
2. $(a + 4)(a + 3)$
3. $(a + 2)(a + 1)$
4. $(x - 2)(2x + 1)$
5. $(2x - 3)(x + 2)$
6. $(3a + b)(2a - 2b)$

7. $(4a - b)(a + 2b)$
8. $(3a - 2b)(2a - 3b)$
9. $(4a - 3b)(2a - 3b)$
10. $(4 - 3b)(5 + 2b)$
11. $(2a - b)(3a - b)$
12. $(7a + 3b)(2a + b)$
13. $(a + 2)^2$

14. $(4x - 3y)^2$
15. $(3x - y)^2$
16. $(a - 2)(a + 2)$
17. $(3a + b)(3a - b)$
18. $(5x - 2y)(5x + 2y)$
19. $(4a + 3b)(a - b)$
20. $(5a + 4b)(2a - b)$

Multiply out the brackets.

1. $(x + 1)(x + 3)$
2. $(a + 3)(a + 3)$
3. $(a + 2)(a + 1)$
4. $(x - 2)(x + 1)$
5. $(5x - 3y)(x + 2y)$
6. $(3a + b)(a - 2b)$
7. $(a - b)(3a + 2b)$
8. $(5a - 2)(a - 3)$
9. $(7 - 3b)(1 - 3b)$
10. $(a - 3b)(2a + b)$
11. $(6a - b)(3a - 2b)$
12. $(a + 3b)(2a + b)$
13. $(a - b)^2$
14. $(2x - 1)^2$
15. $(x + 4)^2$
16. $(a - 5)(a + 5)$
17. $(2a + b)(2a - b)$
18. $(3x - 2y)(3x + 2y)$
19. $(5a + 4b)(2a - b)$
20. $(7a - 5b)(a - b)$

Simplifying expressions using indices

Remember that: $\quad a \times a \times a = a^3$ and $a \times a \times a \times a \times a = a^5$.
This can be extended: $a^3 \times a^5 = (a \times a \times a) \times (a \times a \times a \times a \times a) = a^8$
which is the same as: $\quad a^3 \times a^5 = a^{3+5} = a^8$

This suggests a general rule for indices.

$$a^m \times a^n = a^{m+n}$$

Similarly: $a^5 \div a^3 = (a \times a \times a \times a \times a) \div (a \times a \times a)$

$$= \frac{a \times a \times a \times a \times a}{a \times a \times a} = a \times a = a^2 \qquad \boxed{\text{Cancelling } a \times a \times a \text{ top and bottom.}}$$

This is the same as $a^5 \div a^3 = a^{5-3} = a^2$
which suggests another general rule for indices.

$$a^m \div a^n = a^{m-n}$$

Now $(a^2)^3 = a^2 \times a^2 \times a^2 = a^6$ $\qquad \boxed{\text{By the first rule.}}$

This is the same as $(a^2)^3 = a^{2 \times 3} = a^6$
and this suggests yet another rule.

$$(a^n)^m = a^{n \times m}$$

$a^3 \div a^3 = a^{3-3} = a^0 \qquad\qquad$ but $\qquad a^3 \div a^3 = 1$

This gives another rule.

$$a^0 = 1$$

You can use these rules, together with the algebra you have already learnt, to simplify a number of different algebraic expressions.

Example 9

Simplify these.

(a) $3a^2 \times 4a^3$ (b) $\dfrac{6a^5}{2a^3}$ (c) $(a^3)^4 \times a^3 \div a^5$

(a) $3a^2 \times 4a^3 = 12a^5$ The numbers are just multiplied and the indices are added.

(b) $\dfrac{6a^5}{2a^3} = 3a^2$ The numbers are divided and the indices are subtracted.

(c) $(a^3)^4 \times a^3 \div a^5 = a^{12} \times a^3 \div a^5$
$$= a^{15} \div a^5$$
$$= a^{10} \quad \text{Use the three rules.}$$

Example 10

Simplify where possible.

(a) $4a^2b^3 \times 3ab^2$ (b) $\dfrac{12ab^3 \times 3a^2b}{2a^3b^2}$ (c) $4a^2 + 3a^3$

(a) $4a^2b^3 \times 3ab^2 = 12a^3b^5$ The numbers are multiplied and the indices are added for each letter. Note that a is the same as a^1, so $a^2 \times a = a^{2+1} = a^3$.

(b) $\dfrac{12ab^3 \times 3a^2b}{2a^3b^2} = 18b^2$ The numbers combine as $12 \times 3 \div 2 = 18$, $a \times a^2 \div a^3 = a^{1+2-3} = a^0 = 1$, $b^3 \times b \div b^2 = b^{3+1-2} = b^2$.

(c) $4a^2 + 3a^3$ This cannot be simplified. The two terms are different and cannot be added.

Exercise 17.4a

Simplify where possible.

1. $3a^2 \times 4a^3$

2. $\dfrac{12a^5}{6a^3}$

3. $(3a^3)^2$

4. $2a^2b \times 3a^3b^2$

5. $4a^2b - 2ab^2$

6. $\dfrac{15a^2b^3 \times 3a^2b}{9a^3b^2}$

7. $\dfrac{9p^2q \times (2p^3q)^2}{12p^5q^3}$

8. $\dfrac{4abc \times 3a^2bc^3}{6a^2bc^2}$

9. $\dfrac{12t^3}{(2t)^2}$

10. $2a^2b \times 3ab^2 - 4a^3b^3$

Exercise 17.4b

Simplify where possible.

1. $\dfrac{a^5 \times a^3}{a^6}$

2. $3a^2 \times 4a^2$

3. $(2c)^3$

4. $3a^2b^3 \times 2a^3b^4$

5. $12a^2 \times 3b^2$

6. $2a^2 + 3a^3$

7. $8a^2b^3 \times 2a^3b \div 4a^4b^2$

8. $\dfrac{(3a^2b^2)^3}{(a^3b)^2}$

9. $4a^2 \times 2b^3 - a \times 3b \times ab^2$

10. $6a^2 \times (2ab^2)^2 \div 12b^2a$

Finding the *n*th term of a sequence

The *n*th term of a sequence, where the rule is linear, can be found by looking at the first differences, which are constant.

If the rule is not linear, for example a quadratic containing n^2, the first differences will not be constant. Find the second differences.

Examiner's tip

Check that the *n*th term is correct by trying the first few terms, putting $n = 1, 2, 3$.

Example Work out the *n*th term for this sequence.

4   7   12   19   28

Look at the differences.

Sequence:	4		7		12		19		28
First differences:		3		5		7		9	
Second differences:			2		2		2		

The second differences are constant.

> The rule for the *n*th term in sequences like this one will involve n^2 and the coefficient of n^2 will be half the second difference.

In the sequence above, dividing 2 by 2 gives 1, so the *n*th term must include $1n^2$ or just n^2.

Now compare the original sequence with the sequence n^2.

Sequence:	4	7	12	19	28
n^2:	1	4	9	16	25
Subtract n^2:	3	3	3	3	3

The *n*th term is $n^2 + 3$.

Example Work out the *n*th term for this sequence.

2   11   26   47   74

Look at the differences.

Sequence:	2		11		26		47		74
First differences:		9		15		21		27	
Second differences:			6		6		6		

Dividing 6 by 2 gives 3, so the *n*th term will include $3n^2$.
Subtract $3n^2$ from each term.

Sequence:	2	11	26	47	74
$3n^2$:	3	12	27	48	75
Subtract:	−1	−1	−1	−1	−1

So the *n*th term is $3n^2 - 1$.

Most of the sequences you will have to deal with will be like the two above, but you may be asked to find the nth term of sequences that include n^2, n and a number.

Example E

Work out the nth term for this sequence.

2   7   14   23   34

Sequence:		2		7		14		23		34
First differences:			5		7		9		11	
Second differences:				2		2		2		

Dividing 2 by 2 gives 1 so the nth term will include n^2.

Subtract n^2 from the sequence.

Sequence:	2	7	14	23	34
n^2:	1	4	9	16	25
Subtract	1	3	5	7	9

This sequence is like those you dealt with in Chapter 3.

The difference is 2 so the nth term will include $2n$.

When $n = 1$, $2n$ is 2, but the term is 1 so the nth term is $2n - 1$ for this sequence.

The nth term for the main sequence is therefore $n^2 + 2n - 1$.

Exercise 17.5a

Find the nth term of each of these sequences.

1.	2	5	10	17	26
2.	1	7	17	31	49
3.	5	8	13	20	29
4.	6	15	30	51	78

5.	−1	2	7	14	23
6.	7	16	31	52	79
7.	1	13	33	61	97
8.	4	9	16	25	36
9.	2	8	16	26	38
10.	2	6	12	20	30

Exercise 17.5b

Find the nth term of each of these sequences.

1.	3	6	11	18	27
2.	4	13	28	49	76
3.	6	9	14	21	30
4.	6	12	22	36	54

5.	1	10	25	46	73
6.	8	14	24	38	56
7.	1	16	41	76	121
8.	1	5	11	19	29
9.	0	5	12	21	32
10.	−2	−2	0	4	10

Factorising algebraic expressions

Factors are numbers or letters which will divide into an expression.

The factors of 6 are 1, 2, 3 and 6.

The factors of b^3 are 1, b, b^2 and b^3.

> Remember that multiplying or dividing by 1 leaves a number unchanged, so 1 is not a useful factor and it is ignored.

To factorise an expression, look for common factors, for example, the common factors of $2a^2$ and $6a$ are 2, a and $2a$.

Example 14

Factorise these fully.

(a) $4p + 6$ (b) $2a^2 - 3a$ (c) $15ab^2 + 10a^2b^2$ (d) $2a - 10a^2 + 6a^3$

(a) $4p + 6 = 2(2p + 3)$ The only common factor is 2 and $2 \times 2p = 4p$, $2 \times 3 = 6$.

(b) $2a^2 - 3a = a(2a - 3)$ The only common factor is a and $a \times 2a = 2a^2$, $a \times -3 = -3a$.

(c) $15ab^2 + 10a^2b^2 = 5ab^2(3 + 2a)$ 5, a and b^2 are common factors and $5ab^2 \times 3 = 15ab^2$, $5ab^2 \times +2a = 10a^2b^2$.

(d) $2a - 10a^2 + 6a^3 = 2a(1 - 5a + 3a^2)$ 2 and a are common factors and $2a \times 1 = 2a$, $2a \times -5a = -10a^2$ and $2a \times 3a^2 = 6a^3$.

Exercise 17.6a

Factorise these fully.

1. $2a + 8$
2. $3a + 5a^2$
3. $2ab - 6ac$
4. $5a^2b + 10ab^2$
5. $2x^2y^2 - 3x^3y$
6. $3a^2b - 6ab^2$
7. $12x - 6y + 8z$
8. $9ab + 6b^2$
9. $4a^2c - 2ac^2$
10. $15xy - 5y$
11. $6a^3 - 4a^2 + 2a$
12. $3a^2b - 9a^3b^2$
13. $5a^2b^2c^2 - 10abc$
14. $2a^2b - 3a^2b^3 + 7a^4b$
15. $4abc - 3ac^2 + 2a^2b$

Examiner's tip

Make sure that you have found all the common factors. Check that the expression in the bracket will not factorise further.

Exercise 17.6b

Factorise these fully.

1. $3x - 12$
2. $4a + 5ab$
3. $4ab - 2a^2$
4. $3ab - 2ac + 3ad$
5. $5x^2 - 15x + 15$
6. $4a^2b - 3ab^2$
7. $9x^2y - 6xy^2$
8. $14a^2 - 8a^3$
9. $21x^2 - 14y^2$
10. $12x^2y + 8xy - 4xy^2$
11. $14s^2t - 7st^2$
12. $10z^3 - 15z^2 + 5z$
13. $5abc - 15a^2b^2c^2$
14. $3a^2bc - 6ab^2c + 9abc^2$
15. $7a^3b^3c^2 - 14a^2b^3c^3$

Factorising expressions of the type $x^2 + ax + b$

Expressions where the last sign is positive

The expression $(x + 2)(x + 3)$ can be simplified to give $x^2 + 5x + 6$. Therefore, $x^2 + 5x + 6$ can be factorised as a product of two brackets, by reversing the process. Look at your answers for Exercises 17.3.

Example Factorise $x^2 + 7x + 12$.

This will factorise into two brackets with x as the first term in each.

$x^2 + 7x + 12 = (x \quad)(x \quad)$

As both the signs are positive, both the numbers will be positive.
You need to find two numbers that multiply to give 12 and add to give 7.

These will be $+3$ and $+4$.

So $x^2 + 7x + 12 = (x + 3)(x + 4)$ or
$x^2 + 7x + 12 = (x + 4)(x + 3)$

The order you write the brackets does not matter.

If the middle sign is negative and the last sign is positive, the two numbers will be negative.

Example Factorise $x^2 - 3x + 2$.

You need to find two negative numbers that multiply to give $+2$ and add to -3.
They are -2 and -1.

$x^2 - 3x + 2 = (x - 2)(x - 1)$.

Examiner's tip

If the last sign is positive (+), both the signs in the brackets must be the same as the sign before the x term.

Exercise 17.7a

Factorise these expressions.

1. $x^2 + 5x + 6$
2. $x^2 + 6x + 5$
3. $x^2 + 6x + 8$
4. $x^2 + 5x + 4$
5. $x^2 + 2x + 1$
6. $x^2 - 7x + 6$
7. $x^2 - 7x + 10$
8. $x^2 - 4x + 3$
9. $y^2 - 9y + 14$
10. $x^2 - 6x + 8$
11. $a^2 + 8a + 12$
12. $a^2 - 6a + 9$
13. $b^2 - 12b + 32$
14. $x^2 + 11x + 24$
15. $x^2 - 9x + 20$

Exercise 17.7b

Factorise these expressions.

1. $x^2 + 7x + 10$
2. $x^2 + 4x + 3$
3. $x^2 + 8x + 15$
4. $x^2 + 9x + 20$
5. $x^2 + 7x + 6$
6. $x^2 - 9x + 18$
7. $x^2 - 7x + 12$
8. $a^2 - 2a + 1$
9. $b^2 - 10b + 24$
10. $c^2 - 4c + 3$
11. $a^2 + 15a + 36$
12. $x^2 - 12x + 27$
13. $b^2 - 10b + 25$
14. $x^2 + 14x + 24$
15. $x^2 - 15x + 56$

Expressions where the last sign is negative

Example

Factorise $x^2 - 3x - 10$.

As the last sign is negative, you need two numbers, with opposite signs, that multiply to give -10 and add to give -3.
The numbers are -5 and $+2$.

$x^2 - 3x - 10 = (x - 5)(x + 2)$

Example

Factorise $x^2 + 4x - 12$.

The last sign is negative, so you need two numbers, with opposite signs, that multiply to give -12 and add to give $+4$.
The numbers are $+6$ and -2.

$x^2 + 4x - 12 = (x + 6)(x - 2)$

Examiner's tip

Remember that if the last sign is negative the two numbers have different signs and the larger number has the sign of the x term.
It is easy to make a mistake when factorising. Always check by multiplying out the brackets.

Exercise 17.8a

Factorise these expressions.

1. $x^2 - 2x - 8$
2. $x^2 + 4x - 5$
3. $x^2 - x - 6$
4. $x^2 + 5x - 6$
5. $x^2 + 2x - 3$
6. $x^2 - 3x - 18$
7. $x^2 - 9x - 10$
8. $x^2 + 9x + 14$
9. $y^2 + 9y - 22$
10. $x^2 + x - 12$
11. $a^2 + 8a - 20$
12. $a^2 - 6a - 27$
13. $b^2 + 12b + 20$
14. $x^2 - 25$
15. $x^2 - 49$

Exercise 17.8b

Factorise these expressions.

1. $x^2 + 2x - 3$
2. $x^2 + 3x - 10$
3. $x^2 - x - 12$
4. $x^2 + 5x - 14$
5. $x^2 - 2x - 15$
6. $x^2 - 3x - 28$
7. $x^2 - 17x + 30$
8. $x^2 + 4x - 32$
9. $a^2 + 9a - 36$
10. $x^2 + x - 20$
11. $y^2 + 19y + 48$
12. $a^2 - 6a - 16$
13. $b^2 - 15b + 36$
14. $x^2 - 4$
15. $x^2 - 144$

Rearranging formulae

Formulae can be treated in the same way as equations. This means they can be rearranged, to change the subject.

Example 19

Rearrange these formulae to make the letter in brackets the subject.

(a) $a = b + c$ (b)
(b) $a = b + cx^2$ (x)
(c) $n = m - 3s$ (s)
(d) $p = r + a(a - r)$ (r)
(e) $s = 3(a + b)$ (b)

(a) $a = b + c$

$a - c = b$	Subtract c from both sides.
$b = a - c$	Reverse to get b on the left.

(b) $a - b = cx^2$ Subtract b from both sides.

$x^2 = \dfrac{a - b}{c}$ Swap and divide both sides by c.

$x = \sqrt{\dfrac{a - b}{c}}$ Take square roots of both sides.

(c) $n = m - 3s$

$n + 3s = m$	Add $3s$ to both sides.
$3s = m - n$	Subtract n from both sides.
$s = \dfrac{m - n}{3}$	Divide both sides by 3.

Example 15
continued

(d) $p = r + a(a - r)$

$p = r + a^2 - ar$ — Multiply out the bracket.

$p - a^2 = r(1 - a)$ — Rearrange and take out r.

$r(1 - a) = p - a^2$ — Reverse to get r on the left.

$r = \dfrac{p - a^2}{1 - a}$ — Divide both sides by $(1 - a)$.

(e) $s = 3(a + b)$

$s = 3a + 3b$ — Multiply out the brackets.

$s - 3a = 3b$ — Subtract $3a$ from both sides.

$\dfrac{s - 3a}{3} = b$ — Divide both sides by 3.

$b = \dfrac{s - 3a}{3}$ — Reverse to get b on the left.

Examiner's tip

If you have difficulty in rearranging a formula, practise by replacing some of the letters with numbers.

Exercise 17.9a

Rearrange each formula to make the letter in the brackets the subject.

1. $a = b - c$ (b)
2. $3a = wx + xy$ (x)
3. $v = u + at$ (t)
4. $A = \dfrac{T}{H}$ (T)
5. $C = P - 3T$ (T)
6. $P = \dfrac{u + v}{2}$ (u)
7. $C = 2\pi r$ (r)
8. $A = p(q^2 + r)$ (q)
9. The formula for finding the perimeter P of a rectangle with sides of length x and y is $P = 2(x + y)$.

    Rearrange the formula to make x the subject.

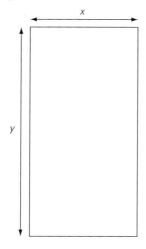

10. The cost (£C) of catering for a wedding reception is given by the formula $C = A + 32n$, where A is the cost of the room and n is the number of guests.

    (a) Rearrange the formula to make n the subject.

    (b) Work out the number of guests when A is £120 and the total cost C is £1912.

Rearrange each formula to make the letter in the brackets the subject.

1. $p = q + 2r$ (q)
2. $B = s + 5r^2$ (r)
3. $s = 2u - t$ (t)
4. $m = \dfrac{pqr}{s}$ (q)
5. $L = 2G - 2FG$ (G)
6. $F = \dfrac{m + 4n}{t}$ (n)
7. $T = \dfrac{S}{2a}$ (S)
8. $A = t(x - 2y)$ (y)
9. The formula for the volume V of a cone of height h and base radius r is $V = \frac{1}{3}\pi r^2 h$.

   Rearrange the formula to make h the subject.

10. The cost (£C) of a minibus to the airport is given by the formula

    $$C = 20 + \dfrac{d}{2}$$

    where d is the distance in miles.

    (a) Rearrange the formula to make d the subject.
    (b) Work out the distance when it costs £65 to go to the airport.

Inequalities

$a < b$ means a is less than b.
$a \leqslant b$ means a is less than or equal to b.
$a > b$ means a is greater than b.
$a \geqslant b$ means a is greater than or equal to b.

Expressions involving these signs are called **inequalities**.

Find the integer (whole number) values of x when:

(a) $-3 < x \leqslant -1$ (b) $1 \leqslant x < 4$.

(a) If $-3 < x \leqslant -1$, then $x = -2$ or -1.

Note that -3 is not included but -1 is.

(b) If $1 \leqslant x < 4$, then $x = 1, 2$ or 3.

Note that 1 is included but 4 is not.

In equations, if you always do the same thing to both sides the equality is still valid.

The same is true for inequalities, except in one case.

Consider the inequality $5 < 7$.

Add 2 to each side:	$7 < 9$	Still true.
Subtract 5 from each side:	$2 < 4$	Still true.
Multiply each side by 3:	$6 < 12$	Still true.
Divide each side by 2:	$3 < 4$	Still true.
Multiply each side by -2:	$-6 < -12$	No longer true.
But reverse the inequality sign:	$-6 > -12$	Now true.

If an inequality is multiplied or divided by a negative number, the **inequality sign must be reversed**.

Otherwise inequalities can be treated in the same way as equations.

 Solve these inequalities.

(a) $3x + 4 < 10$ (b) $2x - 5 \leqslant 4 - 3x$
(c) $x + 4 > 3x - 2$

(a) $3x + 4 < 10$

$3x < 6$	Subtract 4 from each side.
$x < 2$	Divide each side by 2.

(b) $2x - 5 \leqslant 4 - 3x$

$2x \leqslant 9 - 3x$	Add 5 to each side.
$5x \leqslant 9$	Add $3x$ to each side.
$x \leqslant \frac{9}{5} \ (=1\frac{4}{5})$	Divide each side by 5.

(c) $x + 4 > 3x - 2$

$x > 3x - 6$	Subtract 4 from both sides.
$-2x > -6$	Subtract $3x$ from both sides.
$x < 3$	Divide each side by -2 and change the $>$ to $<$ (when dividing by -2).

Exercise 17.10a

1. Write down the integer values of x when:
   (a) $-4 \leqslant x < 0$
   (b) $1 < x \leqslant 5$.

Solve these inequalities.

2. $x - 3 \leqslant 4$
3. $x + 7 > 9$
4. $2x - 3 < 5$
5. $3x + 4 \leqslant 7$
6. $2x \geqslant x + 5$
7. $5x > 3 - x$
8. $2x + 1 < 7$
9. $4x > 2x + 5$
10. $3x - 6 \geqslant x + 2$
11. $2(x + 3) < 1 - 3x$
12. $3(2x - 1) \geqslant 11 - x$
13. $x + 4 > 2x$
14. $2x - 5 < 4x + 1$
15. $3(x - 4) > 5(x + 1)$

Exercise 17.10b

1. Write down two possible values of x for the inequality $x < -2$.
2. Write down the integer values of x when:
   (a) $1 < x \leqslant 4$
   (b) $-5 < x \leqslant -1$.

Solve these inequalities.

3. $x - 2 < 5$
4. $2x + 3 > 6$
5. $3x - 4 \leqslant 8$
6. $3x \geqslant x - 2$
7. $4x + 2 < 3$
8. $5a - 3 > 2a$
9. $2x - 3 < x + 1$
10. $3x + 2 \geqslant x - 1$
11. $x - 2 < 2x + 4$
12. $2x - 1 > x - 4$
13. $3(x + 3) \geqslant 2x - 1$
14. $\frac{1}{2}x + 4 < 5$
15. $3(2x - 4) < 5(x - 6)$

Forming equations and inequalities

Everyday problems can often be solved by forming equations or inequalities and solving them.

Example

The length of a rectangle is $4\,\text{cm}$ longer than its width, which is $x\,\text{cm}$.

(a) Write down an expression in terms of x for the perimeter of the rectangle.

(b) The perimeter is $32\,\text{cm}$.
   (i) Write down an equation in x and solve it.
   (ii) What are the length and width of the rectangle?

(a) The length is $4\,\text{cm}$ longer than the width, so it is $x + 4\,\text{cm}$.

Perimeter $= x + x + 4 + x + x + 4 = 4x + 8$

(b) (i) The perimeter is $32\,\text{cm}$, so:
$$4x + 8 = 32$$
$$4x = 24$$
$$x = 6$$

(ii) The width is $x = 6\,\text{cm}$.
The length is $x + 4 = 10\,\text{cm}$.

Example 22

John is having a party but he has only £60 to spend on it. He has to pay £10 to hire the room and £4 for every person at the party. How many people can he invite to his party?

Let the number of people be n. Write down an inequality involving n and solve it to find the largest number that can go to the party.

Cost of party $\leqslant$ £60.
Therefore $10 + 4n \leqslant 60$
$$4n \leqslant 50$$
$$n \leqslant 12.5$$
So the largest number of people who can be at the party is 12.

Examiner's tip

In these questions you can sometimes work out the answer without forming the equation or inequality but you will lose marks in an examination if you don't write down an equation or inequality.

Exercise 17.11a

1. Erica is x years old and Jayne is three years older than Erica. Their ages add up to 23. Write down an equation in x and solve it to find out their ages.
2. Two angles of a triangle are the same and the other is 15° bigger. Call the two equal angles a. Write down an equation and solve it to find the angles.
3. A man is papering a room. It takes him 30 minutes to prepare his paste and 20 minutes to cut and hang a length of paper. He is working for 4 hours and hangs x lengths.

Write down an inequality in x and solve it to find the largest number of lengths he can hang in the time.

4. In Devonshire School there are 28 more girls than boys. There are 616 pupils in the school. Let the number of boys be x. Write down an equation in x and solve it. Write down the number of boys and girls in the school.

5. To hire a bus the charge is £60 and £2 a mile. The bus company will only hire the bus if they take at least £225. Let the number of miles be x.
   (a) Write down an inequality and solve it for x.
   (b) What is the smallest distance that the bus can be hired to go?

6. It costs £x to hire a bike for an adult, and it is £2 cheaper for a child's bike. Mr Newton hires bikes for two adults and three children.

   (a) Write down an expression in x for the cost of the bikes.
   (b) The cost is £19.
       (i) Write down an equation and solve it to find x.
       (ii) How much did each bike cost?

7. Ameer has 40 metres of fencing, in one-metre lengths that cannot be split. He wants to use as much of it as he can to mark out a rectangle that is twice as long as it is wide.
   (a) Call the width of the rectangle x and write down an inequality.
   (b) Solve it to find the length and width of the biggest rectangle that he can make.

8. Mark, Patrick and Iain all collect matchbox cars. Mark has four more than Patrick. Iain has three more than Mark. They have 41 altogether. Set up an equation and solve it to find how many matchbox cars each boy has.

1. Mrs Pippard and her daughter go shopping. Mrs Pippard spends £x and her daughter spends twice as much. The spend £45 altogether. Set up an equation and solve it to find how much each spends.

2. A pentagon has two angles of 150°, two of x° and one of $(x + 30)$°. The sum of the angles in a pentagon is 540°.
   (a) Write down an equation in x and solve it.
   (b) State the size of each of the angles.

3. Sara is x years old; Mary is ten years older than Sara. The sum of their ages is less than 50.
   (a) Write down an inequality and solve it.
   (b) What is the oldest Sara can be?

4. On a school trip to France, there are 15 more girls than boys. Altogether 53 pupils go.
   (a) If the number of boys is x, write down an equation in x and solve it.
   (b) How many boys and how many girls go on the trip?

5. Paul goes to the shop and buys two chocolate bars at 30p each and x cans of cola at 45p each. He has £2 and wants to buy as many cans as possible.
   (a) Write down an inequality in x and solve it.
   (b) What is the largest number of cans he can buy?

6. In the Oasis café a cup of tea costs x pence; a cup of coffee costs 10 pence more than tea. David bought three teas and two coffees and spent £1.20.
   (a) Write down an equation in x and solve it.
   (b) What do tea and coffee cost at the Oasis café?

7. A firm employs 140 people, of whom x are men. There are ten fewer women than men.
   Use algebra to find how many men and women work for the firm.

8. It costs £5 for each person to go skating. Skates can be hired for £2. Ten friends went skating and n of them hired skates.
   (a) Write down an expression in pounds for the amount they spent.
   (b) They spent £62. Write down an equation in n and solve it to find how many hired skates.

Key points

- Terms can only be added or subtracted if they are of exactly the same type.

  ab and $2ab$ can be added but a^2 and $2a$ cannot.

- When multiplying two brackets , multiply every term in the first bracket by every term in the second.

- When multiplying or dividing algebraic expressions involving powers, add or subtract the indices.

- To find the nth term of a quadratic sequence, look at the second difference and divide by 2 to find the coefficient of the n^2 term.

- When finding common factors make sure you factorise fully.

- To factorise $x^2 + ax + b$:
  - if b is positive find two numbers that multiply to give b and add up to a
  - if b is negative find two numbers that multiply to give b and have a difference of a.

- To solve inequalities treat them like equations, except when multiplying or dividing by a negative number, when you *must reverse* the inequality sign.

- To solve a problem using algebra, set up an equation for the unknown quantity, then solve it.

Revision exercise 17a

1. If $h = 7a - 2bc$, find h when:
   (a) $a = 2$, $b = 3$, $c = -1$
   (b) $a = \frac{1}{4}$, $b = \frac{1}{2}$, $c = \frac{3}{4}$
   (c) $a = 3.6$, $b = 7.4$, $c = 2.5$.

2. The formula for finding the area of the surface of a sphere is $A = 4\pi r^2$.
   Find the area when $r = 4.2$ cm, giving the answer correct to three significant figures.

3. Simplify these by collecting like terms.
   (a) $3a + 4b + 2a - 4b$
   (b) $5ab^2 - 2a^2b + 3a^2b - 4ab^2$
   (c) $2ab + 3ac - 4ad + 2ab + 4ad - ac$
   (d) $x^2 - 2xy + 3yx + 3x^2$

4. Take out the brackets and collect like terms.
   (a) $5(x + 3) + 2(x - 5)$
   (b) $2(3x + 1) - 2(x + 4)$
   (c) $3(a + 4) + 2(2 - a)$
   (d) $3(2x - 2) - 2(x - 5)$
   (e) $2(x + 1) + (2x - 3)$
   (f) $3x(x - 2y) - 4y(3x - y)$

5. Multiply out these brackets.
   (a) $(x + 7)(x + 1)$
   (b) $(a - 3)(a + 5)$
   (c) $(2y - 4)(y + 1)$
   (d) $(x - 5)(2x + 1)$
   (e) $(4a + b)(a - b)$

   (f) $(2 - c)(3 - 2c)$
   (g) $(x - 5)(x + 5)$
   (h) $(x + 2y)^2$
   (i) $(7x - 5y)(2x + 3y)$

6. Simplify these.
   (a) $2a^2 \times a^3$
   (b) $10a^2 \div 2a$
   (c) $(a^3)^2 \times a^3 \div a^4$
   (d) $12a^2b \times 2a^2b^3$
   (e) $6x^2y^2z^2 \div 2xy^2z$
   (f) $\dfrac{8a^2b \times 3abc}{6ab^2}$

7. Find the nth term of the following sequences.
   (a) 4   7   12   19   28
   (b) 3   9   19   33   51
   (c) 4   8   14   22   32

8. Factorise each of these fully.
   (a) $3a + 6b - 12c$
   (b) $2a + 3ab$
   (c) $a^2b - 3ab^2$
   (d) $2x^2y - 6xy$
   (e) $7abc + 14a^2b$
   (f) $9a^2 + 3b^2 - 6c^2$
   (g) $5pq - 10$
   (h) $2a - 4a^2 + 6a^3$
   (i) $100abc - 50ac$

9. Factorise these.
   (a) $x^2 + 5x + 4$
   (b) $x^2 - 6x + 8$
   (c) $x^2 - 10x + 16$
   (d) $x^2 + 8x + 15$
   (e) $x^2 - 6x - 7$
   (f) $x^2 - 3x - 10$
   (g) $x^2 - 8x + 12$
   (h) $x^2 - 2x - 15$
   (i) $x^2 - 3x - 70$
   (j) $x^2 + 16x + 48$
   (k) $x^2 - 7x - 18$
   (l) $16 - x^2$

10. Rearrange the following formulae to make the letter in brackets the subject.
    (a) $x = y - 3b$ (y)
    (b) $t = \dfrac{u + v}{2}$ (u)
    (c) $P = 2b - a$ (a)
    (d) $p = qx + m$ (q)
    (e) $I = \dfrac{PTR}{100}$ (P)
    (f) $v^2 = u^2 + 2as$ (s)

11. Solve the inequalities.
    (a) $2x > 5$
    (b) $x + 3 \leqslant 5$
    (c) $2x - 4 \geqslant x + 2$
    (d) $4x - 3 < 7 - x$
    (e) $4x - 9 \leqslant 2x + 7$
    (f) $2(3x - 1) \leqslant 3x + 5$
    (g) $2x - 3 < 3x - 1$
    (h) $x + 2 > 3x + 1$
    (i) $3(2x - 3) > 2(x - 5)$

12. David has two brothers. One brother is two years younger than him and the other brother is five years older than him.
    (a) Let David be x years old and write down an expression for the sum of their ages.
    (b) The sum of their ages is 39. Write down an equation and solve it to find x.
    (c) What are their ages?

13. Angela has £5 to spend. She spends £3.20 on her lunch and decides to buy as many packets of crisps as possible at 24p each with the rest.
    (a) If the number of packets she buys is x, write down an inequality in x and solve it.
    (b) How many packets of crisps does she buy?

14. A quadrilateral has angles $x°$, $3x°$, $90°$ and $(x + 20)°$.
    (a) Write down an equation in x and solve it.
    (b) What are the sizes of the angles?

Questionnaires and cumulative frequency

> ## You will need to know:
>
> - how to calculate mode, median, mean and range.

Questionnaires

Planning and collecting data

The first part of this unit is about the design of **questionnaires** or **experiments** to test a **hypothesis**.

> A hypothesis is a statement which may be true but for which you have no proof.

Questionnaires are designed to find out, for example, people's opinions. Questionnaires are used a lot. Their findings are often shown on TV or in newspapers, for example:

- 63% of the people asked would vote Labour
- Eight out of ten owners who expressed a preference said their cats preferred fresh fish.

Experiments are tests intended to find out if something is true; for example an experiment could be carried out to see if people are better at catching a ball with their 'writing' hand or with their 'non-writing' hand.

After you have thought of a task, challenge or a question to answer, or had one set for you, there are several important issues for you to consider as you plan how you are going to find the information you need.

Think about and discuss the following three topics.

Hypothesis: Boys are better at mathematics than girls are.

How could you prove it?

You might decide to change the statement into a question which you could then test, for example,

Do boys get higher marks than girls in mathematics tests?

but then you will need to think about issues such as:

- Does age or year group matter? Is it true for boys and girls in year 11, but not for year 7?
- Does the type of test matter?
- Does the topic matter?
  Are the results for algebra the same as for shape and space?
- How will you define 'higher' marks?
  - Will you use the actual highest mark or the average mark?
  - Should the average be the mean, the median or the mode?

- How will you actually get the results?
  - Write a test?
  - Ask your teachers?
  - Use the results from the KS3 tests?
- How will you analyse and show the information?
  - Will the questions you ask help you?
  - Should you be asking different questions?
- Will your results be fair or biased?
  Will you consider a cross-section of students or just those in a particular set?

You must consider bias or fairness when you are trying to find information to prove a hypothesis. If you wanted to find out if people like hamburgers, would you expect to get a fair or unbiased response if you only asked teenagers who were going into or out of a hamburger restaurant?

Activity 2

Survey: What do people eat for breakfast?

You might ask:

'Do you have breakfast?'

What sort of answers might you get?

'yes', 'no', 'sometimes', 'only on a school day' …

- Ask some members of your class and see what they say.

Or you might ask:

'Do you have cereal or toast?'

How would you analyse an answer of 'yes'?

Does it mean toast **or** cereal, or toast **and** cereal?

Are there better questions, or a better way to find out what people had for breakfast?

You might consider setting out your questions like this:

Breakfast survey

Please tick if you have:

Fruit juice	☐	Fruit	☐	Cereal	☐
Tea	☐	Coffee	☐	Toast	☐
Bread	☐	Cooked	☐		

Activity 2 continued

What other questions could you ask?

- Write them down and discuss them.

You might want to ask people other than students at school.

Perhaps you could sort the answers by age, but be careful! It would be tactless to ask, 'How old are you?' so you need to think of an alternative approach. It would be much better to ask people to tick a box from a range of possibilities, for example:

Please tick the appropriate age group (in years):

5–10	☐	11–16	☐	17–20	☐
21–30	☐	31–40	☐	41–60	☐
over 60	☐				

Clearly it is easier to analyse responses to questions that can be ticked or which have a 'yes' or 'no' answer. These are **closed** questions.

Activity 3

How will you know if the questions are good ones?

A quick and simple way is to ask about ten people to answer your questions, and see how they 'work'.

- Can the questions be answered clearly?
- Can you analyse the answers?

This is called a **pilot survey**.

Now think of a question of your own, or choose one of the following:

- Do you think too much homework is set?
- Do tall people have bigger heads than shorter people do?
- Do boys spend more time than girls watching TV?
- Make three sets of ten cards:
  - one set with the numbers 0–9
  - one set with the letters A–J
  - one set with the letters Q–Z

Read the following three questions and then ask your volunteers to sort the cards in each set into order, recording the results appropriately. Shuffle each set before carrying out the experiment, each time.

- Are people quicker to put numbers in order, or letters?
- Are people quicker to put numbers or letters in order using their 'writing' hand or their 'non-writing' hand?
- Are people quicker with the first ten letters or the last ten letters of the alphabet?

Remember:

- write down what questions you will ask
- write down who you will ask, making sure that you ask a cross-section of people.

Try to justify what you do.

- Now carry out the survey or experiment you have chosen.

After you have produced the data you will have to analyse it.

Analysing data

This section suggests one way of analysing data. There are others, including calculating the mean or the mode.

Cumulative frequency tables and diagrams

A plant grower wants to find out if one sort of compost is better than another. He sows equal numbers of seeds, from the same packet, in each compost and measures the height, to the nearest centimetre, of 60 plants which grow in each.

Compost A

22	13	33	31	51	24	37	83	39	28	31	64
23	35	9	34	42	26	68	38	63	34	44	77
37	15	38	54	34	22	47	25	48	38	53	52
35	45	32	31	37	43	37	49	24	17	48	29
57	33	30	36	42	36	43	38	39	48	39	59

Compost B

33	43	17	50	37	59	21	58	45	78	36	34
45	77	52	42	79	38	63	48	47	71	63	49
8	53	47	66	49	69	55	33	54	28	40	68
55	67	36	76	27	86	29	67	57	47	64	55
48	65	58	41	35	57	44	39	59	23	64	36

This gives a total of 120 results (60 for each compost), which is a lot to analyse.

In cases like this it is better to group the results in intervals. A sensible interval for the heights in this case would be 10 cm.

This is like sorting the results into 'bins'.

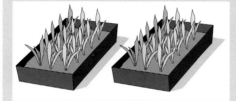

	13	22 24 28	33 31 37 39		
$0 \leq h < 10$	$10 \leq h < 20$	$20 \leq h < 30$	$30 \leq h < 40$	$40 \leq h < 50$	and so on.

Remember that $\leq$ means 'less than or equal to' and $<$ means 'less than' so $30 \leq h < 40$ means all the heights between 30 and 40, including 30 but excluding 40.

Here are the figures for compost A, grouped into a frequency table.

Height h (cm)	Frequency	Height h (cm)	Cumulative frequency	
$0 \leq h < 10$	1	$h < 10$	1	
$10 \leq h < 20$	3	$h < 20$	4	
$20 \leq h < 30$	9	$h < 30$	13	
$30 \leq h < 40$	25	$h < 40$	38	$\leftarrow 38 = 1 + 3 + 9 + 25$
$40 \leq h < 50$	11	$h < 50$	49	
$50 \leq h < 60$	6	$h < 60$	55	
$60 \leq h < 70$	3	$h < 70$	58	
$70 \leq h < 80$	1	$h < 80$	59	
$80 \leq h < 90$	1	$h < 90$	60	

The cumulative frequency in the last column gives the running total. In this case it is the number of plants less than a certain height, for example there are 38 plants less than 40 cm high. Make sure you can see how the cumulative frequency values are obtained.

The values for cumulative frequency can be plotted to give a cumulative frequency diagram.

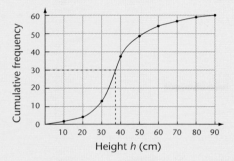

Note that the cumulative frequency values are plotted at the upper value of each interval i.e. at 10, 20, 30 and so on.

You can use a cumulative frequency diagram to estimate the median value.

There are 60 results so the median will be the halfway value, which is the 30th.

Note: If you were using a list of numbers, then for an even number of numbers the median is halfway between the two middle values.

Find 30 on the vertical scale and look across the graph until you reach the curve. Read off the corresponding value on the horizontal scale (see the dotted line on the graph).

The median height is about 37 cm – check you agree.

It is also possible to calculate the **quartiles**. As the name suggests these are quarters – the cumulative frequency is divided into four equal parts. The median is the middle quartile. The lower quartile will be at $\frac{1}{4}$ of 60, which is the 15th value, giving a height of 31 cm. The upper quartile is at $\frac{3}{4}$ of 60, which is the 45th value, giving a height of 45 cm.

The difference between these two values is called the **interquartile range**.

Interquartile range = 45 cm − 31 cm = 14 cm

The interquartile range shows how widely the data are spread out. Half the data are within the interquartile range, and if that range is small then the data are bunched together.

You can also use the cumulative frequency graph to estimate how many plants were taller than a given height, such as 55 cm. From the graph, a height of 55 cm corresponds to cumulative frequency 52, so the number of plants that were taller than 55 cm is 60 − 52 = 8.

- Now construct a grouped frequency table and find the cumulative frequency for the plant heights for compost B.

- Copy the cumulative frequency diagram for compost A. Using the same axes, draw the cumulative frequency diagram for compost B and calculate the median value and the interquartile range.

- Compare the results for the two composts, writing down what you notice.

- Now repeat this process for the results from the survey or experiment you chose to do above.

Examiner's tip

In a cumulative frequency diagram, you can join the points with straight (ruled) line segments instead of a curve.

Writing up your findings

There are several things you need to remember when you are writing up your work. Look back at the table and reread the advice given in the introduction.

You need to:

- show evidence of the planning that you did
- show how you found the information, who you asked and how you chose them
- include ideas for extending the task, for example suggesting other questions that you could ask
- present the information clearly
- explain or state what you wanted to find out
- state what you notice, how your analysis supports this and try to explain why the results occur.

Box-and-whisker plots

As you have seen in this chapter, data which have been ordered and/or presented in a cumulative frequency table or graph can be divided into four equal parts called **quartiles**. This example illustrates the use of quartiles.

Here are the examination results for a class.

80 62 53 76 76 41 59 78 84

66 71 50 79 69 87 64 56 65

58 78 75 60 51 73 74

These can be shown in a frequency table like this...

Mark m	Frequency		Cumulative frequency
$40 \leqslant m < 45$	/	1	1
$45 \leqslant m < 50$			1
$50 \leqslant m < 55$	///	3	4
$55 \leqslant m < 60$	///	3	7
$60 \leqslant m < 65$	///	3	10
$65 \leqslant m < 70$	///	3	13
$70 \leqslant m < 75$	///	3	16
$75 \leqslant m < 80$	⦚⦚⦚⦚ /	6	22
$80 \leqslant m < 85$	//	2	24
$85 \leqslant m < 90$	/	1	25

and plotted like this.

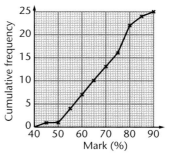

There are 25 values so the median is the middle, the 13th. The lower quartile is one-quarter of the way up, in this case in the $0.25 \times 25 = 6.25$th position. The upper quartile is three-quarters of the way up, in this case in the $0.75 \times 25 = 18.75$th position.

Reading from the graph, the lower quartile is 59, the median is 70 and the upper quartile is 77.

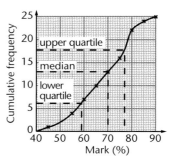

Chapter 18 Box-and-whisker plots

An alternative method of showing the quartiles and the range is with a **box-and-whisker plot.** These plots are produced to allow people to compare data quickly – they are not always useful for accurate work.

The left-hand side of the box is at the point corresponding to the value of the lower quartile. The right-hand side of the box is at a point corresponding to the upper quartile. The median is also drawn in the box.

The left-hand whisker extends from the lower quartile to the minimum value, the right-hand whisker extends from the upper quartile to the maximum value.

The box itself shows the location of the middle 50% of the data, the whiskers show how the data are spread out.

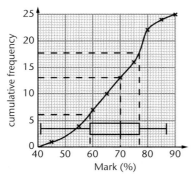

It is not necessary to draw cumulative frequency graphs to find or show quartiles before constructing box-and-whisker plots.

For example, the data below give the numbers of fish caught over a period of 11 days.

0 3 4 5 0 3 5 7 8 7 6

Rearranging these data into order gives:

0 0 3 3 4 5 5 6 7 7 8

With a small amount of data, it is easy to find the quartiles as follows.

- The median is 5.
- The median splits the data into a lower half: 0 0 3 3 4 and an upper half: 5 6 7 7 8.
- Find the median for each of these halves: the lower quartile is 3 and the upper quartile is 7.

You can now draw a box-and-whisker plot, knowing that the lowest value is 0 and the highest is 8.

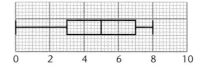

If you have to draw a box and whisker plot for grouped data when you are not given the actual minimum or maximum values the whiskers should be drawn to the minimum value of the lowest group and to the maximum value of the highest group. For example, if you were given a table of examination marks grouped $0 \leqslant M < 10$, $10 \leqslant M < 20$, $20 \leqslant M < 30$, ... $90 \leqslant M < 100$, the whiskers would be drawn to 0 and to 100.

Exercise 18.1a

1. Two policemen did separate traffic surveys at different locations. Each policeman recorded the speeds of 50 cars that passed him. Their findings are recorded in this table.

Speed v (km/h)	Policeman A	Policeman B
$10 \leqslant v < 30$	6	2
$30 \leqslant v < 50$	12	4
$50 \leqslant v < 70$	12	4
$70 \leqslant v < 90$	13	19
$90 \leqslant v < 110$	4	19
$110 \leqslant v < 130$	3	2

(a) Draw a cumulative frequency graph for each set of data and find the median and quartiles.
(b) Show these on box-and-whisker plots.
(c) Compare the results. Can you draw any conclusions about the two locations?

2. Draw box-and-whisker plots for each of these sets of data.
 (a) 8 3 15 8 13 1 20 5 16
 (b) 8.6 3.8 1.5 6.8 4.7 7.6 10.3 5.4 1.6

Exercise 18.1b

1. A sample of 200 potatoes was weighed and sorted into groups, as shown in this table.

Weight w (g)	Frequency	Cumulative frequency
$10 \leqslant w < 30$	12	
$30 \leqslant w < 50$	37	
$50 \leqslant w < 70$	63	
$70 \leqslant w < 90$	52	
$90 \leqslant w < 110$	31	
$110 \leqslant w < 130$	5	

(a) Copy the cumulative frequency table and complete it.
(b) Draw a cumulative frequency graph and use it to find the lower and upper quartiles and the median of the samples.
(c) Draw a box-and-whisker plot for the data.

2. A gardener measured the heights of 170 plants he was growing and produced this table.

Height h (cm)	Frequency	Cumulative frequency
$0 \leqslant h < 6$	5	
$6 \leqslant h < 10$	11	
$10 \leqslant h < 14$	57	
$14 \leqslant h < 18$	52	
$18 \leqslant h < 22$	27	
$22 \leqslant h < 26$	18	

(a) Copy the cumulative frequency table and complete it.
(b) Draw a cumulative frequency graph.
(c) Show the information on a box-and-whisker plot.

Key points

- Questionnaires and experiments can be used to prove or disprove a hypothesis.
- Questionnaires should be designed to avoid bias.
- Closed questions, which have a 'yes' or 'no' answer are easier to analyse for a questionnaire.
- A pilot survey is a small survey carried out with a few people to see if the planned questions 'work'.

- Grouped data are used when there are a lot of results to analyse.
- Cumulative frequency diagrams are useful for estimating the interquartile range and the median of a set of data.
- Findings should be written up clearly, to explain what has been done and why.
- Box-and-whisker plots show the quartiles, the median and the range of a set of data.

Revision exercise 18a

1. In 1970, a small factory employed 200 people. The frequency table shows their annual earnings.

Earnings (£E)	$400 < E \leqslant 600$	$600 < E \leqslant 800$	$800 < E \leqslant 1000$	$1000 < E \leqslant 1200$	$1200 < E \leqslant 1400$
Frequency	50	55	63	27	5

(a) Draw a cumulative frequency curve of the data.
(b) Use your graph to find:
   (i) the median earnings
   (ii) the interquartile range
   (iii) the number of employees who earned more than £900.

2. This table shows the numbers of marks obtained by candidates in an examination.

Mark	10	20	30	40	50	60	70	80	90	100
Number of candidates obtaining less than this mark	7	16	36	64	102	130	151	162	168	170

   (a) Draw a cumulative frequency curve of the data.
   (b) Use your graph to find:
       (i)   the median mark
       (ii) the interquartile range
       (iii) the number of candidates who obtained at least 55 marks
       (iv) the mark achieved by at least 60% of the candidates.

3. A new treatment is tested on 50 young apple trees. When the fruit was picked, these were the results.

Yield y (kg)	$0 \leqslant y < 2$	$2 \leqslant y < 4$	$4 \leqslant y < 6$	$6 \leqslant y < 8$	$8 \leqslant y < 10$	$10 \leqslant y < 12$
Frequency	0	1	2	25	18	4

   (a) Draw a cumulative frequency curve of the data and use your graph to find:
       (i) the median yield
       (ii) the interquartile range.
   These are the individual yields, in kilograms, for 50 similar trees that were not treated.

6	1	8	3	6	5	7	8	4	3
5	7	7	8	7	8	2	9	5	7
8	4	5	6	1	5	3	6	7	5
5	7	8	7	4	6	6	5	8	9
8	6	8	5	5	3	7	7	8	6

   (b) Compare the yields from the two samples. Was the treatment effective?

4. The table shows the age distribution (in millions) for males in England and Wales for two years.

Age	1881	1966
under 15	4.7	5.6
15 and under 30	3.4	4.9
30 and under 45	2.3	4.4
45 and under 60	1.4	4.4
60 and under 75	0.7	0.7
75 and under 90	0.1	0.7

   Draw cumulative frequency diagrams for the two years and use the medians, interquartile ranges and the numbers over 65 to compare the distributions.

5. A local theatre monitored the size of audience on Wednesday and Thursday evenings during one year. The results are shown in the table.

Audience size	50–99	100–199	200–299	300–399	400–499	500–599
Number of Wednesdays	11	20	10	6	4	1
Number of Thursdays	3	3	18	19	5	4

   (a) Draw cumulative frequency graphs to compare the data.
   (b) From these curves draw box-and-whisker plots to show the data.

You will need to know:

- how to interpret significant figures
- how to calculate simple interest
- how to write numbers in standard form.

Checking answers by rounding to one significant figure

It is important to be able to check calculations quickly, without using a calculator. One way to do this is to round the numbers to one significant figure, which was discussed in Chapter 12.

Example 1

Find an approximate answer to the calculation 5.13×3.83.

$5.13 \times 3.83 \approx 5 \times 4$

Rounding 5.13 and 3.83 each to 1 s.f. to give a much simpler calculation

$= 20$

Examiner's tip

In a calculation it may be possible to round one number up and another number down. This might give an answer close to the exact answer.

Exercise 19.1a

1. Find approximate answers to these calculations by rounding each number to one significant figure.
   (a) $31.3 \div 4.85$ (b) 113.5×2.99
   (c) $44.669 \div 8.77$ (d) $3.6 \times 14.9 \times 21.5$
   Now use a calculator to see how close your approximations are to the correct answers.

2. Find approximate answers to these calculations by rounding each number to one significant figure.
   (a) $\dfrac{14.56 \times 22.4}{59.78}$ (b) $\dfrac{4.9^2 \times 49.3}{96.7}$
   (c) $\sqrt{4.9 \times 5.2}$
   Now use a calculator to see how close your approximations are to the correct answers.

3. Find approximate answers to these calculations by rounding each number to one significant figure.
   (a) $(0.35 \times 86.3) \div 7.9$
   (b) $\sqrt{103.5} \div \sqrt{37.2}$
   (c) 9.87×0.0657
   (d) $0.95 \div 4.8$

4. Make up some multiplication and division calculations of your own to test this statement:
   'In multiplication and division calculations, rounding each number to one significant figure will always give an answer which is correct to one significant figure'.

1. Find approximate answers to these calculations by rounding each number to one significant figure.
   (a) 48.67×12.69 (b) 0.89×5.2 (c) 61.33×11.79
   (d) $(1.8 \times 2.9) \div 3.2$
   Now use a calculator to see how close your approximations are to the correct answers.

2. Find approximate answers to these calculations by rounding each number to one significant figure.
   (a) $\dfrac{3.99}{0.8 \times 1.64}$ (b) $198.5 \times 63.1 \times 2.8$ (c) $\dfrac{\sqrt{8.1 \times 1.9}}{1.9}$
   Now use a calculator to see how close your approximations are to the correct answers.

3. Find approximate answers to these calculations by rounding each number to one significant figure.
   (a) $32 \times \sqrt{124}$ (b) $\dfrac{62 \times 9.7}{10.12 \times 5.1}$ (c) 0.246×0.789
   (d) $44.555 \div 0.086$

Compound interest

Compound interest is different from simple interest. At the end of each year the interest is added to the amount. The interest at the end of the next year is calculated on this new figure.

Example 2

Helen puts £500 into the bank, which pays interest at 8% per annum (written as p.a.). How much money will she have at the end of three years?

Interest at the end of 1st year = 8% of £500

$$= 0.08 \times £500$$

$$= £40$$

Amount at the end of 1st year = £500 + £40 = £540

> This amount can be found easily if you remember that the new amount will be $(100 + 8)\% = 108\%$ of the previous amount. £500 $\times$ 1.08 = £540

Amount at the end of the 2nd year = £540 $\times$ 1.08

$$= £583.20$$

Amount at the end of the 3rd year = £583.20 $\times$ 1.08

$$= £629.856$$

Exercise 19.2a

1. Calculate the compound interest on these amounts.
   (a) £300 invested for two years at 5%
   (b) £1000 invested for four years at 4%
   (c) £450 invested for three years at 3%
   (d) £5000 invested for two years at 8%
   (e) £30 000 invested for four years at 7%
2. Is it better to invest £1000 for five years at 8% or for four years at 9%?

Examiner's tip

If you are asked to find the compound interest rather than the total amount, make sure you subtract the original investment from the final balance.

Exercise 19.2b

1. Calculate the compound interest on these amounts.
   (a) £250 invested for two years at 3.5%
   (b) £100 000 invested for three years at 2.5%
   (c) £50 invested for six years at 1.7%
   (d) £2000 invested for four years at 6.1%
   (e) £800 invested for five years at 3.4%
2. Find the difference in interest earned by investing £500 for three years at 12% simple interest or for three years at 10% compound interest.
3. I invest £500 at 7% compound interest. How many years must I leave it, before it doubles in value?

Insurance

Motor insurance

Everyone who drives a car needs to insure it. Each year they will have to pay an insurance company a certain amount, called the **insurance premium**. Then the insurance company will pay for the car to be repaired if it is stolen or if it is damaged in an accident or a fire, and they will also pay some medical expenses for people injured in the crash.

If the car driver doesn't make any claims for a year, the insurance company will give a discount called a no-claim bonus the following year. The discount could be a 20% reduction in the premium for every year (up to three years) without a claim, then from the fourth year a maximum reduction of 65% would apply (see the table on next page). If the driver makes a claim, the no-claim bonus may be reduced. Usually, a driver loses a year's no-claim bonus for making a claim, for example, a driver who makes a claim in the third year would find that the no-claim bonus is reduced from 60% to 40%.

Years with no claim	No-claim bonus
0	0%
1	20%
2	40%
3	60%
more than 3	65%

Example 3

Jackie drives a 1997 Ford Escort 1.8. The basic premium for this type of car is £350. How much would she have to pay if she had completed three years of driving and had never claimed?

From the table, the no-claim bonus = 60%. Therefore she pays 100% − 60% = 40% of the premium.

40% of £350 = 0.4 × £350 = £140

Therefore Jackie must pay £140.

Health insurance

Some people take out health insurance which will give them some money if they are ill and unable to work. Insurance companies work out how much people have to pay depending upon factors such as their age, whether or not they smoke, whether they are male or female.

A simplified table that an insurance company might use is given below.

Age (years)	Cost per month (£)	
	Female	Male
under 16	19.50	20.30
16–25	18.20	18.90
26–35	19.80	20.70
36–45	20.50	21.40
46–55	21.70	22.60
56–65	25.50	26.40
over 65	28.90	31.30
Add 15% for smokers.		

Exercise 19.3a

Refer to the table for motor insurance to calculate how much each of the following motorists will have to pay.
1. Tariq, basic premium £480, driving for four years, never made a claim.
2. Claire, basic premium £530, driving for three years, no claims.

Refer to the table for health insurance to work out the insurance premiums per month for the following people.
3. A 55 year old male who smokes.
4. A 36 year old female, non-smoker.
5. A 66 year old female who smokes.

Exercise 19.3b

Refer to the table for motor insurance to calculate how much each of the following motorists will have to pay.
1. Bill, basic premium £654, driving for three years, one claim.
2. David, basic premium £686, driving for two years, one claim.

Refer to the table for health insurance to work out the insurance premiums per month for the following people.
3. A 22-year-old male who smokes.
4. A married couple:
   (a) the man is aged 29 and a non-smoker
   (b) the woman is aged 25 and a smoker.

Compound measures

Speed

Speed is an example of a compound measure, because it is calculated from two other measurements: distance and time.

For a journey, the average speed is $\dfrac{\text{total distance travelled}}{\text{total time taken}}$.

Speed is written as 'distance per time', for example 30 km/h.

Calculate the average speed of a car that travels 80 km in 2 hours.

Average speed = $\dfrac{\text{total distance travelled}}{\text{total time taken}} = \dfrac{80}{2} = 40$ km/h

The formula for speed can be rearranged to find the distance travelled or the time taken for a journey.

distance = speed × time   time = $\dfrac{\text{distance}}{\text{speed}}$

Density

Another example of a compound measure is **density**, which is linked to mass and volume.

The density of a substance = $\dfrac{\text{mass}}{\text{volume}}$.

It is measured in units such as grams per cubic centimetre (g/cm³).

Example 5   The density of gold is 19.3 g/cm³. Calculate the mass of a gold bar with a volume of 30 cm³.

density = $\dfrac{\text{mass}}{\text{volume}}$

so mass = density × volume

The mass of the gold bar = density × volume = 19.3 × 30 = 579 g.

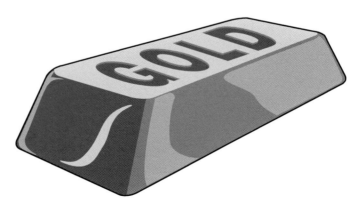

Population density

Population density is another example of a compound measure. It gives an idea of how heavily populated an area is. It is measured as the number of people per square kilometre.

Exercise 19.4a

1. A train covers a distance of 750 metres in a time of 12.4 seconds. Calculate its average speed.
2. Paula jogs at a steady speed of 7 miles per hour. How far does she run in 90 minutes?
3. How long will it take a boat sailing at 6 km/h to travel 45 km?
4. Look at these three different packets of washing powder.

   Calculate the price per kilogram for each size.
5. The density of aluminium is 2.7 g/cm³. What is the volume of a block of aluminium with a mass of 750 g?
6. Calculate the population density of Africa, if the population is 7.43 × 10⁸ and its area is 3.03 × 10⁷ km².

Exercise 19.4b

1. Calculate the density of a 5 cm³ block of copper with a mass of 35 g.
2. Brian's car travels 120 miles in 4 hours. Calculate his average speed.
3. A car is travelling at 25 m/s. Find how long it will take to travel:
   (a) 1 m   (b) 1 km   (c) 1 mile (1 mile is about 1.6 km).
4. The petrol tank of Nick's car has a capacity of 100 litres. She weighs 100 cm³ of petrol and finds it has a mass of 80 g. What is the mass of petrol in the tank when it is full?
5. Which is better value?

6. Calculate the population density of Europe, if the population is 4.95 × 10⁸ and the land area is 4.94 × 10⁶ km².

Working to a reasonable degree of accuracy

Measurements and calculations should always be expressed to a suitable degree of accuracy. For example, it would be silly to say that a car journey took 4 hours 46 minutes and 13 seconds, but reasonable to say that it took four and three-quarter hours, or about five hours. In the same way, saying that the distance the car travelled was 93 kilometres 484 metres and 78 centimetres would be giving the measurement to an unnecessary degree of accuracy. It would more sensibly be stated as 93 km.

As a general rule the answer you give after a calculation should not be given to a greater degree of accuracy than any of the values used in the calculation.

 Example 6

Ben measured the length and width of a table as 1.8 m and 1.3 m. He calculated the area as $1.8 \times 1.3 = 2.34 \,\text{m}^2$. How should he have given the answer?

Ben's answer has two places of decimals (2 d.p.) so it is more accurate than the measurements he took. His answer should be $2.3 \,\text{m}^2$.

Exercise 19.5a

Write down sensible values for each of these measurements.
1. 3 minutes 24.8 seconds to boil an egg.
2. 2 weeks, 5 days, 3 hours and 13 minutes to paint a house.

Work these out and give the answers to a reasonable degree of accuracy.
3. Find the length of the side of a square field with area $33 \,\text{m}^2$.
4. Three friends win £48.32. How much will each receive?
5. It takes 12 hours to fly between two cities, if the aeroplane is travelling at 554 km/h. How far apart are the cities?
6. The length of a strip of card is 2.36 cm and the width is 0.041 cm, each measured to two significant figures. Calculate the area.

Exercise 19.5b

Write down sensible values for each of these measurements.
1. A book weighing 2.853 kg.
2. The height of a door as 2 metres 12 centimetres and 54 millimetres.

Work these out and give the answers to a reasonable degree of accuracy.
3. The length of a field is 92 m correct to two significant figures and the width is 58.36 m correct to four significant figures. Calculate the area of the field.
4. A book has 228 pages and is 18 mm thick. How thick is Chapter 1 which has 35 pages?
5. The total weight of 13 people in a lift is 879 kg. What is their average weight?
6. Last year a delivery driver drove 23 876 miles. Her van travels an average of 27 miles to the gallon. Diesel costs 72p per litre. If one gallon equals 4.55 litres, calculate the cost of the fuel used.

Key points

- Check answers to calculations by rounding numbers to one significant figure.
- Find the amount after compound interest at x% per annum by multiplying by $(100 + x)$% for each year.

- Compound measures such as speed and density are found by dividing one unit by another.
- The degree of accuracy in measurement depends on the purpose of the measurement.

Revision exercise 19a

1. Find approximate answers to these.
   (a) 63.9×14.9
   (b) $\sqrt{143} \times \sqrt{170} \times \sqrt{80}$
   (c) $(6.32 + 5.72) \times (\sqrt{16.1} + \sqrt{48.9})$

2. Calculate the compound interest on £1800 invested for three years at 4.8% p.a.

3. Tom invests £560 in the bank for three years at a rate of interest of 7.5% p.a. How much more interest will he earn than if he had invested the same amount at a rate of 7.5% simple interest?

4. A car travels 158 miles in 3.5 hours. What is its average speed?

5. A car travels an average of 22.6 miles per gallon of fuel. Petrol costs £2.95 per gallon. About how much will fuel for a journey of 500 miles cost?

6. The population of a city is 543 861. The population is expected to grow by 6% next year. What will the population be then?

7. A tap takes 13.5 minutes to fill a 70 litre tank. Calculate the flow rate in litres per minute. How long would it take to fill the same tank from a tap with a flow rate of 18.2 litres/minute?

8. Which of these is better value?

9. The volume of the Earth is approximately $1.4 \times 10^{27}\,\text{m}^3$. The mass of the Earth is approximately $2 \times 10^{30}\,\text{kg}$. Calculate the density of the Earth.

10. The diameter of a circle is 3.5 m correct to two significant figures. Calculate the circumference of the circle. Take $\pi = 3.141\,593$.

11. Light travels at approximately $3 \times 10^5\,\text{km/s}$. If light takes about 8 minutes to reach the Earth from the sun, how far is the Earth from the sun, in kilometres?

12. The mean weight of eight men is 78.7 kg. A ninth man joins them. His weight is 48.6 kg. What is the mean weight of all nine men?

20 Pythagoras' theorem and trigonometry

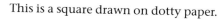

You will need to know how to find:

- the area of a triangle
- squares and square roots on a calculator.

Pythagoras' theorem

This is a square drawn on dotty paper.

Its area is 4 square units.

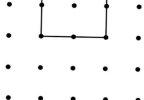

Here is a tilted square.

Calculating its area is more difficult. There are two methods you could use.

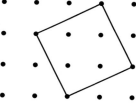

(a) Calculate the area of the large square drawn round the outside and subtract the area of the shaded triangles:

9 − 1 − 1 − 1 − 1 = 5 square units

or

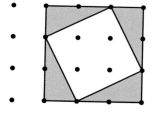

(b) Add together the area of the four shaded triangles and the area of the middle square:

1 + 1 + 1 + 1 + 1 = 5 square units

- Using either method (a) or method (b) calculate the area of the squares in the diagram below.

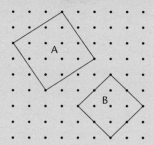

You should have found that the area of square A is 13 square units and the area of square B is 8 square units.

- Draw some more tilted squares of your own on dotty paper and find their areas.

  Look at all the tilted squares.

  Code the tilt by drawing a triangle at the base and writing down the length of its sides, like this.

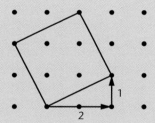

In this diagram the code is (2, 1).

In the diagram on the left, square A has a code of (3, 2) and square B has a code of (2, 2).

Check that you agree.

- Code the squares you have drawn and write the codes and the areas of the squares in a table.

  Include the square from the beginning of this chapter. Its code is (2, 0) and its area is 4 square units. Include all the other squares you have already studied.

Code	Area
2, 0	4
2, 1	5
3, 2	13
2, 2	8

- Look at the codes and their areas. See if you can find a rule linking them together.

You will have found that squaring each code number and then adding the squares together gives the area:

$$3^2 + 2^2 = 9 + 4 = 13$$
$$2^2 + 2^2 = 4 + 4 = 8$$
$$2^2 + 1^2 = 4 + 1 = 5$$

The rule linking them together is called **Pythagoras' theorem**.

Squaring the numbers in the code and adding them is the same as squaring the lengths of the sides of the triangle.

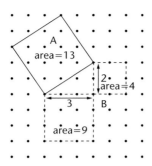

Here is square A again.

Can you see that you can calculate the area of the largest square by adding together the areas of the smaller squares?

The largest square will always be on the longest side of the triangle – this is called the **hypotenuse** of the triangle.

Thus Pythagoras' theorem can be stated as:

> The area of the square on the hypotenuse = the sum of the areas of the squares on the other two sides

but it is normally abbreviated to:

> The square on the hypotenuse = the sum of the squares on the other two sides.

Activity 2

• Calculate the missing area in each of these diagrams.

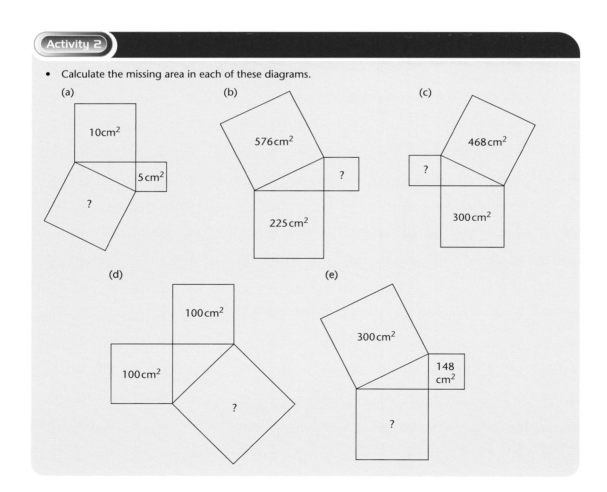

(a)

10 cm²

5 cm²

?

(b)

576 cm²

?

225 cm²

(c)

468 cm²

?

300 cm²

(d)

100 cm²

100 cm²

?

(e)

300 cm²

148 cm²

?

If you know the lengths of two sides of a right-angled triangle you can use Pythagoras' theorem to find the length of the third side.

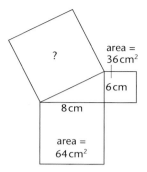

The unknown area = 64 + 36 = 100 cm²

This means that the length of the sides of the unknown square is 10 cm.

When using Pythagoras' theorem, you don't need to draw the squares — you can simply use the rule.

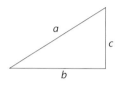

$$a^2 = b^2 + c^2$$

Thus:

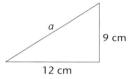

$$a^2 = 9^2 + 12^2 = 81 + 144 = 225$$
$$a = \sqrt{225} = 15\,cm$$

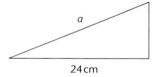

$$a^2 = 7^2 + 24^2 = 49 + 576 = 625$$
$$a = \sqrt{625} = 25\,cm$$

Activity 3

- Find the length of the hypotenuse in each of these triangles:

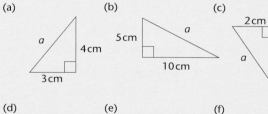

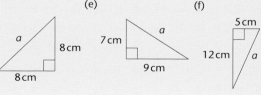

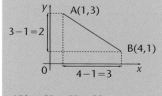
Chapter 20 Pythagoras' theorem

277

If you know the length of the hypotenuse and the length of one other side you can find the length of the third side.

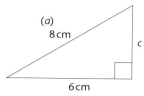

(a)

8 cm

6 cm

c

$a^2 = b^2 + c^2$
$8^2 = 6^2 + c^2$
$64 = 36 + c^2$
$c^2 = 64 - 36 = 28$
$c = \sqrt{28} = 5.29\,\text{cm}$ (to 2 d.p.)

Activity 4

- Calculate the length of the third side in each of these triangles. Give your answers correct to two decimal places.

(a)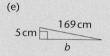
17 cm   15 cm
b

(b)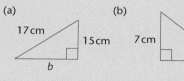
7 cm   9 cm
b

(c)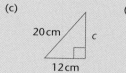
20 cm c
12 cm

(d)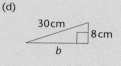
30 cm   8 cm
b

(e)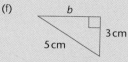
5 cm   169 cm
b

(f)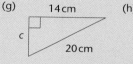
b
5 cm   3 cm

(g)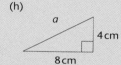
14 cm
c
20 cm

(h)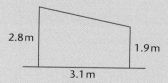
a 4 cm
8 cm

- Solve these problems.

(a) A rectangular field is 225 m long and 110 m wide. Find the length of the diagonal path across it.

(b) A rectangular field is 250 m long. A footpath 38.0 m long crosses the field diagonally. Find the width of the field.

(c) A ladder is 7 m long. It is resting against a wall, with the top of the ladder 5 m above the ground. How far from the wall is the base of the ladder?

(d) Harry is building a kite for his sister. This is his diagram of the kite. The kite is 30 cm wide.

A
26 cm   26 cm
D   B
40 cm   40 cm
C

Harry needs to buy some cane to make the struts AC and DB. What length of cane does he need to buy?

(e) This is the side view of a shed. Find the length of the sloping roof.

2.8 m

1.9 m

3.1 m

Examiner's tip

It is a good idea to draw a sketch if a diagram isn't given. Try to draw it roughly to scale and mark on it any lengths you know. It may help you see any errors in your working.

Trigonometry

You will need a protractor and a ruler for this section.

The hypotenuse is the longest side of a right-angled triangle. It is the side opposite the right angle.

The other sides are named according to the angle under consideration.

For *a*:

hypotenuse   opposite
a
adjacent

For *b*:

hypotenuse   adjacent
b
opposite

In this activity you will need to draw triangles and identify the sides, like this:

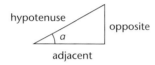

hypotenuse   opposite
a
adjacent

- Draw six different triangles. Use these values for *a* but make the **adjacent** side 5 cm long each time.

  (a) 10°   (b) 20°   (c) 30°   (d) 40°
  (e) 50°   (f) 60°

  a is the angle between the adjacent and the hypotenuse.

  For each triangle, measure the length of the **opposite side** and divide this by the length of the adjacent side (5 cm). Record your results in a table like this one. Round each ratio to one decimal place.

Angle *a* (°)	Adjacent (cm)	Opposite (cm)	opposite / adjacent
10	5		
20	5		
30	5		
40	5		
50	5		
60	5		

Now draw two more triangles. Keep the angle fixed at 30° but change the length of the adjacent side to 10 cm and then 15 cm.

Record your results in a table like this.

Angle *a* (°)	Adjacent (cm)	Opposite (cm)	opposite / adjacent
30	5		
30	10		
30	15		

What do you notice?

What do you think the length of the opposite side would be, if you drew another triangle with *a* = 30° and the adjacent side 20 cm long?

Estimate your answer and then check by drawing and measuring.

- What do you think the length of the adjacent side would be, if you drew another triangle with *a* = 30° and the opposite side 8 cm long?

Estimate your answer and check by drawing and measuring.

The ratio $\frac{\text{opposite}}{\text{adjacent}}$ is called the **tangent** (ratio).

The value of the tangent has been calculated for all angles and is one of the functions on a scientific calculator.

- Now look back at all the triangles you have drawn and for each one measure the length of the hypotenuse.

Copy this table and complete it.

Write the ratios correct to one decimal place.

Angle (°)	Adjacent (cm)	Opposite (cm)	Hypotenuse (cm)	$\frac{\text{opposite}}{\text{hypotenuse}}$	$\frac{\text{adjacent}}{\text{hypotenuse}}$
10	5				
20	5				
30	5				
40	5				
50	5				
60	5				
30	5				
30	10				
30	15				

What do you notice?

The ratio $\frac{\text{opposite}}{\text{hypotenuse}}$ is called the **sine** (ratio)   The ratio $\frac{\text{adjacent}}{\text{hypotenuse}}$ is called the **cosine** (ratio)

Use the tangent ratio to calculate the missing side or angle in a right-angled triangle.

Examiner's tip

Always label the sides of the triangle as adjacent, opposite and hypotenuse (or A, O and H) when you are using the ratios to calculate sides or angles.

Example 1 Calculate the value of x.

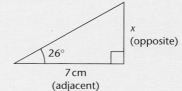

In the triangle: $\tan 26° = \dfrac{\text{opposite}}{\text{adjacent}}$

$$= \dfrac{x}{7}$$

So: $\dfrac{x}{7} = \tan 26°$

$x = 7 \times \tan 26°$

$\quad = 7 \times 0.4877$

$\quad = 3.414$

$x = 3.41$ cm (to 3 s.f.)

Examiner's tip

Only label the sides you are given. This helps you to identify which formula to use. Here, since O and A are labelled, use the tan formula.

Example 2 Find the size of the angle marked a.

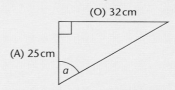

In the triangle: $\tan a = \dfrac{\text{opposite}}{\text{adjacent}}$

$$= \dfrac{32}{25}$$

$\tan a = 1.28$

$a = \tan^{-1} 1.28$

$a = 52.0°$ (to 1 d.p.)

Examiner's tip

Use the $\sin^{-1}$, $\cos^{-1}$ or $\tan^{-1}$ button on your calculator to find the angle.

Chapter 20 Trigonometry

- Find the length of the side marked *x* in each of these triangles.

(a)

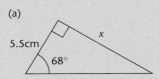

5.5 cm
68°
x

(b)

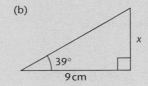

39°
9 cm
x

(c)

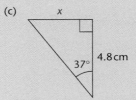

x
37°
4.8 cm

(d)

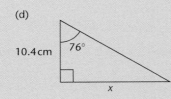

10.4 cm
76°
x

(e)

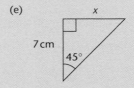

x
7 cm
45°

(f)

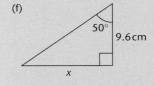

50°
9.6 cm
x

- Find the size of the angle marked *a* in each of these triangles.

(a)

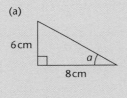

6 cm
a
8 cm

(b)

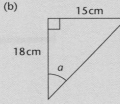

15 cm
18 cm
a

(c)

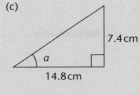

7.4 cm
a
14.8 cm

(d)

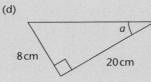

a
8 cm
20 cm

(e)

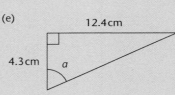

12.4 cm
4.3 cm
a

Use the sine or cosine ratios to calculate the missing side or angle in a right-angled triangle.

Example ③

Calculate the value of *x*.

(H) 5 m
x (O)
34°

In the triangle: $\sin 34° = \dfrac{\text{opposite}}{\text{hypotenuse}}$

So: $\sin 34° = \dfrac{x}{5}$

$x = 5 \times \sin 34°$

$x = 2.80 \text{ cm (to 2 d.p.)}$

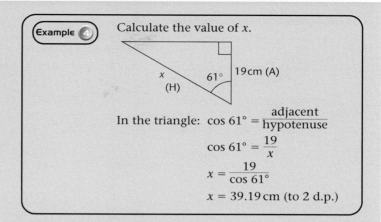

Example 4

Calculate the value of *x*.

19cm (A)

61°

x
(H)

In the triangle: $\cos 61° = \dfrac{\text{adjacent}}{\text{hypotenuse}}$

$\cos 61° = \dfrac{19}{x}$

$x = \dfrac{19}{\cos 61°}$

$x = 39.19\,\text{cm (to 2 d.p.)}$

Examiner's tip

Take care when you are finding the hypotenuse. Remember to multiply by *x* and divide by cos 61°.

- Find the length of the side marked *x* or the angle marked *a* in each of these triangles.

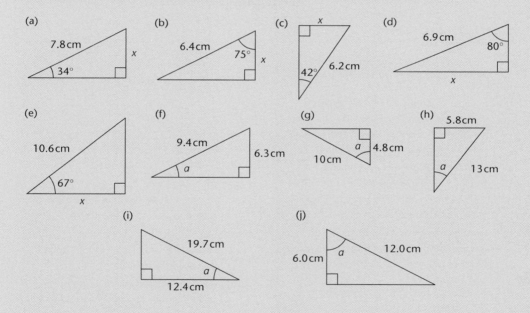

(a) 7.8 cm, 34°, *x*

(b) 6.4 cm, 75°, *x*

(c) *x*, 42°, 6.2 cm

(d) 6.9 cm, 80°, *x*

(e) 10.6 cm, 67°, *x*

(f) 9.4 cm, *a*, 6.3 cm

(g) *a*, 4.8 cm, 10 cm

(h) 5.8 cm, *a*, 13 cm

(i) 19.7 cm, *a*, 12.4 cm

(j) 6.0 cm, *a*, 12.0 cm

Examiner's tip

Follow these three steps.
Step 1 Draw and label a diagram.
Step 2 Label the two appropriate sides as O, A or H.
Step 3 Write down the formula to be used and calculate the answer.

Use trigonometry to solve problems.

Example 5

A kite flies at the end of a string 20 m long. The string is straight and it makes an angle of 47° with the horizontal. How high is the kite from the ground?

Step 1　　　　　　　　　　　　　　　Step 2

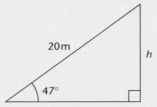　　　　　　　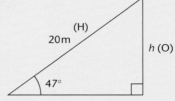

Step 3　In the triangle:　$\sin 47° = \dfrac{O}{H} = \dfrac{h}{20}$

So $h = 20 \times \sin 47°$

$h = 14.6\,\text{m}$ (1 d.p.)

• Use trigonometric ratios to solve these problems.

(a) A ladder of length 4.8 m rests against a vertical wall so that the base of the ladder is 1.8 m from the wall. Calculate the angle between the ladder and the ground.

(b) A ladder of length 5 m rests against a vertical wall. The angle between the ladder and the wall is 62°. How far up the wall does the ladder reach?

(c) From a distance of 25 m, the angle of elevation from the ground to the top of a tower is 37°. How high is the tower?

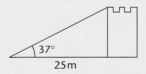

(d) A ship is due south of a lighthouse. It sails on a bearing of 065° until it is due east of the lighthouse.

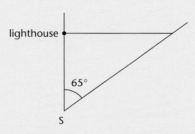

If the ship is now 40 km away from the lighthouse, how far has it sailed?

(e) An isosceles triangle has sides of length 8 cm, 8 cm and 5 cm. Find the angle between the equal sides.

(f) Find the acute angle between the diagonals of a rectangle with sides of 5 cm and 8 cm.

(g) A path slopes up a hill at 12° from the horizontal. The path is 2.8 km long. How high is the hill?

(h) A ship sails for 70 km on a bearing of 130°. How far south and east of its starting point is it?

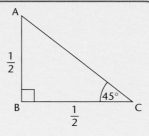

Example 6 Find the length of AC and the value of $\cos 45°$.

$AC^2 = (\frac{1}{2})^2 + (\frac{1}{2})^2 = \frac{1}{4} + \frac{1}{4} = \frac{2}{4}$

$AC = \frac{\sqrt{2}}{2}$

$\cos 45° = \frac{\frac{1}{2}}{\frac{\sqrt{2}}{2}} = \frac{1}{\sqrt{2}}$ Multiply top and bottom by 2.

Expressions like these, which include square roots ($\sqrt{\,}$) are called **surds**.

- Use Pythagoras' theorem and trigonometry to work these out.
  Do not use a calculator. Leave surds in your answer.

  (a) Find the length of the hypotenuse.

  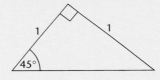

  (b) Find: (i) the length of BD
       (ii) the values of $\sin 30°$ and $\cos 30°$.

  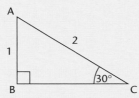

  (c) Find: (i) the length of the hypotenuse
       (ii) the values of $\sin 45°$ and $\cos 45°$.

  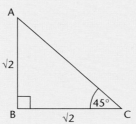

Examiner's tip

You could be asked to use Pythagoras' theorem and trigonometry without a calculator. If so, leave any square roots in your answer.

Key points

- For a right-angled triangle, Pythagoras' theorem states that $a^2 = b^2 + c^2$

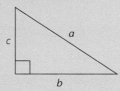

- In a right-angled triangle, for angle x:

$$\tan x = \frac{O}{A}$$

$$\sin x = \frac{O}{H}$$

$$\cos x = \frac{A}{H}$$

Revision exercise 20a

1. Calculate the value of x in each triangle.

   (a) 4 cm, 6 cm, x

   (b) x, 14.9 cm, 8 cm

   (c) 12.1 cm, 10 cm, x

2. Calculate the length of the diagonal of a rectangle with length 22 cm and width 12 cm.

3. A ship sails 20 km due north and then 30 km due west. How far is it from its starting point?

4. Calculate the lengths or angles marked with letters. (All lengths are in cm.)

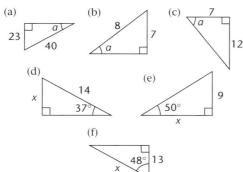

5. A boy is flying a kite with a string of length 45 m. If the string is straight and it makes an angle of 75° with the ground, how high is the kite? Ignore the height of the boy.

6. The sides of a triangle are 5 cm, 5 cm and 7 cm. Calculate the angles of the triangle.

7. A ramp for disabled people must slope at not more than 10°. If the height of the ramp has to be 0.8 m, how long must the ramp be?

8. A man sails for 5 km on a bearing of 285° from a harbour.

   (a) How far north and west of the harbour is he?

   He then sails 3 km due north.

   (b) Find the bearing on which he needs to sail to return to the harbour, and how far he needs to sail.

9. Find the length of PQ.

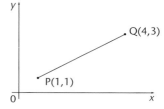

Equations and manipulation 2

Solving harder linear equations

Some linear equations you may be asked to solve may include decimals or have the unknown in the denominator.

Example 1

Solve $2(5x - 4) = 3(x + 2)$.

$$10x - 8 = 3x + 6$$ Multiply out the brackets.

$$[10x - 8 + 8 = 3x + 6 + 8]$$ Add 8 to each side.

$$10x = 3x + 14$$

$$[10x - 3x = 3x + 14 - 3x]$$ Subtract $3x$ from each side.

$$7x = 14$$

$$x = 14 \div 7 = 2$$ Divide by 7.

The lines in square brackets are often missed out.

Example 2

Solve $\frac{x}{3} = 2x - 3$.

$$x = 3(2x - 3)$$ Multiply each side by 3.

$$x = 6x - 9$$ Multiply out the bracket.

$$6x - 9 = x$$ Reverse the equation to put the x-term with the larger positive coefficient on the left.

$$[6x - 9 + 9 = x + 9]$$ Add 9 to each side.

$$6x = x + 9$$

$$[6x - x = x + 9 - x]$$ Subtract x from each side.

$$5x = 9$$

$$x = 9 \div 5 = 1\frac{4}{5}$$ Divide each side by 5.

A common error when multiplying through by a number or letter is to multiply just the first term. Use brackets to make sure.

Another common error in the above example would be to give the answer as $\frac{5}{9}$ rather than $\frac{9}{5}$.

It is helpful to swap the sides if necessary to make the coefficient of x greater on the left-hand side.

Example 3 Solve the equation $\dfrac{400}{x} = 8$.

$400 = 8x$ Multiply each side by x.

$400 \div 8 = x$ Divide each side by 8.

$x = 50$

Example 4 Solve the equation $3.6x = 8.7$.

$x = 8.7 \div 3.6$ Divide each side by 3.6.

$x = 2.416\,666\,6$ Use a calculator and give the
$= 2.42$ answer to three significant figures unless you are told otherwise.

Exercise 21.1a

Solve these equations.

1. $5(x - 2) = 4x$
2. $3(2x + 3) = 9$
3. $4(2x - 3) = 3(x + 1)$
4. $2(4x - 5) = 2x + 6$
5. $10(x + 2) = 3(x - 5)$
6. $3(2x - 1) = 2(x + 4)$
7. $\dfrac{x}{2} = 3x - 10$
8. $\dfrac{2x}{3} = x - 2$

9. $\dfrac{3x}{2} = 7 - 2x$
10. $\dfrac{5x}{3} = 4x - 2$
11. $\dfrac{50}{x} = 2$
12. $\dfrac{300}{x} = 15$
13. $\dfrac{75}{2x} = 3$

Now give the answers to the remaining questions correct to three significant figures.

14. $3.5x = 9.6$
15. $5.2x = 25$
16. $4.6x = 7.5$
17. $\dfrac{x}{1.4} = 2.6$
18. $2.1(x - 3.2) = 4.4$
19. $2.2(2x + 5.1) = 4.9$
20. $\dfrac{2.3}{x} = 4.5$

Exercise 21.1b

Solve these equations.

1. $2(3x - 5) = 14$
2. $4(3x - 1) = 10x$
3. $3(2x + 1) = 7x + 1$
4. $5(2x - 2) = 2(x + 3)$
5. $3(4x + 3) = 2(x + 6)$
6. $5(x + 2) = 3(4 - x)$
7. $5(x + 2) = 3(2x + 1)$
8. $\dfrac{x}{3} = x - 4$
9. $\dfrac{2x}{5} = x - 3$

10. $\dfrac{3x}{5} = 4 - x$
11. $\dfrac{2x}{3} = 4x - 5$
12. $\dfrac{200}{x} = 4$
13. $\dfrac{25}{2x} = 5$
14. $\dfrac{15}{2x} = 3$

Give the answers to the remaining questions correct to three significant figures.

15. $2.4x = 9.7$
16. $22x = 7.55$
17. $4.2x = 9.3$
18. $2.1(3x - 6.4) = 9.2$
19. $\dfrac{x}{3.4} = 2.5$
20. $\dfrac{3.5}{x} = 1.6$

Solving inequalities

The inequalities in earlier chapters had only one answer. These are more complicated.

Example 5

Solve the inequality $x^2 < 9$.

Since $3^2 = 9$, one answer is $x < 3$, but remember that $(-3)^2$ also gives an answer of 9. The inequality has two solutions, since 9 has two possible square roots. $x > -3$ also satisfies the inequality.

Take square roots on both sides but if you take the negative value you must reverse the inequality sign.

$x < 3$ or $x > -3$

This can be written as $-3 < x < 3$.

Example 6

Solve the inequality $x^2 - 3 \geqslant 6$.

$x^2 \geqslant 9$ Add 3 to each side.

$x \leqslant -3$ or $x \geqslant 3$ Again change the inequality for the negative answer. These cannot be combined.

Examiner's tip

Take care with the direction of the inequality when taking the square root. It is useful to check. In Example 6, the answer is $x \leqslant -3$ so check by putting in a value of x such as $x = -4$. This gives $x^2 - 3 = 16 - 3 = 13$ and $13 \geqslant 6$ as required.

Exercise 21.2a

Solve these inequalities.

1. $2x + 3 < 5$
2. $5x - 4 > 10 - 2x$
3. $3(2x - 1) > 15$
4. $4(x - 4) \geqslant x - 1$
5. $\dfrac{x}{2} > 3$

6. $x^2 \geqslant 4$
7. $x^2 < 25$
8. $x^2 \geqslant 1$
9. $x^2 - 2 \leqslant 14$
10. $x^2 + 6 < 22$

Exercise 21.2b

Solve these inequalities.

1. $4n - 2 > 6$
2. $2n + 6 < n + 3$
3. $4n - 9 \geqslant 2n + 1$
4. $3(x - 1) \geqslant 6$
5. $2(3x - 1) > 4x + 6$

6. $x^2 < 1$
7. $x^2 \leqslant 9$
8. $x^2 > 36$
9. $x^2 - 5 \leqslant 4$
10. $x^2 + 8 > 33$

Forming equations and inequalities

Simple problems can be solved using equations and inequalities.

Example 7

The length of a rectangle is a cm, the width is 15 cm shorter. The length is three times the width.

Write down an equation in a and solve it to find the length and width of the rectangle.

$a - 15$

a

If the length = a, the width = $a - 15$

and the length = $3 \times$ width = $3(a - 15)$.

The equation is $a = 3(a - 15)$.

$$a = 3a - 45 \qquad \text{Multiply out the brackets.}$$
$$3a - 45 = a \qquad \text{Swap sides to write them the other way round.}$$
$$[3a - 45 + 45 = a + 45] \qquad \text{Add 45 to each side.}$$
$$3a = a + 45$$
$$[3a - a = a + 45 - a] \qquad \text{Subtract } a \text{ from each side.}$$
$$2a = 45$$
$$a = 22.5$$

So the length = 22.5 cm and the width = 7.5 cm.

Examiner's tip

When you are asked to set up an equation and solve it you will not score any marks if you just give the answer without the equation.

Example 8

£400 was shared by n people and each received £16.

Set up an equation and find how many people there were.

The equation is $\dfrac{400}{n} = 16$

$400 = 16n$ \qquad Multiply each side by n.

$n = 400 \div 16 = 25$

There were 25 people.

Exercise 21.3a

1. Two angles in a triangle are x and $2x - 30$. The first angle is twice the size of the second. Set up an equation and solve it to find the size of the two angles.

2. The width of a rectangle is 3 cm and the length is $x + 4$ cm. The area is 27 cm². Set up an equation and solve it to find x.

3. In a class of 32 pupils, x are girls. There are three times as many girls as boys. Set up an equation and solve it to find out how many boys and how many girls there are.

4. A greengrocer sells potatoes at x pence per kilogram. He paid $\frac{2x}{3}$ per kilogram for them. This is 20p less than x. Set up an equation and solve it to find x.

5. Stephen thinks of a number. If he doubles the number and then subtracts 5, he gets the same answer as if he subtracts 2 from the number and then multiplies by 3. Let the number be n. Set up an equation and solve it to find n.

6. On a bus trip each child pays £p and each adult pays £12 more than each child. There are 28 children and four adults on the bus. The same amount of money is collected from all the children as from all the adults. Set up an equation and solve it to find how much each child and each adult pays.

7. At Joe's Diner one-course meals cost £x. Two-course meals cost £2 more. A group of eight people bought three one-course meals and five two-course meals. They paid £38. Set up an equation and solve it to find the cost of a one-course meal.

8. The square of a number is less than 36. Set up an inequality and find the possible values for the number.

Exercise 21.3b

1. Two angles of a pentagon are $x°$ and the other three are each $(2x - 20)°$. The total of all the angles is 540°. Write down an equation and solve it to find the size of the angles.

2. A triangle has a base of x cm and a height of 5 cm. The area is 30 cm². Set up an equation in x and solve it to find the length of the base.

3. The cost per person of a flight from Sheffield Airport is the charge by the airline plus £40 tax. Four people flew from Sheffield to Cairo and the total they had to pay was £1640. Let the charge by the airline be £x. Write down an equation in x and solve it to find the charge by the airline.

4. A 32-year-old man has three children who are x, $2x$ and $2x + 4$ years old. The man is four times as old as his eldest child. Set up an equation and solve it to find the ages of the children.

5. Jane thinks of a number. Her number divided by three gives the same answer as taking the number away from sixteen. Let the number be n. Set up an equation and solve it to find what the number was.

6. At Deno's Pizza Place, a basic pizza costs £x and extra toppings are 50p each. Bernard and four of his friends each have pizzas with two extra toppings. They pay £25.50. Set up an equation and find the cost of a basic pizza.

7. Ahmed had £x. He spent £4 on books and still had three-fifths of his money left. Write down an equation in x and solve it to find how much he had to start with.

8. Timothy squares a number and gets an answer greater than 49. Set up an inequality and find the values that the number can take.

Simultaneous equations

An equation in two unknowns does not have a unique solution. For example, the graph of the equation $x + y = 4$ is a straight line. Every point on the line will have coordinates that satisfy the equation.

When you are given two equations in two unknowns, such as x and y, they usually have common solutions where the two lines meet in a point. These are called **simultaneous equations**.

Solving by the method of elimination

Example 9

Solve the simultaneous equations $x + y = 4$ and $2x - y = 5$.

$$x + y = 4 \quad \text{①}$$
$$2x - y = 5 \quad \text{②}$$

Write the two equations, one under the other, and label them.

Look to see if either of the unknowns (x or y) has the same coefficient in both equations. In this case there is $1y$ in equation ① and $1y$ in equation ②. As their signs are different, the two y-terms will be eliminated (cancel each other out) if the two equations are added.

$$x + 2x + y + (-y) = 4 + 5 \quad \text{Adding ①+②.}$$
$$3x = 9$$
$$x = 3$$

To find the value of y, substitute $x = 3$ in equation ①.

$$3 + y = 4 \quad \text{Replacing } x \text{ by 3.}$$
$$y = 1$$

So the solution is $x = 3$ and $y = 1$.

Check in equation ②: The left-hand side is $2x - y = 6 - 1 = 5$ which is corrected.

Example 10

Solve the simultaneous equations $2x + 5y = 9$ and $2x - y = 3$.

$$2x + 5y = 9 \quad \text{①} \quad \text{Set out in line.}$$
$$2x - y = 3 \quad \text{②}$$

This time $(+)2x$ appears in each equation, so subtract to eliminate the x-terms.

$$2x - 2x + 5x - (-y) = 9 - 3 \quad \text{①-② Take care with the signs. } 5y - (-y) = 5y + y.$$
$$6y = 6$$
$$y = 1$$
$$2x + 5 = 9 \quad \text{Substitute in ①. } 5y \text{ is replaced by } 5 \times 1 = 5.$$
$$2x = 4$$
$$x = 2$$

The solution is $x = 2$, $y = 1$.

Check in equation ②: The left-hand side is $2x - y = 4 - 1 = 3$ which is correct.

Example 11

Solve the simultaneous equations
$x + 3y = 10$, $3x + 2y = 16$.

$x + 3y = 10$ ① Set out in line.
$3x + 2y = 16$ ②

This time the coefficients of x and y are different in both equations.

Multiply ① by 3 to make the coefficient of x the same as in equation ②.

$3x + 9y = 30$ ③ ① × 3
$3x + 2y = 16$ ②

Now $(+)3x$ appears in both equations, so subtract.

$3x - 3x + 9y - 2y = 30 - 16$ ③ − ②
$$7y = 14$$
$$y = 2$$
$$x + 6 = 10$$ Substitute in ①.
$$x = 4$$

The solution is $x = 4$, $y = 2$.

Check in equation ②: The left-hand side is $3x + 2y = 12 + 4 = 16$ which is correct.

Examiner's tip

When eliminating, if the signs of the letter to be eliminated are the same, subtract. If they are different, add.
When subtracting, take great care with the signs. This is where most errors are made. If your check is wrong, see if you have made an error with any signs.

Examiner's tip

When subtracting equations, you can use ① − ② or ② − ①. It is better to make the letter positive.
Always write down clearly what you are doing.

There is no need to write as much detail as in the last example. The next example shows what is required. The commentary can be omitted.

 Example 12

Solve simultaneously $4x - y = 10$ and $3x + 2y = 13$.

$4x - y = 10$ ① Set out in line.
$3x + 2y = 13$ ②
① × 2: $8x - 2y = 20$ ③ To get $2y$ in each equation.

② + ③: $8x + 3x + 2y + (-2y) = 13 + 20$ To eliminate y.
$$11x = 33$$
$$x = 3$$

Substitute in ①: $12 - y = 10$
$$-y = -2$$
$$y = 2$$

The solution is $x = 3$, $y = 2$.

Check in ②: LHS = $3x + 2y = 9 + 4 = 13$ which is correct.

Exercise 21.4a

Solve these simultaneous equations.

1. $x + y = 5$
   $2x - y = 7$
2. $3x + y = 9$
   $2x + y = 7$
3. $2x + 3y = 11$
   $2x + y = 5$
4. $2x + y = 7$
   $4x - y = 5$
5. $2x + 3y = 13$
   $3x - 3y = 12$
6. $2x + 3y = 14$
   $5x + 3y = 26$
7. $3x + y = 7$
   $2x + 3y = 7$
8. $2x - 3y = 0$
   $3x + y = 11$
9. $2x + 3y = 13$
   $x + 2y = 8$
10. $3x + 2y = 13$
    $x + 3y = 16$
11. $2x + 3y = 7$
    $3x - y = 5$
12. $x + y = 4$
    $4x - 2y = 7$
13. $2x + 2y = 7$
    $4x - 3y = 7$
14. $4x - 2y = 14$
    $3x + y = 8$
15. $2x - 2y = 5$
    $4x - 3y = 11$

Exercise 21.4b

Solve these simultaneous equations.

1. $x + y = 3$
   $2x + y = 4$
2. $2x + y = 6$
   $2x - y = 2$
3. $2x - y = 7$
   $3x + y = 13$
4. $2x + y = 12$
   $2x - 2y = 6$
5. $3x - y = 11$
   $3x - 5y = 7$
6. $2x + y = 6$
   $3x + 2y = 10$
7. $x + 3y = 9$
   $2x - y = 4$
8. $x + 2y = 19$
   $3x - y = 8$
9. $x + 2y = 6$
   $3x - 3y = 9$
10. $2x + y = 14$
    $3x + 2y = 22$
11. $2x + y = 3$
    $3x - 2y = 8$
12. $2x + 4y = 11$
    $x + 3y = 8$
13. $2x - y = 4$
    $4x + 3y = 13$
14. $2x - 4y = 2$
    $x + 3y = -9$
15. $x + y = 0$
    $2x + 4y = 3$

Further simultaneous equations

Sometimes the letters in the equations are not in the same order,
so the first thing to do is to rearrange them.

Example 13

Solve simultaneously the equations $y = 3x - 4$, $x + 2y = -1$.

$$-3x + y = -4 \quad \text{①} \qquad \boxed{\text{Rearrange the equation.}}$$
$$x + 2y = -1 \quad \text{②}$$

This can be solved in two ways, either ① $\times 2$ and subtract or ② $\times 3$ and add.
It is normally easier to add.

$$\text{②} \times 3 \quad 3x + 6y = -3 \quad \text{③}$$
$$-3x + y = -4 \quad \text{①} \qquad \boxed{\text{① is copied down.}}$$

$$\text{③} + \text{①} \quad 3x + (-3x) + 6y + y = -3 + (-4)$$
$$7y = -7$$
$$y = -1$$

Substitute in ①: $-3x - 1 = -4$ $\qquad \boxed{\text{Replace } y \text{ by } -1.}$
$$[-3x - 1 + 1 = -4 + 1]$$
$$-3x = -3$$
$$x = 1$$

The solution is $x = 1$, $y = -1$.

Check in ②: LHS $= x + 2y = 1 - 2 = -1$ which is correct.

Sometimes each of the equations needs to be multiplied by a
different number.

Example 14

Solve the equations $3y = 4 - 4x$, $6x + 2y = 11$.

$$4x + 3y = 4 \quad \text{①} \qquad \boxed{\text{Rearrange the first equation.}}$$
$$6x + 2y = 11 \quad \text{②}$$

To eliminate x multiply ① by 3 and ② by 2 and subtract, or to eliminate y
multiply ① by 2 and ② by 3 and subtract.

$$\text{①} \times 3 \quad 12x + 9y = 12 \quad \text{③} \qquad \boxed{\text{Eliminate } x.}$$
$$\text{②} \times 2 \quad 12x + 4y = 22 \quad \text{④}$$
$$\text{③} - \text{④} \quad 5y = -10$$
$$y = -2$$

Substitute in ①: $4x - 6 = 4$ $\qquad \boxed{3y \text{ is replaced by } -6.}$
$$[4x - 6 + 6 = 4 + 6]$$
$$4x = 10$$
$$x = 2\tfrac{1}{2}$$

The solution is $x = 2\tfrac{1}{2}$ and $y = -2$.

Check in ②: LHS $= 6x + 2y = 15 - 4 = 11$ which is correct.

Exercise 21.5a

Solve these simultaneous equations.

1. $y = 2x - 1$
   $x + 2y = 8$
2. $y = 3 - 2x$
   $3x - 3y = 0$
3. $3y = 11 - x$
   $3x - y = 3$
4. $3x + 2y = 7$
   $2x - 3y = -4$
5. $3x - 2y = 3$
   $2x - y = 4$
6. $2x + 3y = 7$
   $7x - 4y = 10$
7. $3x + 4y = 5$
   $2x + 3y = 4$
8. $4x - 3y = 1$
   $5x + 2y = -16$
9. $3x + 2y = 5$
   $2x + 3y = 10$
10. $4x - 2y = 3$
    $5y = 23 - 3x$

Exercise 21.5b

Solve these simultaneous equations.

1. $3y = 5 - x$
   $2x + y = 5$
2. $5y = x + 1$
   $2x + 2y = 10$
3. $y = 3x - 3$
   $2x + 3y = 13$
4. $4x - y = 2$
   $5x + 3y = 11$
5. $3x - 2y = 11$
   $2x + 3y = 16$
6. $2x - 3y = 5$
   $3x + 4y = 16$
7. $2x + 3y = 4$
   $3x - 2y = -7$
8. $4x + 3y = 1$
   $3x + 2y = 0$
9. $y = x + 2$
   $2x - 4y = -9$
10. $2y = 4x - 5$
    $3x - 5y = 9$

Solving quadratic equations

For any two numbers, if $A \times B = 0$, then either $A = 0$ or $B = 0$.

If $(x - 3)(x - 2) = 0$ then either $(x - 3) = 0$ or $(x - 2) = 0$.

To solve a quadratic equation, factorise it into two brackets and then use this fact.

Remember: to factorise $x^2 + ax + b$:
- if b is positive find two numbers with product b and sum a
  the signs in the bracket are both the same as a
- if b is negative find two numbers with product b and difference a
  the signs in the bracket are different
  the bigger number in the bracket has the same sign as a.

Example 15

Solve the equation $x^2 - 4x + 3 = 0$.

$(x - 3)(x - 1) = 0$ Factorising: both signs are negative, $1 \times 3 = +3$ and $1 + 3 = 4$.

$x - 3 = 0$ or $x - 1 = 0$

The solution is $x = 3$ or $x = 1$.

Example 16

Solve the equation $x^2 + 5x + 6 = 0$.

$(x + 3)(x + 2) = 0$ Factorising: both signs are positive, $2 \times 3 = 6$ and $2 + 3 = 5$.

$x + 3 = 0$ or $x + 2 = 0$

The solution is $x = -3$ or $x = -2$.

Example 17

Solve the equation $x^2 - 3x - 10 = 0$.

$(x - 5)(x + 2) = 0$ Factorising: the signs are different, $5 \times 2 = 10$ and $5 - 2 = 3$.

$x - 5 = 0$ or $x + 2 = 0$

The solution is $x = 5$ or $x = -2$.

If an equation is written as $x^2 - 2x = 15$ or $x^2 = 2x - 15$, first rearrange it so that all three terms are on the same side.

Example 18

Solve the equation $x^2 = 4x - 4$ by factorisation.

$x^2 - 4x + 4 = 0$ Rearrange so that all three terms are on the same side.

$(x - 2)(x - 2) = 0$ Factorising: the signs are both negative, $2 \times 2 = 4$, $2 + 2 = 4$.

$x - 2 = 0$ or $x - 2 = 0$

The solution is $x = 2$ (twice).

There are always two answers, so if they are both the same write 'twice'.

Exercise 21.6a

Solve these equations by factorisation.

1. $x^2 - 5x + 6 = 0$
2. $x^2 - 6x + 5 = 0$
3. $x^2 + 6x + 8 = 0$
4. $x^2 + 5x + 4 = 0$
5. $x^2 + 2x + 1 = 0$
6. $x^2 - 7x + 6 = 0$

7. $x^2 - 7x + 10 = 0$
8. $x^2 - 4x + 3 = 0$
9. $x^2 - 9x + 14 = 0$
10. $x^2 - 6x + 8 = 0$
11. $x^2 - 2x - 8 = 0$
12. $x^2 + 4x - 5 = 0$
13. $x^2 - x - 6 = 0$
14. $x^2 + 5x - 6 = 0$

15. $x^2 + 2x - 3 = 0$
16. $x^2 - 3x - 18 = 0$
17. $x^2 - 9x - 10 = 0$
18. $x^2 + 9x + 14 = 0$
19. $x^2 + 9x - 22 = 0$
20. $x^2 + x - 12 = 0$

Exercise 21.6b

Solve these equations by factorisation.

1. $x^2 - 7x + 10 = 0$
2. $x^2 - 4x + 3 = 0$
3. $x^2 - 8x + 15 = 0$
4. $x^2 + 9x + 20 = 0$
5. $x^2 + 7x + 6 = 0$
6. $x^2 - 9x + 18 = 0$

7. $x^2 + 7x + 12 = 0$
8. $x^2 - 2x + 1 = 0$
9. $x^2 - 10x + 24 = 0$
10. $x^2 + 4x + 3 = 0$
11. $x^2 + 2x - 3 = 0$
12. $x^2 + 3x - 10 = 0$
13. $x^2 - x - 12 = 0$
14. $x^2 + 5x - 14 = 0$

15. $x^2 - 2x - 15 = 0$
16. $x^2 - 3x - 28 = 0$
17. $x^2 - 17x + 30 = 0$
18. $x^2 + 4x - 32 = 0$
19. $x^2 + 9x - 36 = 0$
20. $x^2 + x - 20 = 0$

Graphical methods of solving equations

One way to solve simultaneous linear, quadratic and cubic equations is to use a graph. The point(s) where the lines or curves meet will give the solution.

Example 19

Solve the simultaneous equations

$$y = 2x - 4 \text{ and}$$
$$3y = 12 - 2x$$

graphically. Use values of x from 0 to 6.

Three points for equation ① are $(0, -4)$, $(3, 2)$, $(6, 8)$.

Three points for equation ② are $(0, 4)$, $(3, 2)$, $(6, 0)$.

The two lines cross at $(3, 2)$ so the solution is $x = 3$, $y = 2$.

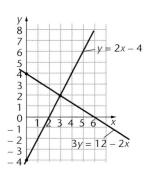

(a) Draw the graph of $y = x^2 - 2x - 8$ for values of x from -3 to $+5$.

(b) Solve the equation $x^2 - 2x - 8 = 0$.

(a)

x	-3	-2	-1	0	1	2	3	4	5
x^2	9	4	1	0	1	4	9	16	25
$-2x$	6	4	2	0	-2	-4	-6	-8	-10
-8	-8	-8	-8	-8	-8	-8	-8	-8	-8
$y = x^2 - 2x - 8$	7	0	-5	-8	-9	-8	-5	0	7

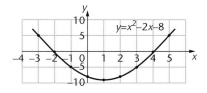

(b) The solution of $x^2 - 2x - 8 = 0$ is when $y = 0$, where the curve cuts the x-axis. The solution is $x = -2$ or $x = 4$.

(a) Draw the graph of $y = x^3 - 4x + 1$, for values of x from -3 to $+3$.

(b) Solve the equation $x^3 - 4x + 1 = 0$.

(a)

x	-3	-2	-1	0	1	2	3
x^3	-27	-8	-1	0	1	8	27
$-4x$	12	8	4	0	-4	-8	-12
1	1	1	1	1	1	1	1
$y = x^3 - 4x + 1$	-14	1	4	1	-2	1	16

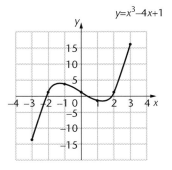

(b) The solution is when $y = 0$, where the curve cuts the x-axis. The solution is $x = -2.2$, $x = 0.2$ or $x = 1.9$.

Exercise 21.7a

Solve the simultaneous equations in questions 1−4 graphically.

1. $y = 2x$ and $y = 8 - 2x$. Use **values of x from** −1 to 4.
2. $y = 3x + 5$ and $y = x + 3$. Use **values of x from** −3 to +2.
3. $y = 5 - x$ and $y = 2x - 7$. Use **values of x from** −1 to +5.
4. $2y = 2x + 1$, $2y + x = 7$. Use **values of x from** 0 to 7.
5. (a) Draw the graph of $y = x^2 - 7x + 10$ for values of x from 0 to 7.
   (b) Solve the equation $x^2 - 7x + 10 = 0$.
6. (a) Draw the graph of $y = x^2 - x - 2$ for values of x from −2 to +3.
   (b) Solve the equation $x^2 - x - 2 = 0$.
7. (a) Draw the graph of $y = x^2 - 8$ for values of x from −3 to +3.
   (b) Solve the equation $x^2 - 8 = 0$.
8. (a) Draw the graph of $y = x^2 + x - 3$ for values of x from −3 to +2.
   (b) Solve the equation $x^2 + x - 3 = 0$.
9. (a) Draw the graph of $y = x^3 - 3x$ for values of x from −3 to +3.
   (b) Solve the equation $x^3 - 3x = 0$.
10. (a) Draw the graph of $y = x^3 - 5x + 3$ for values of x from −3 to +3.
    (b) Solve the equation $x^3 - 5x + 3 = 0$.

Exercise 21.7b

Solve the simultaneous equations in questions 1−4 graphically.

1. $y = 3x$ and $y = 4x - 2$. Use values of x from −1 to +4.
2. $y = 2x + 3$ and $y = 4x + 1$. Use values of x from −2 to +3.
3. $y = x + 4$ and $4x + 3y = 12$. Use values of x from −3 to +3.
4. $y = 2x + 8$ and $y = -2x$. Use values of x from −5 to +1.
5. (a) Draw the graph of $y = x^2 - 4x + 3$ for values of x from −1 to +5.
   (b) Solve the equation $x^2 - 4x + 3 = 0$.
6. (a) Draw the graph of $y = x^2 - 3x$ for values of x from −2 to +5.
   (b) Solve the equation $x^2 - 3x = 0$.
7. (a) Draw the graph of $y = x^2 - 5$ for values of x from −3 to +3.
   (b) Solve the equation $x^2 - 5 = 0$.
8. (a) Draw the graph of $y = x^2 - 3x - 2$ for values of x from −2 to +5.
   (b) Solve the equation $x^2 - 3x - 2 = 0$.
9. (a) Draw the graph of $y = x^3 - 6x$ for values of x from −3 to +3.
   (b) Solve the equation $x^3 - 6x = 0$.
10. (a) Draw the graph of $y = x^3 - 8x - 2$ for values of x from −3 to +3.
    (b) Solve the equation $x^3 - 8x - 2 = 0$.

Solving cubic equations by trial and improvement

Questions 9 and 10 in the previous exercises show that solving a cubic equation by a graphical method is not very accurate. In fact, it is difficult to be accurate even to one decimal place. A more accurate method is **trial and improvement**.

Example 22

A solution of the equation $x^3 - 4x + 1 = 0$ lies between 1 and 2. Use trial and improvement to find the solution correct to one decimal place.

For the first trial try 1:	$1^3 - 4 \times 1 + 1 = -2$	Too small.
Try 1.5:	$1.5^3 - 4 \times 1.5 + 1 = -1.625$	Too small, try solutions between 1.5 and 2.0.
Try 1.8:	$1.8^3 - 4 \times 1.8 + 1 = -0.368$	Too small.
Try 1.9:	$1.9^3 - 4 \times 1.9 + 1 = 0.259$	Too big.

The solution lies between 1.8 and 1.9.

To find which is nearer, try 1.85:

$1.85^3 - 4 \times 1.85 + 1 = -0.0684$ Too small.

So the solution lies between 1.85 and 1.9. It is nearer to 1.9 than 1.8.

The solution is $x = 1.9$ correct to one decimal place.

You should be able to find the solution within about five or six trials.

Example 23

A solution of $x^3 - 3x = 6$ lies between 2 and 3. Find it correct to one decimal place.

Try 2:	$2^3 - 3 \times 2 = 2$	Too small.
Try 2.5:	$2.5^3 - 3 \times 2.5 = 8.125$	Too big, try solutions between 2 and 2.5.
Try 2.3:	$2.3^3 - 3 \times 2.3 = 5.267$	Too small, try solutions between 2.3 and 2.5.
Try 2.4:	$2.4^3 - 3 \times 2.4 = 6.624$	Too big, try solutions between 2.3 and 2.4.
Try 2.35:	$2.35^3 - 3 \times 2.35 = 5.928$	Too small, the solution is between 2.35 and 2.4.

The solution is $x = 2.4$ correct to one decimal place.

Find the two values between which the answer lies to the required degree of accuracy and then try midway values.

Exercise 21.8a

Use trial and improvement to find the solutions.

1. A solution of $x^3 = 5$ lies between 1 and 2. Find it correct to one decimal place.
2. A solution of $x^3 - 8x = 0$ lies between 2 and 3. Find it correct to one decimal place.
3. A solution of $x^3 - 5x = 8$ lies between 2 and 3. Find it correct to one decimal place.
4. A solution of $x^3 - x = 90$ lies between 4 and 5. Find it correct to one decimal place.
5. A solution of $x^3 - x^2 = 30$ lies between 3 and 4. Find it correct to one decimal place.
6. A solution of $x^3 = 12$ lies between 2 and 3. Find it correct to two decimal places.
7. A solution of $x^3 + 50 = 0$ lies between −4 and −3. Find it correct to one decimal place.
8. A solution of $x^3 + 4x + 25 = 0$ lies between −3 and −2. Find it correct to one decimal place.
9. A solution of $x^3 - 2x^2 = 4$ lies between 2 and 3. Find it correct to two decimal places.
10. A solution of $x^3 + 3x^2 + x = 0$ lies between −3 and −2. Find it correct to two decimal places.

Exercise 21.8b

Use trial and improvement to find the solutions.

1. A solution of $x^3 = 15$ lies between 2 and 3. Find it correct to one decimal place.
2. A solution of $x^3 - 2x = 0$ lies between 1 and 2. Find it correct to one decimal place.
3. A solution of $x^3 - 7x = 25$ lies between 3 and 4. Find it correct to one decimal place.
4. A solution of $x^3 + 2x = 2$ lies between 0 and 1. Find it correct to one decimal place.
5. A solution of $x^3 - x^2 = 1$ lies between 1 and 2. Find it correct to one decimal place.
6. A solution of $x^3 = 56$ lies between 3 and 4. Find it correct to two decimal places.
7. A solution of $x^3 + 12 = 0$ lies between −3 and −2. Find it correct to one decimal place.
8. A solution of $x^3 - 2x + 6 = 0$ lies between −3 and −2. Find it correct to one decimal place.
9. A solution of $x^3 - 4x^2 + 9 = 0$ lies between 2 and 3. Find it correct to one decimal place.
10. A solution of $x^3 - 5x^2 + 2x = 0$ lies between 0 and 1. Find it correct to two decimal places.

Problems that lead to simultaneous or quadratic equations

 Example 24

In a café two cups of tea and three cups of coffee cost £5.30. Three cups of tea and a cup of coffee cost £4.10.

Let a cup of tea cost t pence and a cup of coffee cost c pence.

(a) Write down two equations in t and c.

(b) Solve them to find the cost of a cup of tea and a cup of coffee.

(a) $2t + 3c = 530$ ①   Working in pence.
    $3t + c = 410$ ②

(b) ②× 3 $9t + 3c = 1230$ ③
                  $2t + 3c = 530$ ①
    ③−① $7t = 700$
                      $t = 100$
  Substitute in ①: $200 + 3c = 530$
                     $3c = 330$
                     $c = 110$

So tea costs £1 a cup, coffee costs £1.10 a cup.

Check in the problem: 2 teas + 3 coffees cost £2 + £3.30 = £5.30
                           and 3 teas + 1 coffee cost £3 + £1.10 = £4.10.

 Example 24

A rectangle has width x cm and length $x + 4$ cm. The area is 21 cm².

Write down an equation in x and solve it to find the dimensions of the rectangle.

$x(x + 4) = 21$ Form the equation.

$x^2 + 4x = 21$ Expand the brackets.

$x^2 + 4x - 21 = 0$ Take all the non-zero terms to the left-hand side.

$(x + 7)(x - 3) = 0$ Solve the equation.

The solution is $x = -7$ or $x = 3$.   From $x + 7 = 0$ or $x - 3 = 0$.

Lengths must be positive, so the width is 3 cm, the length is $3 + 4 = 7$ cm.
Check: $3 \times 7 = 21$.

Exercise 21.9a

1. Two numbers x and y have a sum of 47 and a difference of 9.
   (a) Write down two equations in x and y.
   (b) Solve them to find the numbers.

2. Cassettes cost £c each and compact discs cost £d each. John bought two cassettes and three discs and paid £27.50. Shahida bought three cassettes and one disc and paid £18.50.
   (a) Write down two equations in c and d.
   (b) Solve them to find the cost of a cassette and a disc.

3. Paint is sold in small and large tins. Peter needs 13 litres and he buys one small and two large tins.
   Gamel needs 11 litres and he buys two small and one large tin. Both have exactly the correct amount.
   Let the small tin hold s litres and the large tin hold b litres.
   (a) Write down two equations in s and b.
   (b) Solve them to find the amount each tin holds.

4. A coach journey cost each adult £a and each child £c. Tickets for one adult and two children cost £31. Tickets for two adults and three children cost £54. Use algebra to find the cost of each ticket.

5. Two consecutive odd numbers x and $x + 2$ have a product of 63. Set up a quadratic equation and solve it to find the two numbers.

Exercise 21.9b

1. Two numbers x and y have a sum of 86 and a difference of 16.
   (a) Write down two equations in x and y.
   (b) Solve them to find the two numbers.

2. At Turner's corner shop beans cost b pence a tin and spaghetti costs s pence a tin. Three tins of beans and two tins of spaghetti cost £1.37. Two tins of beans and a tin of spaghetti cost 81p.
   (a) Write down two equations in b and s.
   (b) Solve them to find the cost of each tin.

3. Orange juice is sold in cans and bottles. Cans hold c ml and bottles hold b ml. Three cans and four bottles contain 475 cl altogether. Four cans and three bottles hold 400 cl altogether. Use algebra to find how much each holds.

4. The entry fees for Barford museum are £a for adults and £c for children. Mr Ekebussi paid £25 for two adults and five children. Mrs Taylor paid £14 for one adult and three children. Use algebra to work out the cost of each ticket.

5. Joan is x years old and her mother is 25 years older. The product of their ages is 306.
   (a) Write down a quadratic equation in x.
   (b) Solve the equation to find Joan's age.

Showing regions on graphs

It is often possible to show the area on a graph that satisfies an inequality.

Example 26 Write down the inequality that describes the region shaded in each graph.

(a)

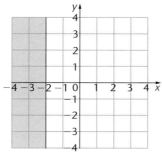

(b)

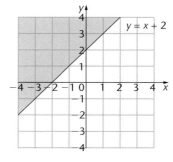

(a) $x < -2$ — The line drawn is $x = -2$. This line divides the graph into two regions $x < -2$ and $x > -2$. The shaded region is $x < -2$. Check by testing any point in the region.

(b) $y > x + 2$ — The line is $y = x + 2$ and it divides the graph into two regions $y < x + 2$ and $y > x + 2$. To decide which side is shaded choose any point not on the line and test it, for example, $(0, 0)$. Here $x + 2 = 2$, and $y = 0$, so $y < x + 2$ at $(0, 0)$ and $(0, 0)$ is not in the region. So the shaded region is $y > x + 2$.

Example 27 On separate grids shade the regions: (a) $y > 2$ (b) $y < 2x - 3$.

(a)

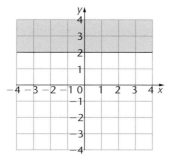

It is clear that $y > 2$ is above the line $y = 2$.

(b)

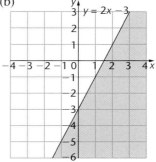

Draw the line $y = 2x - 3$. Then the two regions are $y > 2x - 3$ and $y < 2x - 3$. To test which side is wanted, choose any point not on the line, for example $(0, 0)$. Here $y = 0$ and $2x - 3 = -3$, so $y > 2x - 3$ at $(0, 0)$. Therefore $(0, 0)$ is not in the region required. Shade the other region.

In the previous examples the region required has been shaded. If more that one region is required, then it is best to shade out the regions not required and leave blank the required region.

Examiner's tip

When testing a region, if possible use (0, 0). If the line goes through (0, 0) choose a positive number to test the region.

Examiner's tip

Either shading in or shading out is acceptable, but indicate the required region by labelling it clearly.

Example 29

Show by shading the region where $x > 0$, $y > 0$ and $x + 2y < 6$.

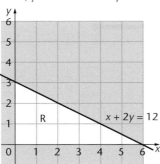

Shade out the regions $x < 0$ and $y < 0$.
Draw the line $x + 2y = 6$ and test (0, 0);
$x + 2y = 0 < 6$ so (0, 0) is in the region
$x + 2y < 6$, which is the required region.
So shade out the region not containing (0, 0).
R is the region required.

Exercise 21.10a

For questions 1–4, write down the inequality that describes the region shaded.

1.

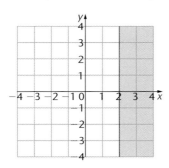

3.

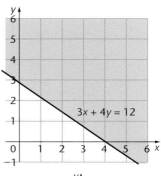

2.

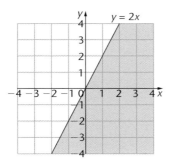

4.

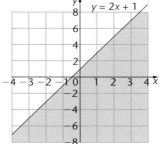

5. Draw a set of axes and label them from −4 to +4 for x and y. Shade the region $y > -3$.

6. Draw a set of axes and label them from −3 to +6 for x and from −3 to +5 for y. Shade the region $2x + 5y < 10$.

7. Draw a set of axes and label them from 0 to 5 for x and y. Show by shading the region where $y > 0$, $x > 0$ and $3x + 5y < 15$.

8. Draw a set of axes and label them from −3 to +3 for x and from −6 to +10 for y. Show by shading the region where $x > 0$, $y < 8$ and $y > 2x$.

Exercise 21.10b

For questions 1−4, write down the inequality that describes the region shaded.

1.
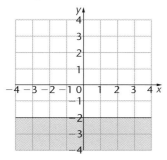

2.
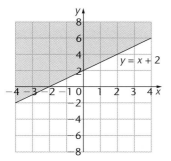

3.
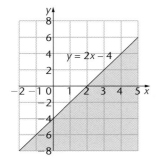

4.
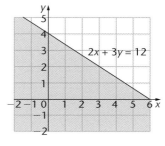

5. Draw a set of axes and label them from −4 to +4 for x and y. Shade the region $x > -1$.

6. Draw a set of axes and label them from −2 to 6 for x and from −2 to +5 for y. Shade the region $4x + 5y < 20$.

7. Draw a set of axes and label them from 0 to 12 for x and from 0 to 8 for y.
Show by shading the region where $y > 0$, $x > 0$ and $3x + 5y < 30$.

8. Draw a set of axes and label them from −3 to +3 for x and from −6 to +10 for y.
Show by shading the region where $x > 0$, $3x + 8y > 24$ and $5x + 4y < 20$.

Rearranging formulae

All the formulae that have been studied so far included the new subject only once, and the subject was not raised to a power. The following examples show how to deal with these situations.

Example 29

Rearrange the formula $A = \pi r^2$ to make r the subject.

$$A = \pi r^2$$

$$\frac{A}{\pi} = r^2 \qquad \text{Divide both sides by } \pi.$$

$$r^2 = \frac{A}{\pi} \qquad \text{Change sides to get terms involving } r \text{ on the left-hand side.}$$

$$r = \sqrt{\frac{A}{\pi}} \qquad \text{Take the square root of both sides.}$$

Example 30

Rearrange the formula $V = \frac{4}{3}\pi r^3$ to make r the subject.

$$V = \frac{4}{3}\pi r^3$$

$$3V = 4\pi r^3 \qquad \text{Multiply both sides by 3.}$$

$$r^3 = \frac{3V}{4\pi} \qquad \text{Swap sides and divide both sides by } 4\pi.$$

$$r = \sqrt[3]{\frac{3V}{4\pi}} \qquad \text{Take the cube root of both sides.}$$

Example 31

Rearrange the formula $a = x + \frac{cx}{d}$ to make x the subject.

$$a = x + \frac{cx}{d}$$

$$ad = dx + cx \qquad \text{Multiply both sides by } d.$$

$$dx + cx = ad \qquad \text{Change sides to get all the } x\text{-terms to the left-hand side.}$$

$$x(d + c) = ad \qquad \text{Factorise the left-hand side, taking out the factor } x.$$

$$x = \frac{ad}{d + c} \qquad \text{Divide both sides by the term in the bracket, } (d + c).$$

Example 32

Rearrange the equation $ax + by = cy - ad$ to make a the subject.

$$ax + by = cy - ad$$

$$ax + ad = cy - by \qquad \text{Rearrange to get all the terms involving } a \text{ on the left-hand side and all the other terms on the right-hand side (adding } ad \text{ to both sides, subtracting } by \text{ from both sides).}$$

$$a(x + d) = cy - by \qquad \text{Factorise the left-hand side, taking out the factor } a.$$

$$a = \frac{cy - by}{x + d} \qquad \text{Divide both sides by the term in the bracket } (x + d).$$

Exercise 21.11a

For each question, make the letter in brackets the subject.

1. $s = at + 2bt$ (t)

2. $P = t - \dfrac{at}{b}$ (t)

3. $A = \pi r^2$ (r)

4. $ab - cd = ac$ (a)

5. $ab - cd = ac$ (c)

6. $s - 2ax = b(x - s)$ (s)

7. $s - 2ax = b(x - s)$ (x)

8. $a = \dfrac{t}{b} - st$ (t)

9. $a = b + c^2$ (c)

10. $A = P + \dfrac{PRT}{100}$ (P)

Exercise 21.11b

For each question, make the letter in brackets the subject.

1. $s = ab - bc$ (b)

2. $v^2 = u^2 - 2as$ (u)

3. $3(a + y) = by + 7$ (y)

4. $2(a - 1) = b(1 - 2a)$ (a)

5. $\dfrac{a}{b} - 2a = b$ (a)

6. $s = 2r^2 - 1$ (r)

7. $a(b + d) = c(b - d)$ (d)

8. $a(b + d) = c(b - d)$ (b)

9. $V = 5ab^2 + 3c^2$ (c)

10. $s = \dfrac{uv}{u + v}$ (v)

Key points

- If a problem asks you to use algebra to solve it, you must start with an equation.
- To solve simultaneous equations, make the coefficient (number) of one of the letters the same in both equations. If they are the same sign, subtract. If different signs, add.
- To solve quadratic equations, factorise and then put each bracket equal to zero.
- To solve linear simultaneous equations graphically, draw the lines on a graph and find where they cross.

- To solve quadratic or cubic equations graphically, draw the curve and find where it cuts the x-axis.
- To solve a cubic equation by trial and improvement, find the two values between which it lies, to the required degree of accuracy and then test the midway point.
- When representing inequalities on a graph, if there is more than one region it is best to shade out the regions not required, leaving the region required clear.

1. Solve these equations.
   (a) $3(x - 2) = x$
   (b) $5(2x + 3) = 55$
   (c) $4(x - 3) = 3(x - 2)$
   (d) $2(3x - 4) = 4(x + 1)$
   (e) $\frac{x}{2} = 3x - 10$
   (f) $\frac{x}{3} = 3 - 2x$
   (g) $\frac{500}{x} = 20$
   (h) $\frac{300}{x} = 60$

2. Solve these inequalities.
   (a) $2x - 1 < 5$
   (b) $3x + 4 \leq 16$
   (c) $5x - 2 > 3 + 4x$
   (d) $x^2 \leq 49$
   (e) $x^2 \geq 16$
   (f) $x^2 - 5 < 31$

3. A number x divided by 3 is the same as 3 times the number minus 24. Write down an equation and solve it to find the number.

4. An ice-lolly costs x pence and an ice-cream costs 20 pence more.
   (a) Write down the cost of an ice-cream in terms of x.
   Jon buys three ice-lollies and two ice-creams and pays £3.40.
   (b) Write down an equation in x and solve it to find the cost of an ice-lolly and of an ice-cream.

5. Marcia is x cm tall and her friend Carole is 25 cm shorter.
   (a) Write down Carole's height in terms of x. Carole is $\frac{4}{5}$ as tall as Marcia.
   (b) Write down an equation in x and solve it to find Marcia's height.

6. Solve these simultaneous equations.
   (a) $x + y = 15$
       $2x + y = 22$
   (b) $2x + 3y = 13$
       $3x - y = 3$
   (c) $2x - 3y = 3$
       $4x + 5y = 17$
   (d) $3x - 6y = 3$
       $2x + 3y = 16$

   (e) $x + 2y = 3$
       $3x + 3y = 3$
   (f) $y = x + 5$
       $2x + 3y = 5$
   (g) $2x + 3y = 8$
       $5x - 2y = 1$
   (h) $x + y = 3$
       $5x + 3y = 10$
   (i) $6x + 5y = -2$
       $4x - 3y = 5$

7. Solve the quadratic equations.
   (a) $x^2 - 6x + 8 = 0$
   (b) $x^2 + 5x + 6 = 0$
   (c) $x^2 - 2x - 3 = 0$
   (d) $x^2 - 3x - 10 = 0$
   (e) $x^2 - 5x + 4 = 0$
   (f) $x^2 + 7x + 10 = 0$
   (g) $x^2 - 5x - 14 = 0$
   (h) $x^2 + 17x + 30 = 0$
   (i) $x^2 - 9x + 20 = 0$
   (j) $x^2 + 4x + 3 = 0$
   (k) $x^2 - 9x - 36 = 0$
   (l) $x^2 + 7x - 18 = 0$

8. Solve these simultaneous equations graphically.
   (a) $y = x + 3$ and $y = 6 - 2x$. Use values of x from -1 to $+3$.
   (b) $y = 2x - 1$ and $3x + 2y = 12$. Use values of x from 0 to 4.

9. (a) Draw the graph of $y = x^2 + 2x$ for values of x from -4 to $+2$.
   (b) Solve the equation $x^2 + 2x = 0$ from your graph.

10. (a) Draw the graph of $y = x^2 - 5x + 5$ for values of x from 0 to $+5$.
    (b) Solve the equation $x^2 - 5x + 5 = 0$ from your graph.

11. (a) Draw the graph of $y = x^3 - 7x$ for values of x from -3 to $+3$.
    (b) Solve the equation $x^3 - 7x = 0$ from your graph.

12. Find a solution for each of these by trial and improvement, giving your answer correct to one decimal place.
    (a) $x^3 - 7x = 0$ between 2 and 3
    (b) $x^3 - 35 = 0$ between 3 and 4
    (c) $x^3 - 2x = 5$ between 2 and 3
    (d) $x^3 + 40 = 0$ between -4 and -3

13. A packet of crisps costs c pence and a can of apple juice costs a pence.

    Three packets of crisps and two cans of apple juice cost £1.39.

    Two packets of crisps and one can of apple juice cost 81p.

    Write down two equations in c and a and solve them to find the cost of the crisps and the cans.

14. Packets of detergent are sold in medium and large packets. A medium packet holds x grams and a large holds y grams. A large packet holds 200 grams more than a medium.

    Joan buys three medium and two large and has 2400 grams of detergent.

    Write down two equations in x and y and solve them to find how much each holds.

15. Write down the inequality satisfied by the shaded region in each diagram.

    (a)

    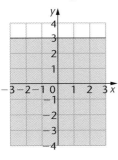

(b)

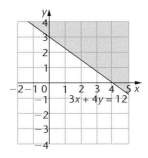

16. Draw sets of axes and label them from -4 to $+4$ for x and y. Shade these regions.

    (a) $x < 1$

    (b) $2y < 3x + 2$

17. Draw a set of axes and label them from -1 to $+8$ for x and y. Show, by shading, the region where $x > 0$, $y > 0$, $x < 8$, $y < x - 1$.

18. Draw a set of axes and label x from -1 to $+8$ and y from -1 to 6. Show, by shading, the region where $x > 0$, $y > 0$, $y < x + 2$, $3x + 7y < 21$.

Measurement and compound units

Estimating measurements

Sometimes you may be asked to estimate measures with which you are not familiar. For example, you may know how much a bag of sugar weighs, but what about a hen's egg? Would it be 5 g, 50 g or 500 g?

Here are some ideas.

- Know your own height, in metric and imperial units, such as 5 ft 7 in or 170 cm.
- Know how much you weigh, in metric and imperial units, such as 9 st 10 lb or 62 kg.
- Occasionally pick up a kilogram bag of sugar to remind yourself how heavy it feels.
- Measure your handspan. Knowing what it is (for example, 20 cm) enables you to measure the width of a table quickly in handspans, for instance, and then estimate the width in centimetres.
- Draw a 10 cm line and look at it carefully, to see how long it is. Practise drawing lines of given lengths without measuring them and see how close you can get to the correct length.
- For a 100 metre distance, think of the length of the 100 m race on an athletics track.

When comparing, ask yourself questions such as:

- Is it the same as … ?
- Is it twice as long as … ?
- Is it much heavier than … ?
- How many of these would weigh the same as … ?
- How many of these laid end to end would be the same length as … ?

Example ❶

Choose the most suitable value, from this list, for the mass of the telephone directory for Guildford and West Surrey.

10 g   100 g   1000 g   10 000 g

A telephone directory is quite a large book. A mass of 10 g is quite small. Change the larger masses in the list to kilograms.

1000 g = 1 kg
10 000 g = 10 kg

A mass of 10 kg is too much to carry easily, but 1 kg is about right.

> Checking with the smaller masses, 100 g is a small pack of cheese, for instance, and this would be too light in comparison.

So the answer is 1000 g.

Exercise 22.1a

1. Estimate the length of this line.

_____

2. Estimate the size of this angle.

3. Estimate the height of this tower.

4. Which of these is closest to the mass of a tablespoon of sugar?
   2.5 g   25 g   250 g

5. Which of the masses below is likely to be how much Sarah is carrying in one of these shopping bags?
   40 g   400 g   4 kg   40 kg

1. Estimate the length of this line.

2. Estimate the size of this angle.

3. Estimate the length of this trailer.

4. Choose the most appropriate of these measurements for the length of a sports hall.

   7 m   70 m   700 m   7000 m

5. When a large jug is full, it holds enough water to fill six tumblers. Choose the most likely measurement from the list for the capacity of the jug.

   0.02 litres   0.2 litres   2 litres   20 litres

Discrete and continuous measures

Discrete measures can be counted. They can only take particular values.

Continuous measures include length, time and mass. They cannot be measured exactly.

Look at this table of some data for a bicycle.

Number of wheels	2
Number of gears	15
Diameter of wheel	66 cm
Frame size	66 cm
Price	£99.99

In the table, the discrete measures are: number of wheels
number of gears
price

The continuous measures are: diameter of wheel
frame size.

Accuracy

Although the table above does not make this clear, the frame size is given to the nearest centimetre. A less accurate measurement, such as to the nearest 10 cm, would not give enough information about the size of the bicycle. Someone wanting to buy a bicycle would not be able to tell whether it would be the right size for them. A more accurate measurement would be unnecessary in the context.

Similarly, giving a person's height to the nearest millimetre would not be sensible, since a person's height varies by more than a millimetre during the day.

When solving problems, the accuracy of your answer is limited by the accuracy of the data available. The answer cannot be accurate to more significant figures than the data. It is often accurate to one fewer significant figure.

> **Examiner's tip**
>
> When giving the answers to practical situations, think what would be an appropriate degree of accuracy to use. Where there are no practical considerations, answers requiring rounding are usually given to three significant figures.

Example 2

Bryn is asked to calculate the hypotenuse of a right-angled triangle, for which the other two sides are given as 2.8 cm and 5.1 cm. To what accuracy should he give his answer?

The data are given to two significant figures, so his answer should be accurate within one or two significant figures. An accuracy of one significant figure would give the answer to the nearest centimetre, which may not be sufficiently accurate. He should round to two significant figures, which is to the nearest millimetre.

Exercise 22.2a

In questions 1 and 2, look at the descriptions from catalogues. For each measurement, identify whether the data are discrete or continuous.

1. Prestige 20 cm polyester golf bag, 6-way graphite-friendly top, 2 accessory pockets
2. Black attaché case, 2 folio compartments, 3 pen holders, size (H) 31.5 cm, (W) 44.5 cm, (D) 11.5 cm

Read this extract from a newspaper article.

> Andy James has now scored 108 goals in just 167 games, making him the Town's most prolific scorer ever. In Saturday's game a penalty brought his first goal after 30 minutes, with Pete Jeffreys having been fouled. Six minutes later, James volleyed into the net again, after a flick on from Neil Matty, five yards outside the penalty box.

3. Give two examples of discrete data in the article above.
4. Give three examples of continuous data in the article above.

5. Write a description including two discrete measurements and three continuous measurements.
6. What is the appropriate degree of accuracy for the length of a line that could be drawn on this page?
   (a) to the nearest cm
   (b) to the nearest 10 cm
   (c) to the nearest mm
   (d) to the nearest 0.1 mm
7. To what degree of accuracy would a field be measured?
8. What is the usual degree of accuracy on road signs:
   (a) on motorways
   (b) on country lane signposts?
9. For teenagers, what is the usual degree of accuracy for giving their age?
10. What is the usual degree of accuracy for measuring flour in cake recipes?

Exercise 22.2b

In questions 1 and 2, look at these descriptions from catalogues. For each measurement, identify whether it is discrete or continuous.

1. 16 piece dinner set, 4 dinner plates (dia. 24.5 cm), side plates and bowls
2. Food blender, 1.5 litre working capacity, 3 speed settings, 400 watt

Read this extract from a newspaper article.

> Lightning killed two people in Hyde Park yesterday as storms swept the south-east, where 1.75 in of rain fell in 48 hours. In Pagham, winds of up to 120 mph damaged more than 50 houses and bungalows and several boats. One catamaran was flung 100 ft into the air and landed in a tree.

3. Give two examples of discrete data in the article above.
4. Give three examples of continuous data in the article above.

5. Write a description including three discrete measurements and two continuous measurements.
6. What is the appropriate degree of accuracy for a waist measurement?
   (a) to the nearest cm
   (b) to the nearest 10 cm
   (c) to the nearest mm
   (d) to the nearest 0.1 mm
7. To what degree of accuracy is the length of a garden usually given?
8. To what degree of accuracy is: (a) a baby's (b) an adult's mass usually given?
9. To what degree of accuracy would the time for the winner of a 50 m swimming race normally be given?
10. To what degree of accuracy is body temperature usually given?

Bounds of measurement

Suppose a measurement is given as 26 cm 'to the nearest centimetre'. This means the next possible measurements on either side are 25 cm and 27 cm. Where does the boundary between these measurements lie?

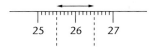

Any measurement that is nearer to 26 cm than to 25 cm or 27 cm will be counted as 26 cm. This is the marked interval on the number line above.

The boundaries of this interval are 25.5 cm and 26.5 cm. These values are exactly halfway between one measurement and the next. Usually when rounding to a given number of decimal places or significant figures, you would round 25.5 up to 26 and 26.5 up to 27.

So this gives:

> the interval for 26 cm to the nearest centimetre is m cm where
> $25.5 \leqslant m < 26.5$.

25.5 cm is called the **lower bound** of the interval

26.5 is called the **upper bound** of the interval (although it is not actually included in the interval).

Example **3**

Simon won the 200 m race in his year in a time of 24.2 s, to the nearest tenth of a second.
Complete the sentence below:
Simon's time was between ... s and ... s.

As the measurement is stated to the nearest tenth of a second, the next possible times are 24.1 s and 24.3 s.

Simon's time was between 24.15 s and 24.25 s.

Chapter 22 Bounds of measurement

Exercise 22.3a

Give the upper and lower bounds of the measurements in questions 1–5.

1. Given to the nearest centimetre:
   (a) 27 cm   (b) 30 cm   (c) 128 cm
2. Given to the nearest 10 cm:
   (a) 10 cm   (b) 30 cm   (c) 150 cm
3. Given to the nearest millimetre:
   (a) 5.6 cm   (b) 0.8 cm   (c) 12.0 cm
4. Given to the nearest centimetre:
   (a) 1.23 m   (b) 0.45 m   (c) 9.08 m
5. Given to the nearest hundredth of a second:
   (a) 10.62 s   (b) 9.81 s   (c) 48.10 s

Complete the sentences in questions 6–10.

6. A mass given as 57 kg to the nearest kilogram is between … kg and … kg.
7. A height given as 4.7 m to two significant figures is between … m and … m.
8. A volume given as 468 ml (to the nearest ml) is between … ml and … ml.
9. A winning time given as 34.91 s to the nearest hundredth of a second is between … s and … s.
10. A mass given as 0.634 kg to the nearest gram is between … kg and … kg.

Exercise 22.3b

Give the upper and lower bounds of the measurements in questions 1–5.

1. Given to the nearest centimetre:
   (a) 34 cm   (b) 92 cm   (c) 210 cm
2. Given to the nearest 10 cm:
   (a) 20 cm   (b) 60 cm   (c) 210 cm
3. Given to the nearest millimetre:
   (a) 2.7 cm   (b) 0.2 cm   (c) 18.0 cm
4. Given to the nearest centimetre:
   (a) 8.17 m   (b) 0.36 m   (c) 2.04 m
5. Given to the nearest hundredth of a second:
   (a) 15.61 s   (b) 12.10 s   (c) 54.07 s

Complete the sentences in questions 6–10.

6. A mass given as 57 kg to the nearest kilogram is between … kg and … kg.
7. A height given as 8.3 m to two significant figures is between … m and … m.
8. A volume given as 234 ml (to the nearest ml) is between … ml and … ml.
9. A winning time given as 27.94 s to the nearest hundredth of a second is between … s and … s.
10. A mass given as 0.256 kg to the nearest gram is between … kg and … kg.

Compound units

Some measures depend on others, which means you need to multiply or divide other measures.

One important example of this is:

$$\text{average speed} = \frac{\text{total distance travelled}}{\text{total time taken}}$$

The units of your answer will depend on the units you begin with.

Example 4 Find the average speed of an athlete who runs 100 m in 20 s.

$$\text{average speed} = \frac{100\,\text{m}}{20\,\text{s}} = 5\,\text{m/s}$$

Example 5 Find the average speed of a delivery driver who travelled 45 km in 30 minutes.

The average speed $= \dfrac{45 \text{ km}}{30 \text{ minutes}} = 1.5 \text{ km/minute}$

However, the speed here is more likely to be needed in kilometres per hour. To find this, first change the time into hours.

So the average speed $= \dfrac{45 \text{ km}}{0.5 \text{ h}} = 90 \text{ km/h}$

You may also be able to see other ways of obtaining this result.

Other examples of compound units are:

- density $= \dfrac{\text{mass}}{\text{volume}}$ with units such as g/cm^3.

- population density $= \dfrac{\text{population}}{\text{area}}$ with units such as number of people/km^2.

Exercise 22.4a

1. Find the average speed of a car which travels 75 miles in one and a half hours.
2. Find the average speed of a runner who covers 180 m in 40 s.
3. Calculate the density of a stone of mass 350 g and volume 40 cm^3.
4. Waring has a population of 60 000 in an area of 8 square kilometres. Calculate its population density.
5. A motorbike travels 1 mile in 3 minutes. Calculate its average speed, in miles per hour.

6. A bus travels at 5 m/s on average. How many kilometres per hour is this?
7. A foam plastic ball with volume 20 cm^3 has density 0.3 g/cm^3. What is its mass?
8. A town has a population of 200 000. Its population density is 10 000 people per square mile. What is the area of the town?
9. A runner's average speed in a 80 m race is 7 m/s. Find the time he takes for the race, to the nearest 0.1 seconds.
10. A car travels 15 km in 12 minutes. What is the average speed in km/h?

Exercise 22.4b

1. Find the average speed of a car which travels 63 miles in one and a half hours.
2. Find the average speed of a runner who goes 180 m in 48 s.
3. Calculate the density of a stone of mass 690 g and volume 74 cm^3. Give your answer to a suitable degree of accuracy.
4. Trenton has a population of 65 000 in an area of 5.8 square kilometres. Calculate its population density, correct to two significant figures.
5. A cyclist rides 0.6 mile in 3 minutes. Calculate her average speed, in miles per hour.

6. A bus travels at 6.1 m/s on average. How many kilometres per hour is this?
7. A rubber ball with volume 28.3 cm^3 has density 0.7 g/cm^3. What is its mass?
8. A town has a population of 276 300. Its population density is 9800 people per square mile. What is the area of the town?
9. A runner's average speed in a 200 m race is 5.3 m/s. Find the time he takes for the race, to the nearest 0.1 seconds.
10. A car travels 15 km in 14 minutes. What is the average speed, in km/h?

Key points

- When estimating measures in unfamiliar contexts, try to compare them with measures you do know.
- Discrete measures can be counted. They can only take particular values.
- Continuous measures include length, time, mass etc. They cannot be measured exactly.
- When giving the answers to practical situations, think what is an appropriate degree of accuracy to use. Where there are no practical considerations, answers requiring rounding are usually given to three significant figures.
- A time of 5.47 s to the nearest one hundredth of a second lies between 5.465 s and 5.475 s.
- Some compound units are average speed (such as m/s), density (such as g/cm³) and population density (such as population/km²).

Revision exercise 22a

1. Estimate the height of this work-top.

2. Walking at a normal pace, it takes Yasmin five minutes to walk from home to school.
   Estimate how far her school is from her home.

3. Name: (a) three discrete (b) three continuous measurements that could be used in describing a car.

4. Jeni and Suni were trying to draw a line 10 cm long. Jenny said hers was 0.102 m long. Suni said his was 9.68 cm long.
   (a) Whose line was more accurate?
   (b) Comment on the units and degree of accuracy they were using.

5. To what degree of accuracy are the lengths of CD tracks usually given?

6. The dimensions of a picture frame were given as 17 cm by 28 cm. Assuming these were to the nearest centimetre, what is the least these dimensions can be?

7. A length is stated as being between 6.805 cm and 6.815 cm.
   (a) What measurement would be recorded?
   (b) What is its degree of accuracy?

8. Sasha runs a 100 m race in 13.58 s. Calculate her average speed. Give your answer to a suitable degree of accuracy.

9. A cyclist travels 5 km in 20 minutes. Calculate her speed in kilometres per hour.

10. A metal weight has mass 200 g and density 25 g/cm³. What is its volume?

You will need to know

- how to draw straight line graphs
- how to draw curved graphs.

Story graphs

Some graphs tell a story – they show what happened in an event. To find out what is happening, first look at the labels on the axes. They tell you what the graph is about.

Look for important features on the graph. For instance, does it increase or decrease at a steady rate (a straight line) or is it curved?

Examiner's tip

When drawing a graph, don't forget to label the axes.

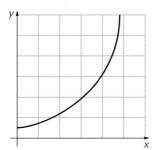

The rate of change is increasing.

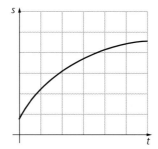

The rate of change is decreasing.

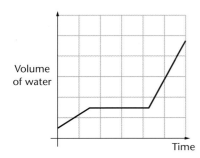

A flat part of the graph – no change for the variable on the vertical axis.

Example 1

This graph shows the noise levels at a football stadium one afternoon.

The boxes describe what may have caused the change in shape of the graph at certain points.

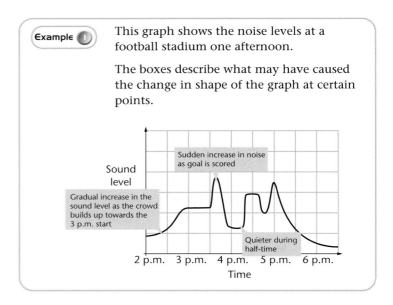

Sound level

Sudden increase in noise as goal is scored

Gradual increase in the sound level as the crowd builds up towards the 3 p.m. start

Quieter during half-time

2 p.m.   3 p.m.   4 p.m.   5 p.m.   6 p.m.

Time

Example 2

John ran the first two miles to school at a speed of 8 mph. He then waited 5 minutes for his friend. They walked the last mile to school together, taking 20 minutes.

The graph for this story has been started. Finish the graph. (The different line on the graph shows where this has been done.)

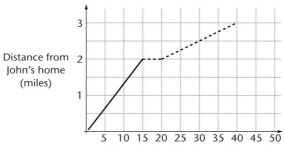

Distance from John's home (miles)

Time after leaving home (minutes)

The first part of this graph is steeper than the last part. This shows that John went faster in the first 15 minutes than he did in the last 20 minutes. The flat part of the graph shows where John stayed in the same place for 5 minutes.

1. This graph shows the volume of water in a bath.

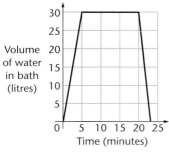

(a) How long did the bath take to fill?

(b) How much water was in the bath when the taps were turned off?

(c) How many litres per minute went down the plughole when the bath was emptied?

2. This graph shows the number of people at a theme park one bank holiday.

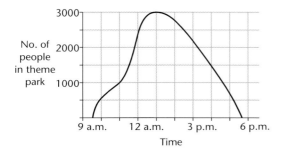

(a) When did the park open?

(b) During which hour did most people go into the park?

3. At a rock concert, the gates opened at 5 p.m. People came in fairly slowly at first, but then quite steadily from 5.45 until the start at 7 p.m. There were then 50 000 people in the stadium. The concert lasted until 10 p.m. At the end people left quickly and the stadium was almost empty by 10.30 p.m.

Sketch a graph to show how the number of people in the stadium for this rock concert changed.

4. This graph shows the cost of hiring a car for a day.

(a) Pedro travelled 150 miles. How much did he pay for his car hire?

(b) Jane paid £48 for her car hire. How many miles did she travel?

(c) What was:
   (i) the basic hire charge
   (ii) the charge per mile?

5. Water is poured at a steady rate into this conical glass until it is full, Sketch a graph to show how the depth of water in the glass changes with time.

1. This graph shows the amount of fuel in a car's petrol tank.

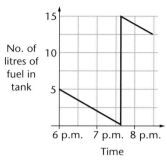

(a) How many litres were used between 6 and 7 p.m?

(b) Describe what happened between 7.30 and 8 p.m.

2. The speed of a car at the start of a journey is shown on this graph.

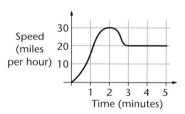

(a) What is happening on the flat portion of the graph?

(b) Between which times is the car slowing down?

3. Jane ran hot water into a bath for 4 minutes at a rate of 15 litres per minute. She then turned on the cold tap too so that the bath filled at 20 litres a minute for another 2 minutes.

(a) Draw a graph to show how the volume of water in the bath changed.

(b) How much water was there in the bath at the end of this time?

4. This graph shows the monthly bill for a mobile phone for different amounts of minutes used.

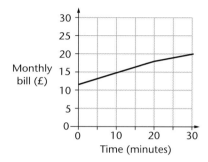

(a) How many minutes have been used if the bill is £15?

(b) There are two line segments on the graph. What do they show?

5. A water company charges £8 each quarter for a meter, then 50p per cubic metre for the first 100 cubic metres used, and 70p per cubic metre for water used above this amount.

Draw a graph to show the total bill for different amounts of water used, up to 200 cubic metres.

Examiner's tip

If you are asked to describe a story graph, try to include numerical information. For example, instead of 'stopped' write 'stopped at 10.14 p.m. for 6 minutes'.

Gradient

The **gradient** of a graph is the mathematical way of measuring its **steepness** or **rate of change**.

$$\text{gradient} = \frac{\text{increase in } y}{\text{increase in } x}$$

To find the gradient of a line, mark two points on the graph, then draw in the horizontal and the vertical to form a triangle as shown.

$$\text{gradient} = \frac{6}{2} = 3$$

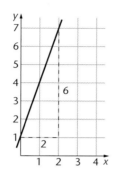

Examiner's tip

Choose two points far apart on the graph, so that the *x*-distance between them is an integer. If possible choose points where the graph crosses gridlines. This makes reading values and dividing easier.

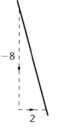

Here the gradient $= \dfrac{-8}{2}$ or $\dfrac{8}{-2}$.

Both give the answer -4.

Lines with a **positive gradient** slope forwards /.

Lines with a **negative gradient** slope backwards \.

Flat lines — have a gradient of zero.

Examiner's tip

Check you have the correct sign, positive or negative, for the slope of the line.

Example 3

Find the gradient of the line joining the points (3, 5) and (8, 7).

Increase in $x = 5$ Subtract $8 - 3 = 5$.

Increase in $y = 2$ Subtract $7 - 5 = 2$. Remember to subtract in the same order.

Gradient $= \frac{2}{5} = 0.4$

You can do this type of example without drawing a diagram, as shown above, but draw a sketch to help you, if you prefer, so that you can see the triangle.

When interpreting graphs about physical situations, the gradient tells you the rate of change.

When calculating gradients from a graph, count the number of units on the axes, not the number of squares on the grid.

Examiner's tip

Use the units on the axes to help you to recognise what the rate of change represents.

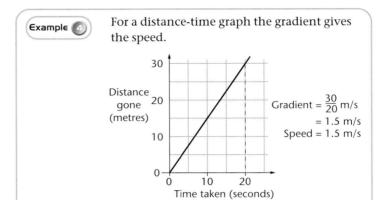

Example 4 For a distance-time graph the gradient gives the speed.

Gradient = $\frac{30}{20}$ m/s
= 1.5 m/s
Speed = 1.5 m/s

Exercise 23.2a

1. Find the gradient of each of these lines.

(a)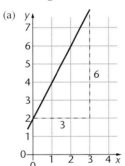

(b)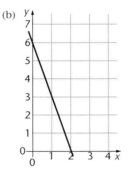

(c)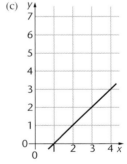

2. Find the gradient of each of these lines.

(a)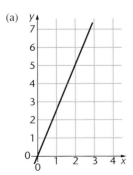

(b)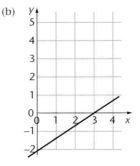

(c)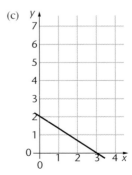

3. Calculate the gradients of the lines joining each of these pairs of points.
   (a) (3, 2) and (4, 8)
   (b) (5, 3) and (7, 3)
   (c) (0, 4) and (2, −6)
   (d) (−1, 1) and (3, 2)

4. Calculate the gradients of the lines joining each of these pairs of points.
   (a) (1, 8) and (5, 6)
   (b) (−3, 0) and (−1, 5)
   (c) (3, −1) and (−1, −5)
   (d) (2.5, 4) and (3.7, 4.9)

5. A ball bearing rolls in a straight groove. The graph shows its distance from a point P in the groove. Find the gradient of the line in this graph. What information does it give?

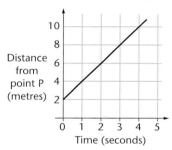

6. Find the gradient of each of the sides of triangle ABC.

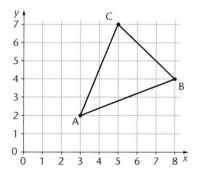

7. The table shows the cost of x minutes of calls on a mobile phone.

Number of minutes (x)	5	12	20	23
Cost (£C)		1.30	3.12	5.20	5.98

Find the gradient of the graph of C against x, and say what this gradient represents.

8. Draw, on the same diagram, the graphs of:
   (a) $y = 3x$
   (b) $y = 3x + 2$ and find their gradients.

9. Draw a graph for each of these straight lines and find their gradients.
   (a) $y = 2x + 1$
   (b) $y = 5x - 2$
   (c) $y = 4x + 3$

10. Draw a graph for each of these straight lines and find their gradients.
    (a) $y = -2x + 1$
    (b) $y = -3x + 2$
    (c) $y = -x$

1. Find the gradient of each of these lines.

(a)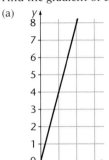

(b)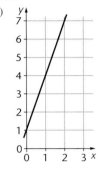

(c)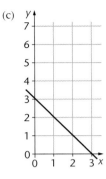

2. Find the gradient of each of these lines.

(a)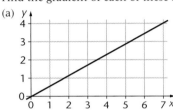

(b)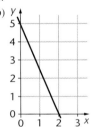

(c)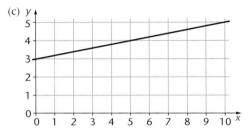

3. Calculate the gradient of the lines joining each of these pairs of points.
   (a) (4, 0) and (6, 8)   (b) (−1, 4) and (7, 2)
   (c) (1, 5) and (3, 5)   (d) (−2, 6) and (0, 4)

4. Calculate the gradients of the lines joining each of these pairs of points.
   (a) (2, 10) and (10, 30)
   (b) (−3, 6) and (−1, −2)
   (c) (0.6, 3) and (3.6, −9)
   (d) (2.5, 7) and (4, 2.2)

5. Find the gradient of the line in this graph. What information does it give?

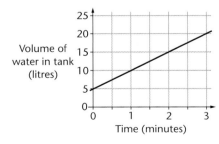

6. Find the gradient of each of the sides of triangle ABC.

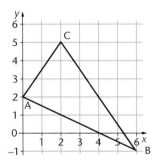

7. The table gives the cost when x metres of ribbon are sold.

Number of metres (x)	0.25	0.5	1.75	3.00
Cost (C pence)		21	42	147	252

Find the gradient of the graph of C against x and say what this gradient represents.

8. Draw, on the same diagram, the graphs of:
   (a) $y = 2x$ (b) $y = 2x + 1$
   and find their gradients.

9. Draw a graph for each of these straight lines
   and find their gradients.
   (a) $y = x + 1$ (b) $y = 2x - 3$
   (c) $y = 4x$

10. Draw a graph for each of these straight lines
    and find their gradients.
    (a) $y = -x + 3$ (b) $y = -3x$
    (c) $y = -2x - 5$

Straight-line graphs

If you did the last two questions in the exercises on gradients,
you may have noticed a connection between the equation of a
line and its gradient.

> When the equation is written in the form $y = mx + c$, where
> m and c are numbers, then:
> m is the gradient of the line
> c is the value of y where the graph crosses the y-axis. In other words,
> the graph passes through $(0, c)$.

Using these facts means that:

- you can work out the equation of a line from its graph
- if you know the equation of a line you can easily find its
  gradient and where it crosses the y-axis.

 Example 5   Find the equation of this straight line.

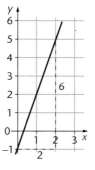

The gradient is $\dfrac{6}{2} = 3$.

The line passes through $(0, -1)$.

So the equation is $y = 3x - 1$.

Example 6

The equation of a straight line is $5x + 2y = 10$. Find its gradient.

Rearranging the equation: $2y = -5x + 10$
$$y = -2.5x + 5$$

So the gradient is -2.5.

Equations of curved graphs

You are expected to know the shapes of the graphs of some types of equation and be able to sketch them. These include curved graphs, and you have already plotted some of them.

Here is a reminder.

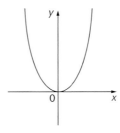

$y = ax^2$
a is positive

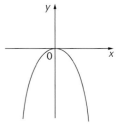

$y = ax^2$
a is negative

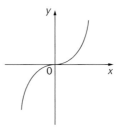

$y = ax^3$
a is positive

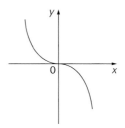

$y = ax^3$
a is negative

Examiner's tip

These last two graphs each have two separate curves, since any number divided by zero is infinity.

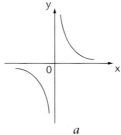

$y = \dfrac{a}{x}$
a is positive

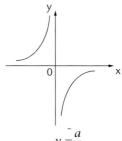

$y = \dfrac{^-a}{x}$
a is negative

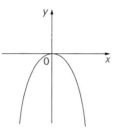

Example 7 Sketch the graph of $y = -2x^2$.

Since x^2 is multiplied by a negative number, the graph opens downwards.

Exercise 23.3a

1. Write down the equations of the straight lines:
   (a) with gradient 3 and passing through (0, 2)   (b) with gradient -1 and passing through (0, 4)
   (c) with gradient 5 and passing through (0, 0).
2. Find the equations of these lines.

(a)

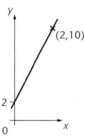

(b)

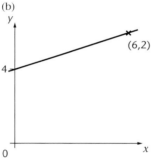

(c)

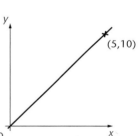

3. Find the equations of these lines.

(a)

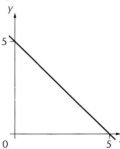

(b)

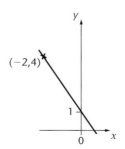

(c)

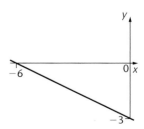

4. Find the gradient of these lines and where they cross the y-axis.
   (a) $y = 3x - 2$
   (b) $y = 2 + 5x$
   (c) $y = 7 - 2x$

5. Find the gradient of these lines and where they cross the y-axis.
   (a) $y + 2x = 5$
   (b) $4x + 2y = 7$
   (c) $6x + 5y = 10$

6. Find an equation for the cost (£C) of travelling m miles for the car hire in question 4, Exercise 23.1a.

7. The table shows the cost of x minutes of calls on a mobile phone (as in Exercise 23.2a, question 7).

Number of minutes (x)	5	12	20	23
Cost (£C)		1.30	3.12	5.20	5.98

   Find an equation connecting x and C.

8. On the same diagram, sketch the graphs of these three equations.
   (a) $y = 2x + 1$
   (b) $y = 2x - 3$
   (c) $y = -4x + 1$

9. Match each of these sketch graphs with the correct equation.

(a) 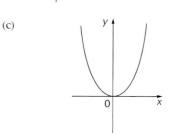 (b)

(c)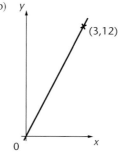

   (i) $y = 3x$
   (ii) $y = -\dfrac{3}{x}$
   (iii) $y = 3x^2$

10. Sketch these graphs.
   (a) $y = 3x - 2$
   (b) $y = -\dfrac{2}{x}$
   (c) $y = -x^2$
   (d) $y = 3x^3$

Exercise 23.3b

1. Write down the equations of the straight lines:
   (a) with gradient 4 and passing through $(0, -1)$
   (b) with gradient -2 and passing through $(0, 5)$
   (c) with gradient 3 and passing through the origin.

2. Find the equations of these lines.
   (a)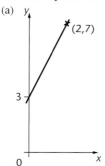

(b)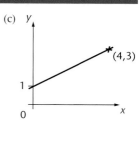

(c)

Chapter 23 Interpreting graphs

3. Find the equations of these lines.

(a)

(b)

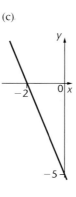

$(-3,15)$

(c)

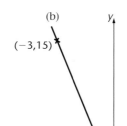

4. Find the gradient of these lines and where they cross the y-axis.
   (a) $y = 5x - 3$ (b) $y = 7 + 2x$
   (c) $y = 9 - 3x$

5. Find the gradient of these lines and where they cross the y-axis.
   (a) $y - 5x = 1$ (b) $3x + 2y = 8$
   (c) $2x + 5y = 15$

6. Using the diagram for question 6 in Exercise 23.2b, find the equations of the sides of triangle ABC.

7. The table gives the cost when x metres of ribbon are sold (as in Exercise 23.2b question 7).

Number of metres (x)	0.25	0.5	1.75	3.00
Cost (C pence)	21	42	147	252

   Find an equation connecting x and C.

8. On the same diagram, sketch the graphs of these questions:
   (a) $y = 3x + 2$ (b) $y = 3x - 2$
   (c) $y = -x + 2$

9. Match each of these sketch graphs with the correct equation.

   (a)

   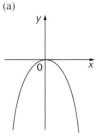

   (b)

   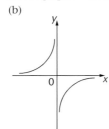

   (c)

   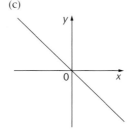

   (i) $y = -2x$ (ii) $y = -2x^2$ (iii) $y = -\dfrac{3}{x}$

10. Sketch these graphs.
    (a) $y = 4x - 1$ (b) $y = -\dfrac{2}{x}$ (c) $y = 3x^2$ (d) $y = -2x^3$

Key points

Story graphs
- The labels on the axes tell you what a graph is about.
- Each feature on the graph is part of the story. For instance, does it increase or decrease at a steady rate (a straight line) or is it curved?

Gradient of straight lines
- Gradient $= \dfrac{\text{increase in } y}{\text{increase in } x}$
- Lines with a positive gradient slope forwards /.
- Lines with a negative gradient slope backwards \.

- Flat lines —— have a gradient of zero.
- Gradient gives the rate of change in graphs about physical situations.

Equation of a straight-line graph
- The equation of a line can be written as $y = mx + c$, where m and c are numbers. m is the gradient of the line and c is the value of y where the graph crosses the y-axis. In other words, the graph passes through $(0, c)$.

Shapes of graphs
- Learn the shapes of graphs such as $y = x^2$, $y = x^3$, $y = \dfrac{1}{x}$.

Revision exercise 23a

1. The distance travelled by a train between two stations is shown on this graph.

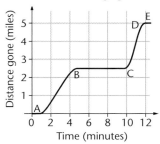

(a) How far is it between the stations?
(b) What was happening on section BC of the graph?
(c) What was happening on section DE of the graph?
(d) On which section did the train travel at the greatest speed?

2. A kite was launched and gained height, slowly at first but then more quickly, until it was 30 m up after about 10 s. It flew at this height for 30 s, then came down 20 m very quickly. It descended the remaining 10 m more gently, landing 50 s after it started.
Draw a graph to show this information.

3. Find the gradient of the lines joining these pairs of points.
(a) (2, 4) and (4, 9)
(b) (2, 4) and (6, 0)
(c) (−1, 2) and (5, 2)

4. Plot and join the points A(3, 1), B(−3, 4) and C(5, 6). Calculate the gradients of the sides of triangle ABC.

5. State the gradients of these lines, and the coordinates of their intersection with the y-axis.
(a) $y = 2x - 3$
(b) $x = 2y$
(c) $3y = x + 2$
(d) $2x + 5y = 10$

6. (a) Plot a graph for these data for the distance (d km) of a car from a motorway junction at time t minutes.

t	2	4	8	15
d	5.0	7.4	12.2	20.6

(b) What was the speed of the car, in km per minute?
(c) How far was the car from the junction when t was zero?
(d) Write an equation connecting d and t.

7. Find an equation for each of these lines.

(a)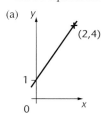

(b)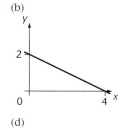

(c)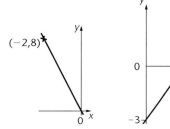

(d)

8. Sketch the graphs of these straight lines.
   (a) $y = 3x - 2$
   (b) $y = -3x + 1$
   (c) $x = 2y$
   (d) $x = 2$

9. Sketch the graphs of these curves.
   (a) $y = x^3$
   (b) $y = -\dfrac{12}{x}$
   (c) $y = 4x^2$

10. Match each of these graphs with the correct equation.

(a)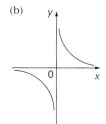

(b)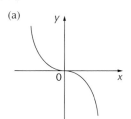

(c)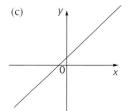

(d)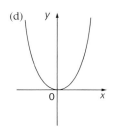

(i) $y = 2x^2$　　(ii) $y = 2x + 1$

(iii) $y = -2x^3$　　(iv) $y = \dfrac{2}{x}$

Probability 2

Covering all the possibilities

To find the probability of an event happening it is essential to consider all the possible outcomes. There are various ways to set out the work, to ensure that none of the outcomes is overlooked.

Each method is introduced in the context of tossing two coins to see whether they come up heads or tails.

List of equally likely outcomes

When looking at the outcomes for tossing two coins it is tempting to say that the possible outcomes are:

two heads   two tails   one of each.

However, the simple practical experiment of tossing two coins a large number of times should show that 'one of each' comes up approximately twice as often as 'two heads' or 'two tails'. This is because to throw either 'two heads' or 'two tails' both coins must come up the same, whereas to throw 'one of each' it does not matter which way the first coin comes up as long as the second one comes up the other way.

So the correct list of equally likely outcomes is:

head, head	head, tail	tail, head	tail, tail

and therefore $P(\text{2 heads}) = \frac{1}{4}$ $P(\text{2 tails}) = \frac{1}{4}$
but $P(\text{1 head and 1 tail}) = \frac{2}{4} = \frac{1}{2}$

The main disadvantage of this method is immediately obvious. It is all too easy to miss possible outcomes.

Another disadvantage is that listing outcomes can be very time-consuming, especially if there are very many possible outcomes. For example, if you are throwing two dice, there are 36 possible outcomes since for every one of six possibilities for the first die there are six possibilities for the second.

Table of outcomes

This method is useful for organising the list of outcomes, and reduces the chance of leaving some out. The table for the above example is on the right. The probabilities can be found exactly as before.

The disadvantage of a table is that again it can be time-consuming if there are many possible outcomes.

Showing outcomes on a grid (a possibility space)

In this method the outcomes for the first trial are listed on the x-axis and the outcomes for the second trial are listed on the y-axis. The outcomes for the combination of trials are then represented by crosses on the grid. The outcomes for two coins are shown like this. Again the probabilities can now easily be found.

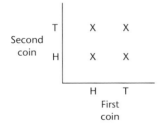

The advantages of this method are:

- it is a very quick way of showing the outcomes when there are very many possibilities, for example it is a very good way to show all 36 outcomes for throwing two dice
- if the outcomes are numerical, the crosses can be replaced by the numerical result. For example, if you are interested in the total score when you throw two dice, you can replace the X for (3, 4) by 7 and the X for (5, 6) by 11 etc.

A disadvantage is that it cannot be extended for a third or subsequent trial. It is difficult to draw a three-dimensional graph but it is easy to add a third column to a table.

Examiner's tip

When filling in a table of outcomes, always try to work logically changing one thing at a time. If you choose outcomes haphazardly you may miss some or repeat some.

First coin	Second coin
head	head
head	tail
tail	head
tail	tail

Tree diagrams

In this method 'branches' are drawn from a starting point to show the possibilities for the first trial. From the end of each of the first branches, further branches are drawn showing each of the possibilities for the second trial, and so on.

The tree diagram for tossing two coins looks like this.

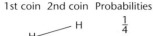
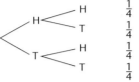

The advantages of a tree diagram are:

- it can easily be extended for a third and subsequent trials
- it can also be used when the outcomes are not equally likely. (This is covered later in the chapter.)

The main disadvantage of a tree diagram is that it can look very messy if there are too many possibilities. For example it is difficult to organise for the example of throwing two dice. However, if there are only two or three possible outcomes for each trial it is often the best method.

Examiner's tip

When drawing a tree diagram:
1. allow plenty of space on the page
2. always line up the possibilities for each trial underneath each other

3. for the first trial, draw the branches to points approximately $\frac{1}{4}$ way from the top of your space to $\frac{1}{4}$ way from the bottom of your space.

Example 1

Fatima throws an ordinary die and tosses a coin. Show the possible outcomes in a table and find the probability that:

(a) Fatima scores a head and a 6

(b) Fatima scores a tail and an odd number.

(a) P(head and 6) $= \frac{1}{12}$

(b) P(tail and odd number) = P[(T, 1) or (T, 3) or (T, 5)] $= \frac{3}{12} = \frac{1}{4}$

This question could also have been answered using a grid.

Coin	T	X	X	X	X	X	X
	H	X	X	X	X	X	X
		1	2	3	4	5	6
				Die			

Choose the outcomes logically, changing the number on the die first.

Die	Coin
1	head
2	head
3	head
4	head
5	head
6	head
1	tail
2	tail
3	tail
4	tail
5	tail
6	tail

Therefore there are 12 possible outcomes.

Example 2 — Gareth tosses two dice. Show the outcomes on a grid (possibility space) and use the grid to find the probability that:

(a) Gareth scores a double

(b) Gareth scores a total of 11.

From the grid, there are 36 possible outcomes.

(a) P(double) $= \frac{6}{36} = \frac{1}{6}$

(b) P(score of 11) $= \frac{2}{36} = \frac{1}{18}$

The grid could also have been drawn with the total scores replacing the Xs.

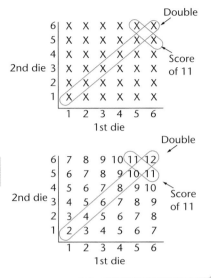

Example 3 — Rachel is selecting a main course and a sweet from this menu.

MENU

Main course	Sweet
Sausage & Chips	Apple Pie
Ham Salad	Fruit Salad
Vegetable Lasagne	

Draw a tree diagram to show Rachel's possible selections. If she is equally likely to select any of the choices, what is the probability that she selects Vegetable Lasagne and Apple Pie?

Therefore there are six possible outcomes.

P(Vegetable Lasagne and Apple Pie) $= \frac{1}{6}$

*Rachel's Selection

1. Jamil has brown socks, red socks and green socks in his drawer. He selects two socks at random from the drawer. Copy the table and complete it to show the possible outcomes.

First sock	Second sock

2. In tennis one player must win, a draw is not possible. Alex plays three games of tennis against Meiling. Copy the table and complete it to show the possible winner of each game. The first entry has been done for you

First game	Second game	Third game
Alex	Alex	Alex

3. Mr and Mrs Green plan to have three children. Draw a table to show the possible sexes for the first, second and third child. Assuming that all the outcomes are equally likely, what is the probability that Mr and Mrs Green will have:
   (a) all girls   (b) two boys and a girl?

4. The picture shows a fair spinner. Claire spins the spinner twice and records the total score.

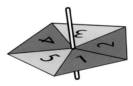

Draw a grid, with each of the axes marked 1−5, to show the possible outcomes.
Find the probability that Claire scored:
   (a) 10   (b) 5.

5. There are four suits in a set of playing cards: hearts (H), spades (S), diamonds (D) and clubs (C). There are equal numbers of each. Hearts and diamonds are red, spades and clubs are black.
   David chooses a card from a set of playing cards. He records the suit, replaces the card and then chooses another. Draw a grid with the axes labelled H, S, D and C to show the possible outcomes for his two cards.
   What is the probability that David chooses:
   (a) two spades   (b) two red cards?

6. Soraya chooses a card from a pack of playing cards and records its suit. She then throws a fair die. Draw a grid showing the possible outcomes. What is the probability that Soraya scores:
   (a) a club and a 5
   (b) a red card and a 6
   (c) a heart and an even number?

7. Fiona chooses a playing card, records its suit and tosses a coin. Copy the tree diagram below and complete it.

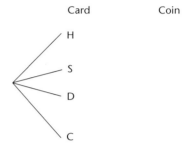

Find the probability that Fiona scores:
   (a) a club and a head
   (b) a red card and a tail.

8. Lisa is choosing from this menu.
   Draw a tree diagram to show her possible choices.
   You could use the initial letters e.g. OB to save you writing the names out in full.
   Assuming she is equally likely to choose any of the items, find the probability that she chooses Onion Bhaji and Lamb Madras.

> ### STARTER AND MAIN COURSE £6
> #### MENU
> | Starters | Main Course |
> |---|---|
> | Onion Bhaji | Chicken Tikka |
> | Sheek Kebab | Lamb Madras |
> | | Vegetable Bhuna |

1. Robbie tosses three coins together. Copy the table and complete it to show all the possible outcomes. The first has been done for you.

First coin	Second coin	Third coin
head	head	head

Find the probability that Robbie tosses:
(a) two heads and a tail
(b) at least one head.

2. Anne plays a game. She tosses a coin to see whether to pick up a number or a letter card at random. If she tosses a head she picks a number card: 1, 2 or 3. If she tosses a tail she picks a letter card: A, B or C. Copy the table and complete it to show her possible outcomes.

Coin	Card

What is the probability that she picks:
(a) card C
(b) a card with an odd number on it?

3. In a game, Bobbie spins both of the spinners. Make a table showing all the possible outcomes. What is the probability of getting a B and a 3?

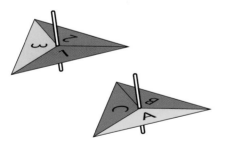

4. Draw a grid showing all the outcomes when two dice are thrown. What is the probability of scoring:
(a) a double six   (b) a total score of 9?

5. In the game of Monopoly, you throw two dice and your score is the total. Use the grid you drew for question 4 to answer these questions.

(a) To buy Park Lane, Hamish needs to score 11. What is the probability that Hamish can buy Park Lane after his next go?
(b) To get out of jail, Sylvia needs to throw a double. What is the probability that Sylvia gets out of jail on her next go?
(c) If Sanjay scores 7, he lands on Regent Street. What is the probability that Sanjay does **not** land on Regent Street?

6. Mr Ahmed is choosing his new company car. He can choose a Rover or a Peugeot. He can choose red, blue or black. Copy the tree diagram and complete it to show his possible choices.

Make   Colour

Rover

Peugeot

If he chooses completely randomly, what is the probability that he will choose:
(a) a blue Peugeot   (b) a black car?

7. Nicola and John are doing their Maths coursework. They can choose 'Billiard tables', 'Stacking cans' or 'Number trees'. Copy and complete the tree diagram to show their possible choices.

Nicola   John

BT

SC

NT

8. Gary and Salma are going out on Saturday night. They can go bowling, to the cinema or to the fair. After that they can either go for a meal or go dancing. Since they cannot agree what to do they decide to choose randomly.
Draw a tree diagram to show their possible choices. Find the probability that they go bowling and then on for a meal.

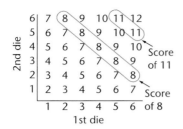

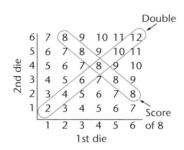

Probability of event A or event B happening

Look again at the grid for throwing two dice.

Suppose Louise needs a score of 8 or 11. Out of a total of 36 possible outcomes, there are seven that give 8 or 11.

The probability of scoring 8 or 11 is therefore $\frac{7}{36}$.

But the probability of scoring 8 is $\frac{5}{36}$ and the probability of scoring 11 is $\frac{2}{36}$

and $\frac{5}{36} + \frac{2}{36} = \frac{7}{36}$

So $P(8 \textbf{ or } 11) = P(8) + P(11)$

If the two events are 'scoring a double' or 'scoring 8' the situation is different.

There are ten outcomes that give a double or a score of 8 and therefore:

$P(\text{double } \textbf{or } 8) = \frac{10}{36}$ $P(\text{double}) = \frac{6}{36}$ $P(8) = \frac{5}{36}$

and $\frac{6}{36} + \frac{5}{36}$ does not equal $\frac{10}{36}$.

So $P(\text{double } \textbf{or } 8)$ does **not** equal $P(\text{double}) + P(8)$.

This is because the events 'scoring a double' and 'scoring 8' are not mutually exclusive events. It is possible to do both by throwing a double 4.

The addition rule only applies to mutually exclusive events.

> If events A and B are mutually exclusive then $P(A \textbf{ or } B) = P(A) + P(B)$

Independent events

If two coins are tossed, the way the first one lands cannot possibly affect the way the second one lands.

Similarly, if two dice are thrown, the way the first one lands cannot possibly affect the way the second one lands.

If there are six red balls and four black balls in a bag, and one is selected and replaced before a second one is selected, the probability of getting a red ball is exactly the same on the second choice as on the first: $\frac{6}{10}$.

When an event is unaffected by what has happened in another event, the events are said to be **independent**.

In the example of six red balls and four black ones, if the first ball is not replaced then the probability of getting a red ball on the second draw is no longer $\frac{6}{10}$ as there are fewer balls in the bag. In this case the events are **dependent**.

Probability of event A and event B happening

In Example 1 Fatima tossed a coin and threw a die. Since there were 12 equally likely outcomes, and scoring a head and a 6 was one of them, it was concluded that:

P(head and a 6) $= \frac{1}{12}$

Now P(head) $= \frac{1}{2}$ and P(6) $= \frac{1}{6}$

But $\frac{1}{2} \times \frac{1}{6} = \frac{1}{12}$ so P(head and a 6) = P(head) $\times$ P(6)

In the first example in this chapter, tossing two coins, it was concluded that:

P(2 heads) $= \frac{1}{4}$

Now P(head) $= \frac{1}{2}$

But $\frac{1}{2} \times \frac{1}{2} = \frac{1}{4}$ so P(2 heads) = P(head) $\times$ P(head)

These results are only true because the events are independent. If they were dependent events, the second probability in the multiplication sum would be different.

> For independent events P(A **and** B) = P(A) $\times$ P(B)

Clearly it is more of a coincidence to throw two heads than one, so it is to be expected that the probability will be less. Multiplying fractions and decimals less than one gives a smaller answer, whereas adding them gives a bigger answer.

The result for events A and B extends to more than two events. For example in Exercise 10b, Question 1 you should have found that when tossing three coins, the probability of getting all three heads $= \frac{1}{8}$.

P(head) $\times$ P(head) $\times$ P(head) $= \frac{1}{2} \times \frac{1}{2} \times \frac{1}{2} = \frac{1}{8}$

So included in the multiply rule are words such as 'both' and 'all'.

Examiner's tip

It is very common for examination candidates to add probabilities when they should have multiplied. If you get an answer to a probability question that is more than 1 you have almost certainly added instead of multiplied.

Example 4

The probability that the school hockey team will win their next match is 0.4. The probability that they draw their next match is 0.3. What is the probability that they will win or draw their next match?

The events are mutually exclusive, since they cannot both win and draw their next match, so:

P(win **or** draw) = P(win) + P(draw) = 0.3 + 0.4 = 0.7

Example 5

Matt spins the fair spinner shown in the picture twice.

What is the probability that Matt scores a 4 on both his spins?

The events are independent, since the second spin cannot be affected by the first.

$P(4 \text{ and } 4) = P(4) \times P(4) = \frac{1}{4} \times \frac{1}{4} = \frac{1}{16}$

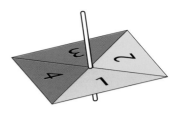

Example 6

There are six red balls and four black balls in a bag. Gina selects a ball, notes its colour and replaces it. She then selects another ball. What is the probability that Gina selects:

(a) two red balls

(b) one of each colour?

Since Gina replaces her first ball before choosing the second the events are independent.

(a) $P(2 \text{ reds}) = P(\text{red}) \times P(\text{red}) = \frac{6}{10} \times \frac{6}{10} = \frac{36}{100} = \frac{9}{25}$

(or in decimals $0.6 \times 0.6 = 0.36$)

(b) Before doing this question it is important to think about what the outcomes are.

Gina requires   first ball red **and** second ball black
or   first ball black **and** second ball red.

Both the add and multiply rules are needed.

$P(\text{one of each colour}) = (\frac{4}{10} \times \frac{6}{10}) + (\frac{6}{10} \times \frac{4}{10}) = \frac{24}{100} + \frac{24}{100} = \frac{48}{100} = \frac{12}{25}$

(or in decimals $(0.4 \times 0.6) + (0.6 \times 0.4) = 0.24 + 0.24 = 0.48$)

Questions like part (b) of example 6, which require both rules, are clearly more difficult. Later in the chapter, you will see that these can often be more easily tackled using tree diagrams.

Exercise 24.2a

1. There are five green balls, three red balls and two yellow balls in a bag. If a ball is selected at random, find the probability that it is green or red.

2. Craig is choosing his next holiday. The probability that he will choose Ibiza is 0.4, the probability that he will chose Corfu is 0.35 and the probability that he will choose Tenerife is 0.25. Find the probability that Craig chooses Ibiza or Corfu.

3. There are four kings and four queens in a pack of 52 playing cards. Salim chooses a card at random from the pack. What is the probability that it is a king or queen?

4. There are five green balls, three red balls and two yellow balls in a bag. Ian chooses a ball at random, notes its colour and puts it back in the bag. He then does this a second time. Find the probability that both Ian's choices are red.

5. The probability that I take sandwiches for dinner is 0.4. The probability that I have a school lunch is 0.6. Assuming the events are independent, what is the probability that I have sandwiches on Monday and a school lunch on Tuesday?

6. What is the probability that I get a multiple of 3 when I throw a single fair die?
   If I throw the die twice, what is the probability that both throws give a multiple of 3?

7. There are four kings in a pack of 52 playing cards. Roger selects a card at random from the pack, returns it to the pack, shuffles the pack and then selects another. Find the probability that both Roger's selections were kings.

8. Alice and Carol are choosing clothes to go out. The probability that Alice chooses jeans is 0.6. The probability that Carol chooses jeans is 0.5. Assuming that their choices are independent, find the probability that they both choose jeans. Explain why this assumption may not be true.

Exercise 24.2b

1. The probability that the school hockey team will win their next game is 0.3. The probability that they draw the next game is 0.45. What is the probability that they will not lose their next game? (That is, they win or draw the game.)

2. If the results of the hockey team are independent, use the probability given in question 1 to find the probability they win both their next two games.

3. Janine travels to school by bus, cycle or car. She says that the probability that she travels by bus is 0.25, by cycle is 0.1 and by car is 0.6. Why must she be incorrect?

4. Rachel is selecting a main course and a sweet from this menu.

> ### MENU
>
> **Main course**
> Sausage & Chips (0.35)
> Ham Salad (0.4)
> Vegetable Lasagne (0.25)
>
> **Sweet**
> Apple Pie (0.4)
> Fruit Salad (0.6)

   The numbers next to the items are the probabilities that Rachel chooses those items.
   (a) Find the probability that Rachel chooses Ham Salad or Vegetable Lasagne for her main course.
   (b) Assuming her choices are independent, find the probability that Rachel chooses Vegetable Lasagne and Fruit Salad.

5. There are 12 picture cards in a pack of 52 playing cards. Ubaid picks a card at random. He then replaces the card and chooses another.
   (a) Find the probability, as a fraction in its lowest terms, that Ubaid's first card is a picture card.
   (b) Find the probability that both Ubaid's cards are picture cards.

6. The weather forecast says 'there is a 40% chance of rain tomorrow'.
   (a) Write 40% as a decimal.
   (b) Assuming the probability that it rains on any day is independent of whether it rained or not the previous day, find the probability that it rains on two successive days.
   (c) State why the assumption made in (b) is unlikely to be correct.

7. There is an equal likelihood that someone is born on any day of the week. What is the probability that Gary and Rushna were both born on a Monday?

8. Sally spins this five-sided spinner three times.

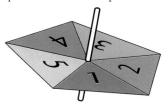

   What is the probability that all Sally's spins landed on 1?

Using tree diagrams for unequal probabilities

In the first section, tree diagrams were used as a way of organising work on probability when the outcomes were equally likely. It is possible to use them when outcomes are not equally likely.

Look again at Rachel's choices on the menu, from question 4 in the last exercise. These can be shown on a tree diagram with the probabilities written on the branches.

MENU

Main course
Sausage & Chips (0.35)
Ham Salad (0.4)
Vegetable Lasagne (0.25)

Sweet
Apple Pie (0.4)
Fruit Salad (0.6)

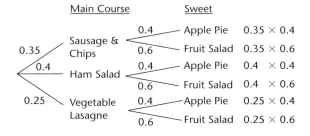

So the probability of choosing Sausage & Chips and Fruit Salad
= 0.35 × 0.6 = 0.21

and the probability of choosing Ham Salad and Apple Pie is
0.4 × 0.4

and so on.

As you go along the 'branches' of any route through the tree, **multiply** the probabilities. Now look at Example 6 in a different way.

Examiner's tip

If you are going along the 'branches' of a tree diagram **multiply** the probabilities. At the end, if you want more than one route through the tree, **add** the probabilities.

Example 7

There are six red balls and four black balls in a bag. Gina selects a ball, notes its colour and replaces it. She then selects another ball. What is the probability that Gina selects:

(a) two red balls

(b) one of each colour?

A tree diagram can be drawn to show this information.

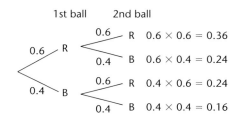

Notice that at each stage the probabilities add up to 1 and at the end all four probabilities add up to 1.

(a) Probability of red followed by red = 0.6 × 0.6 = 0.36.

(b) For one of each colour, Gina needs either the second route **or** the third route through the tree diagram.

So P(one of each colour) = (0.6 × 0.4) + (0.4 × 0.6) = 0.24 + 0.24 = 0.48.

Exercise 24.3a

1. There are seven red balls and three yellow balls in a bag. Lee chooses a ball at random, notes its colour and replaces it. He then chooses another. Copy and complete the tree diagram to show Lee's choices.

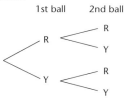

1st ball   2nd ball

What is the probability that Lee chooses:
(a) two red balls
(b) a red ball and then a yellow ball
(c) a yellow ball and then a red ball
(d) a red ball and a yellow ball in either order?

2. Li is choosing a starter and main course from this menu. The probabilities of each of her choices are in brackets next to the items.

> ### MENU
> **Starter**
> Soup (0.3)
> Spring Rolls (0.7)
> **Main course**
> Chicken Fried Rice (0.3)
> Beef Satay (0.2)
> Sweet & Sour Pork (0.5)

(a) Draw a tree diagram to show Li's choices.
(b) Calculate the probability that Li chooses:
   (i) Spring Roll and Beef Satay
   (ii) Soup and Sweet & Sour Pork.

3. The probability that Aftab wakes up when his alarm goes off is 0.8. Copy the tree diagram and complete it for the first two days of the week.

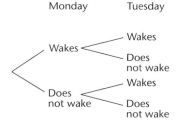

Monday   Tuesday

Calculate the probability that Aftab:
(a) wakes on both days
(b) wakes on one of the two days.

4. There are five red balls, two blue balls and three yellow balls in a bag. Susan chooses a ball at random, notes its colour and replaces it. She then chooses another.
(a) Draw a tree diagram to show the results of Susan's choices.
(b) Calculate the probability that Susan chooses:
   (i) two red balls
   (ii) two balls of the same colour.

5. Mr and Mrs Jones plan to have three children. Assuming there is an equal chance of a boy and girl, draw a tree diagram to show the possible sexes of the three children.
Calculate the probability that Mr and Mrs Jones have:
(a) three girls
(b) two girls and a boy.

Exercise 24.3b

1. On any day the probability that Sarah's bus is late is 0.2. Copy the tree diagram and complete it for two days.

1st day   2nd day

Calculate the probability that Sarah's bus is:
(a) late on both days
(b) late on one of the two days.

2. In an experiment a drawing pin falls point up 300 times in 500 throws.

(a) Write down, as a fraction in its lowest terms, the probability of the pin landing point up.

(b) Draw a tree diagram to show the result of two throws, and the pin landing point up or point down.

(c) Find the probability that the pin lands point up on:
  (i) both throws
  (ii) one of the two throws.

3. Extend the tree diagram you drew for question 2 to show the results of three throws.
Find the probability that the pin lands point up on:
(a) all three throws
(b) one of the three throws.

4. There are ten red balls, three blue balls and seven yellow balls in a bag. Waseem chooses a ball at random, notes its colour and replaces it. He then chooses another.
(a) Draw a tree diagram to show the results of Waseem's choices.
(b) Calculate the probability that Waseem chooses:

  (i) two blue balls
  (ii) two balls of the same colour
  (iii) two balls of different colours. (Look for the quick way of doing it.)

5. Brian drew this tree diagram for the results of choosing coloured balls from a bag.

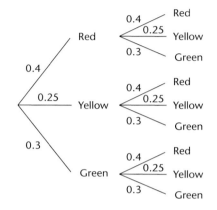

Explain why Brian must have made a mistake.

• Events can be shown using:

Tables	Advantage	Easy to read for probabilities of equally likely events	
	Disadvantages	Lengthy, cannot be used if events are not equally likely	
Grids	Advantage	Easy to read for probabilities of equally likely events	
	Disadvantages	Lengthy, cannot be used if events are not equally likely	
		Cannot be used for more than two successive events.	
Tree diagrams	Advantages	Can be used for more than two successive events	
		Can be used when events are not equally likely	
	Disadvantage	Can be messy when there are more than three outcomes.	

• If events A and B are mutually exclusive then P(A or B) = P(A) + P(B)
• For independent events P(A and B) = P(A) × P(B)
• The multiply rule should also be used for words like 'both' and 'all'.

1. The probability that Brenda goes to school by bus is 0.4. The probability that she goes by car is 0.15. What is the probability that she goes by bus or by car?

2. If Brenda's choices of travel are independent, use the probabilities in question 1 to find the probability that she travels by bus on Monday and by car on Tuesday.

3. A fairground game offers a bottle of champagne if the fair spinner lands on the shaded section twice in succession. What is the probability of winning the champagne?

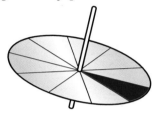

4. The probability of the school football team winning any of their games is 0.2. What is the probability that they do **not** win either of their next two games?

5. Salma throws two dice and records the result of multiplying the two scores together.
   Draw a grid to show all the possible outcomes. Find the probability that Salma's result is:
   (a) 36   (b) 12   (c) 4.

6. Colin throws two dice. He is not interested in all the individual results, just whether he gets a six or not.
   Copy the tree diagram and complete it for the results of Colin's two throws.

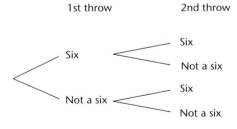

Find the probability that Colin throws:
(a) two sixes   (b) at least one six.

7. The probability that the school netball team will win any match is 0.4. The probability that they draw any match is 0.1. Draw a tree diagram to show the outcomes of their next two matches. Find the probability that the team:
   (a) loses both matches
   (b) does not lose both matches
   (c) wins one of the two matches and draws the other.

8. The probability that it rains on 15 July is 0.1. The probability that it rains on 16 July is also 0.1.
   Find the probability that it:
   (a) rains on both days
   (b) rains on one of the two days.

9. The whole of year 9 take tests in English, Maths and Science. The probability that a randomly-chosen pupil passes English is 0.8, Maths is 0.7 and Science is 0.9.
   Copy the tree diagram and complete it for the three subjects. Assume these events are independent.

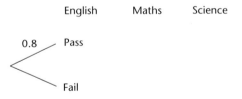

Calculate the probability that a randomly chosen Year 9 pupil:
(a) passes all three subjects
(b) passes two subjects.

Length, area and volume 2

Area of a parallelogram

A parallelogram may be cut up and rearranged to form a rectangle or two congruent triangles.

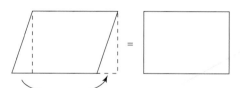

Area of a rectangle = base × height

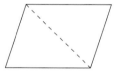

Area of a triangle = $\frac{1}{2}$ × base × height

Both these ways of splitting a parallelogram show:

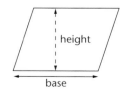

Area of a parallelogram = base × height

Examiner's tip

Make sure you use the perpendicular height and not the sloping edge when finding the area of a parallelogram.

Example 1   Find the area of this parallelogram.

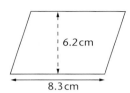

Area of a parallelogram = base × height
= 8.3 × 6.2 cm²
= 51.46 cm²
= 51.5 cm² to three significant
figures

Examiner's tip

Don't forget to give your final answer to a suitable degree of accuracy, but don't use rounded answers in your working.

Exercise 25.1a

1. Find the area of each of these parallelograms.

(a)

(b)

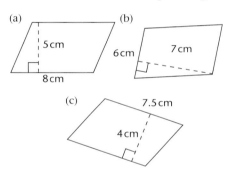

(c)

2. Find the area of each of these parallelograms. The lengths are in centimetres.

(a)

(b)

(c)

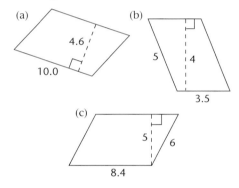

3. Measure the base and height and calculate the area of each of these parallelograms.

(a)   (b)

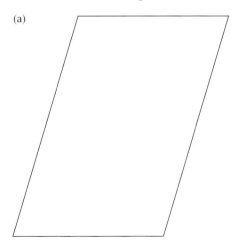

Chapter **25** Area of a parallelogram

(c)

4. Find the values of x, y and z.

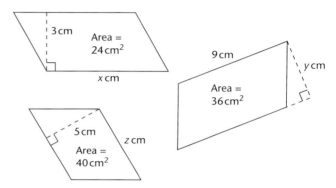

5. This rectangle and parallelogram have the same area. Calculate the height of the parallelogram.

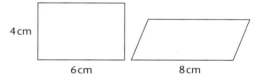

Exercise 25.Ib

1. Find the area of each of these parallelograms.

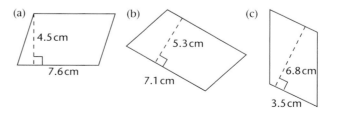

2. Find the area of each of these parallelograms. The lengths are in centimetres.

(a)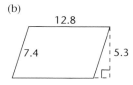

(b)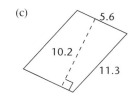

(c)

3. Measure the base and height and calculate the area of each of these parallelograms.

(a)

(b)

(c)

4. Find the values of x, y and z.

(a)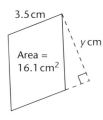
Area = 25.2 cm² 4.2 cm
x cm

(b) 3.5 cm
y cm
Area = 16.1 cm²

(c)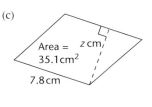
Area = 35.1 cm² z cm
7.8 cm

5. This triangle and parallelogram have the same area. Calculate the height of the parallelogram.

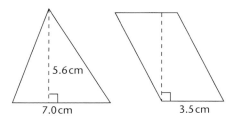

5.6 cm
7.0 cm

3.5 cm

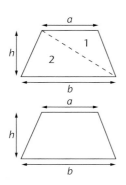

Area of a trapezium

A trapezium has one pair of opposite sides parallel. A trapezium can also be split into two triangles.

Area of triangle 1 $= \frac{1}{2} \times a \times h$

Area of triangle 2 $= \frac{1}{2} \times b \times h$

Area of trapezium $= \frac{1}{2} \times a \times h + \frac{1}{2} \times b \times h = \frac{1}{2} \times (a + b) \times h$

Area of a trapezium $= \frac{1}{2} \times (a + b)h$

In words, a useful formula to remember is:

Area of a trapezium = half the sum of the parallel sides $\times$ the height.

Examiner's tip

When finding the area of a parallelogram or a trapezium, don't try to split the shape up. Instead, learn the area formulae and use them as it is quicker.

Example 2

Calculate the area of this trapezium.

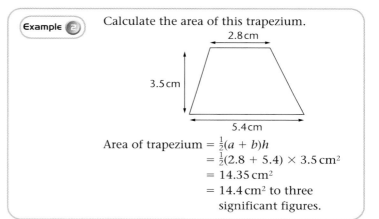

Area of trapezium $= \frac{1}{2}(a + b)h$
$= \frac{1}{2}(2.8 + 5.4) \times 3.5 \text{ cm}^2$
$= 14.35 \text{ cm}^2$
$= 14.4 \text{ cm}^2$ to three significant figures.

Example 3

Without using a calculator, find the area of this trapezium.

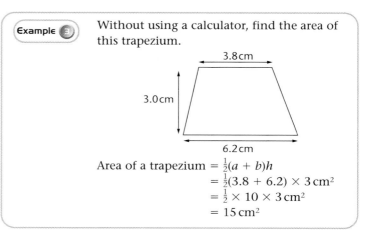

Area of a trapezium $= \frac{1}{2}(a + b)h$
$= \frac{1}{2}(3.8 + 6.2) \times 3 \text{ cm}^2$
$= \frac{1}{2} \times 10 \times 3 \text{ cm}^2$
$= 15 \text{ cm}^2$

Examiner's tip

When finding the area of a trapezium, an efficient method is to use the brackets function on your calculator. Without a calculator, work out the brackets first.

1. Find the area of each of these trapezia.

   (a)

   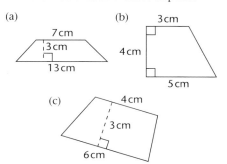

   (b)

   (c)

2. Find the area of each of these trapezia.

   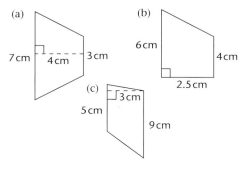

   (a)

   (b)

   (c)

3. Measure the lengths you need and calculate the area of each of these trapezia.

   (a)

   (b)

   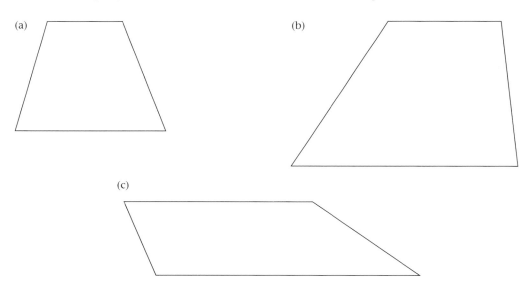

   (c)

4. Find the values of a, b and c in these trapezia.

   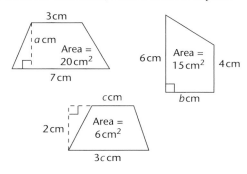

5. A trapezium has height 4 cm and area 28 cm². One of its parallel sides is 5 cm long. How long is the other parallel side?

1. Find the area of each of these trapezia.

(a)

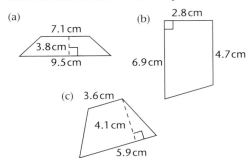

7.1 cm
3.8 cm
9.5 cm

(b)
2.8 cm
6.9 cm
4.7 cm

(c) 3.6 cm
4.1 cm
5.9 cm

2. Find the area of each of these trapezia.

(a)

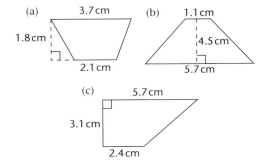

3.7 cm
1.8 cm
2.1 cm

(b)
1.1 cm
4.5 cm
5.7 cm

(c)
5.7 cm
3.1 cm
2.4 cm

3. Measure the lengths you need and calculate the area of each of these trapezia.

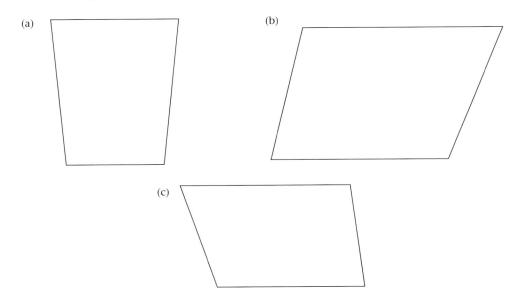

(a)

(b)

(c)

4. Find the values of a, b and c in these trapezia.

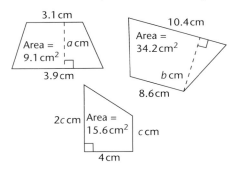

3.1 cm
Area = 9.1 cm²
a cm
3.9 cm

10.4 cm
Area = 34.2 cm²
b cm
8.6 cm

$2c$ cm
Area = 15.6 cm²
c cm
4 cm

5. A trapezium has height 6.6 cm and area 42.9 cm². One of its parallel sides is 5 cm long. How long is the other parallel side?

Volume of a prism

You should remember how to find the volume of a cuboid.

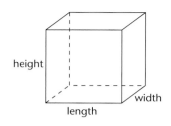

> Volume of a cuboid = length × width × height

It may also be thought of as:

> Volume of a cuboid = area of cross-section × height

This is an example of a general formula for the volume of a prism. When laid on its side, along its length:

> Volume of a prism = area of cross-section × length

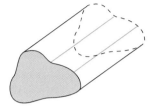

Another important prism is the cylinder. Its cross-section is a circle, which has area πr^2.

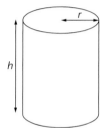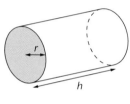

> Volume of a cylinder = $\pi r^2 h$

Example 4

Calculate the volume of a cylinder with base diameter 15 cm and height 10 cm.

Radius of base = $\frac{15}{2}$ = 7.5 cm

Volume of a cylinder = $\pi r^2 h$
$$= \pi \times 7.5^2 \times 10 \text{ cm}^3$$
$$= 1770 \text{ cm}^3 \text{ to three significant figures}$$

Example 5

A chocolate box is a prism with a trapezium as cross-section, as shown. Calculate the volume of the prism.

Area of a trapezium = $\frac{1}{2}(a + b)h$
$$= \frac{1}{2}(20 + 16) \times 6 \text{ cm}^2$$
$$= 108 \text{ cm}^2$$

Volume of a prism = area of cross-section × length
$$= 108 \times 25 \text{ cm}^3$$
$$= 2700 \text{ cm}^3$$

Example 6

A cylinder has volume 100 cm³ and is 4.2 cm high. Find the radius of its base. Give your answer to the nearest millimetre.

$$\text{Volume of cylinder} = \pi r^2 h$$
$$100 = \pi \times r^2 \times 4.2$$
$$r^2 = \frac{100}{\pi \times 4.2}$$
$$= 7.578...$$
$$r = \sqrt{7.578...}$$
$$= 2.8 \text{ cm to the nearest mm}$$

Exercise 25.3a

1. Calculate the volume of a cylinder with base radius 5.6 cm and height 8.5 cm.

2. A cylindrical stick of rock is 12 cm long and has radius 2.4 cm. Find its volume.

3. A cylinder has diameter 8 cm and height 8 cm. Calculate its volume.

4. Calculate the volume of prisms each 15 cm long, with these cross-sections.

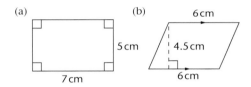

(a)

7 cm, 5 cm

(b)

6 cm, 4.5 cm, 6 cm

(c)

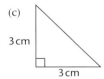

3 cm, 3 cm

5. A chocolate bar is in the shape of a triangular prism. Calculate its volume.

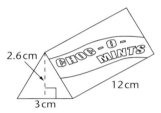

2.6 cm, CHOC-O-MINTS, 12 cm, 3 cm

6. A pencil-box is a prism with a trapezium as its cross-section, as shown. Calculate the volume of the box.

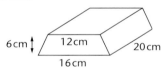

6 cm, 12 cm, 20 cm, 16 cm

7. The area of cross-section of a prism is 75 cm². Its volume is 1200 cm³. Calculate its length.

8. The volume of a cylinder is 800 cm³. Its radius is 5.3 cm. Calculate its length.

9. A cylinder has volume 570 cm³ and height 7 cm. Find its base radius. Give your answer to the nearest millimetre.

10. The volume of a cylindrical tank is 600 m³. Its height is 4.6 m. Calculate the radius of its base.

1. Calculate the volume of a cylinder with base radius 4.3 cm and height 9.7 cm.
2. A cylindrical water tank is 4.2 m high and has radius 3.6 m. Find its volume.
3. A cylinder has diameter 9 cm and height 12 cm. Calculate its volume.
4. Calculate the volume of prisms each 12 cm long, with these cross-sections.

(a)

(b)

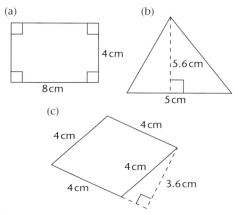

(c)

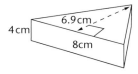

5. A gift box is a prism with a triangular base. Calculate its volume.

6. A vase is a prism with a trapezium as its base. The internal measurements are as shown. How much water can the vase hold? Give your answer in litres. (1 litre = 1000 cm³)

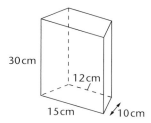

7. The area of cross-section of a prism is 90 cm². Its volume is 1503 cm³. Calculate its length.
8. The volume of a cylinder is 1500 cm³. Its radius is 7.5 cm. Calculate its length. Give your answer to the nearest millimetre.
9. A cylinder has volume 620 cm³ and height 8 cm. Find its base radius. Give your answer to the nearest millimetre.
10. The volume of a cylinder is 1100 cm³. Its length is 10.8 cm. Calculate its radius, giving your answer to the nearest millimetre.

Dimensions

You can tell whether a formula gives length, area or volume by looking at its dimensions.

The area of a circle is πr^2...

...or is it $2\pi r$?

$$\begin{array}{ll}
\text{number} \times \text{length} = \text{length} & \text{[1 dimension]} \\
\text{length} + \text{length} = \text{length} & \text{[1 dimension]} \\
\text{length} \times \text{length} = \text{area} & \text{[2 dimensions]} \\
\text{length} \times \text{length} \times \text{length} = \text{volume} & \text{[3 dimensions]}
\end{array}$$

So which circle formula is for area?

$2\pi r$ = number × length = length   [the circumference of a circle]

πr^2 = number × length × length
    = length × length
    = area [the area of a circle]

Thinking about the number of dimensions also helps you to sort out what units you should be using. For example:

Examiner's tip

For practice, check the dimensions of formulae you know.

$$\begin{array}{ll}
\text{length} = \text{m} & \text{[1 dimension]} \\
\text{area} = \text{m}^2 & \text{[2 dimensions]} \\
\text{volume} = \text{m}^3 & \text{[3 dimensions]}
\end{array}$$

Example 7

If a, b and h are lengths, does the expression $\frac{1}{2}(a + b)h$ represent a length, area or volume, or none of these?

$a + b$ = length

so $\frac{1}{2}(a + b)h$ = number × length × length = length × length = area

Exercise 25.4a

Throughout this exercise, letters in algebraic expressions represent lengths.

1. State whether each of these expressions represents a length, area or volume.
   (a) $r + h$ (b) rh (c) $2\pi rh$
2. Which of these expressions represents a length?
   (a) $\frac{1}{2}bh$ (b) $3b$ (c) $b + 2h$
3. Which of these expressions represents an area?
   (a) xy (b) xy^2 (c) $x(x + y)$

4. Which of these expressions represents a volume?
   (a) r^3 (b) $\pi r^2 h$ (c) $r^2(r + h)$
5. State whether each expression represents length, area, volume or none of these.
   (a) $r(r^2 + h)$ (b) $(3 + \pi)h$ (c) $4\pi r^2$

Exercise 25.4b

Throughout this exercise, letters in algebraic expressions represent lengths.

1. State whether each of these expressions represents a length, area or volume.
   (a) $a + 2b$ (b) $2ab$ (c) $a^2 b$
2. Which of these expressions represents a length?
   (a) $a + 2b + c$ (b) $3a + 2a^2$ (c) $a(2a + b)$

3. Which of these expressions represents an area?
   (a) $4a^2$ (b) $x(x + 2y)$ (c) $\pi r^2 + 2\pi rh$
4. Which of these expressions represents a volume?
   (a) πab (b) $\frac{4}{3}\pi r^3$ (c) $h^2(a + b)$
5. State whether each expression represents length, area, volume or none of these.
   (a) $\frac{1}{3}\pi r^2 h$ (b) $2a^2 b(a + b)$ (c) $a(3 + \pi)$

Key points

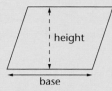

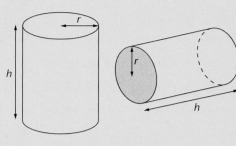

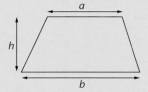

- Area of a parallelogram = base × height

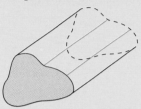

- Area of trapezium = $\frac{1}{2}(a + b)h$

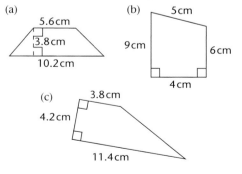

- Volume of a cylinder = $\pi r^2 h$
- Dimensions help to distinguish between formulae for length, area and volume:

  number × length
  = length   [1 dimension]

  length + length
  = length   [1 dimension]

  length × length = area   [2 dimensions]

  length × length × length
  = volume   [3 dimensions]

- Volume of a prism
  = area of cross-section × length

Revision exercise 25a

1. Calculate the area of each of these parallelograms.

   (a)

   3.7 cm
   7.2 cm

   (b)

   5 cm
   9 cm
   3 cm

   (c)

   7.1 cm
   4.2 cm

2. Calculate the area of each trapezium.

   (a)

   5.6 cm
   3.8 cm
   10.2 cm

   (b)

   5 cm
   9 cm
   6 cm
   4 cm

   (c)

   3.8 cm
   4.2 cm
   11.4 cm

3. A prism, 25 cm long, has this L-shape as its cross-section. Calculate the volume of the prism.

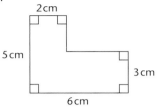

2 cm

5 cm

3 cm

6 cm

4. A cylindrical vase has internal radius 5.6 cm and height 22.5 cm. Calculate how many litres of water this vase can hold.

5. A large cylindrical can of baked beans has volume 3000 cm³ and base radius 7.9 cm. Calculate the height of the can.

6. In these expressions, r and h are lengths. State which of length, area and volume is represented by each of these expressions.
(a) $\pi rh + \pi r^2$
(b) $\frac{1}{2}(r + h)$
(c) $3r^2h$

7. Find the missing powers in these formulae.
(a) volume $= \frac{1}{3}\pi r^2 h$
(b) area $= 6r^2$
(c) length $= \dfrac{r^?}{h^2}$

8. The diagram shows a full-size net for a triangular prism. Use measurements from the drawing to calculate:
(a) the surface area
(b) the volume of the prism.

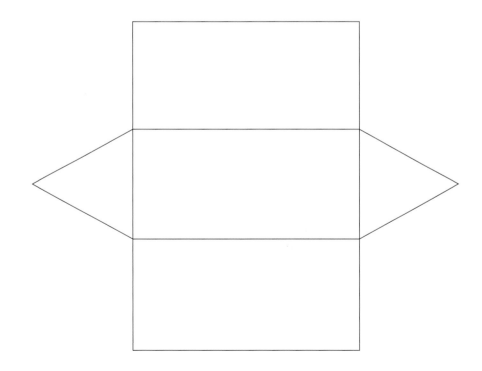

9. Calculate the areas of this parallelogram and trapezium.

(a)

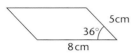

(b)

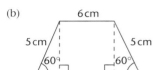

10. The diagram shows the cross-section of a prism which is 8 cm long. Calculate:
    (a) the height of the trapezium
    (b) the volume of the prism.

(c)

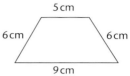

Properties of transformations and similar shapes

26

Properties of transformations

Patterns from transformations

Triangle A is reflected in the mirror line *m* to give triangle B,

triangle B is reflected in the mirror line *n* to give triangle C,

triangle C is reflected in the mirror line *o* to give triangle D,

and so on to form a continuous pattern.

Within patterns like this, there are often other transformations. For example, triangle A maps onto triangle C by a translation.

 Example 1

Triangle P has vertices at (1, 1), (2, 1) and (1, 3).

Triangle P is reflected in the line $x = 3$ to give triangle Q.

Triangle Q is reflected in the line $x = 7$ to give triangle R.

(a) Draw triangles P, Q and R on the same diagram and label them.

(b) Describe fully the transformation that will map triangle P onto triangle R.

Example 1
continued

(a)

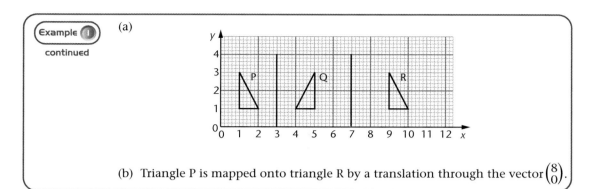

(b) Triangle P is mapped onto triangle R by a translation through the vector $\binom{8}{0}$.

Congruence

If two shapes are identical in shape and size they are **congruent**.

The shapes need not be the same way round. If a tracing of one can be placed exactly over the other, even if it has to be turned over, the shapes are still congruent.

If a shape is rotated, reflected or translated, the new shape is congruent to the original shape, since all these transformations leave the original size and shape unchanged.

Example 2

Explain why triangle ABC is congruent to triangle DEF.

Since the triangle DEF is a reflection of the triangle ABC in the mirror line, m, the two triangles are congruent.

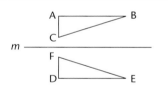

Examiner's tip

When describing congruent shapes using the letters labelling the vertices, always put the points in corresponding order.
For instance, in Example 2, having put A first in the first triangle, you should put D first in the second triangle because it corresponds to A. Similarly, having put B in second place in the first triangle, you should put the corresponding point, E, second in the second triangle and so on. You can spot which points are corresponding by looking at the angles, for example angle A = angle D = 90°.

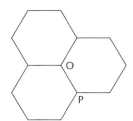

Tessellations

Look at the diagram, which is made up of three regular hexagons.

The interior angle of a regular hexagon is 120°.

Since 360° is exactly divisible by 120°, giving an answer of 3, the hexagons fit together exactly at the point O.

The pattern can be continued by joining another hexagon at the point P and so on.

When shapes fit together exactly to fill up the plane they are said to **tessellate**.

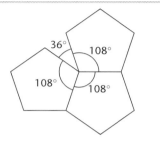

Example 3

With the help of a diagram, explain why regular pentagons will not tessellate.

The interior angle of a regular pentagon is 108°.

Since $3 \times 108 = 324$, there is a gap of 36° that is not filled by the pentagons.

This means that regular pentagons will not tessellate.

Exercise 26.1a

1. Draw a set of axes with x-values from 0 to 14 and y-values from 0 to 4.
   Repeat Example 1, starting with the same triangle P but this time using the mirror lines $x = 4$ and $x = 10$.
   Look at the answers to this question and Example 1. What conclusions can you draw?

   For Questions 2, 3 and 4, refer to the diagram below.

   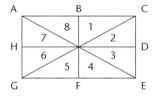

2. Describe fully the transformations that will map triangle:
   (a) 1 onto 2   (b) 2 onto 3   (c) 3 onto 4
   (d) 4 onto 5   (e) 5 onto 6   (f) 6 onto 7
   (g) 7 onto 8.

3. Describe fully the transformations that will map triangle:
   (a) 2 onto 6   (b) 2 onto 5
   (c) 2 onto triangle CEG.

4. Explain why triangle 1 is congruent to:
   (a) triangle 4   (b) triangle 6.

5. Explain why quadrilaterals A, B and C are all congruent to each other.

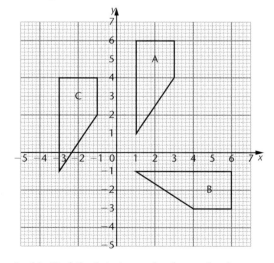

6. (a) Find the interior angle of a regular decagon.
   (b) With the help of a diagram, explain why regular decagons will not tessellate.

1. Draw a set of axes with *x*-values from –6 to +6 and *y*-values from –6 to +6.
   (i) (a) Draw the triangle with vertices at (0, 0), (2, 2) and (6, 0). Label it A.
   (b) Reflect triangle A in the *x*-axis. Label the new triangle B.
   (c) Rotate triangle B through 90° clockwise about the origin. Label the new triangle C.
   (d) Rotate triangle C through 180° about the origin. Label the new triangle D.
   (e) Reflect triangle D in the line $y = -x$. Label the new triangle E.
   (f) Rotate triangle E through 90° anti-clockwise about the origin. Label the new triangle F.
   (ii) Describe fully the transformations that will map F onto G and G onto H to complete a shape with rotational symmetry order 4.

2. Explain why these shapes are not congruent.

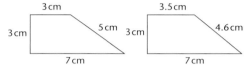

3. Explain why triangles A, B and C are congruent to each other.

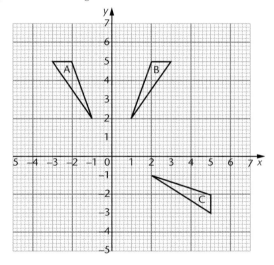

4. With the help of a diagram, explain why trapeziums of this shape will tessellate.

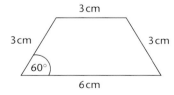

5. (a) Calculate the interior angle of a regular octagon.
   (b) With the help of a diagram, explain why regular octagons will not tessellate.
   (c) Suggest another shape which could be combined with regular octagons to make a tessellating pattern. Draw a diagram which shows how this tessellation would work. You only need a few of the shapes.

Enlargements

Scale drawings

Example 4

The diagram shows a garden shed.

(a) Make a scale drawing of the front of the shed, using a scale factor of $\frac{1}{20}$.

(b) Find the length of the sloping side of the roof.

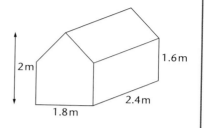

(a) First divide each of the lengths by 20.
$1.6\,m = 160\,cm$, $160 \div 20 = 8\,cm$
$1.8\,m = 180\,cm$, $180 \div 20 = 9\,cm$
$2\,m = 200\,cm$, $200 \div 20 = 10\,cm$

Now draw the shape.

Mark the distance 9 cm along the bottom. Then draw the two sides 8 cm high, at right angles to the first line. To find the top point, mark the middle of the base line and measure up 10 cm at right angles to the base line (the dotted line in the diagram). Then draw the two sloping sides.

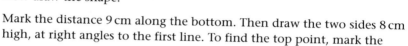

(b) The sloping side measures 4.9 cm.

On the actual shed this length will be 4.9×20 $= 98\,cm$ or $0.98\,m$.

When making a scale drawing of a larger object, the first thing to do is decide on a scale factor. In examination questions this will usually be given to you.

If the units are the same, this scale factor can be given as a simple fraction or ratio.

The scale factor used above could be written $\frac{1}{20}$ or 1 to 20 or $1 : 20$.

Another way of writing the same scale is 1 cm to 0.2 m since 20 cm = 0.2 m.

Once the scale factor has been established, all the lengths on the scale drawing can be calculated. Then the drawing can be made. It is, of course, important to make sure that the angles are the same on the scale drawing as on the actual object.

Maps

Maps are scale drawings of what is actually on the ground. Because the area of the land is very large compared with the area of the piece of paper on which the map is drawn, the scale factor is usually a very small fraction.

Ordnance Survey Outdoor Leisure maps are drawn to a scale of $\frac{1}{25\,000}$ or $1 : 25\,000$.

 Example 5

Find the distance on an Ordnance Survey Outdoor Leisure map which represents a distance of 1 km on the ground.

1 km = 1000 m = 100 000 cm

100 000 ÷ 25 000 = 4

So the distance on the map is 4 cm.

Another way of writing the scale would be **4 cm to 1 km.**

Similar shapes

In Mathematics the word 'similar' has a very exact meaning. It does **not** mean 'roughly the same' or 'alike'.

For two shapes to be similar each shape must be an exact enlargement of the other. For example look at these two rectangles.

The first rectangle is 2 cm wide by 4 cm long.

The second rectangle is 4 cm wide by 7 cm long.

The rectangles are **not** similar because although the width of the large one is twice the width of the small one, the length of the large one is **not** twice the length of the small one. If the length of the large one were 8 cm then the rectangles would be similar.

Now look at these two shapes.

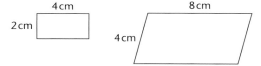

Although the scale factor for both pairs of sides is 2, the shapes are not similar because corresponding angles are not the same.

> For two shapes to be similar:
> * all corresponding sides must have proportional lengths
> * all corresponding angles must be equal.

Similar triangles

If all the sides of a triangle are known the shape of the triangle is fixed. You can demonstrate this, using drinking straws. Take three straws and make a triangle by squeezing the end of one straw and pushing it into the open end of the next. Even if the straws are only loosely jointed, the triangle will be rigid.

This is not the case with a quadrilateral or other polygon, as the angles may change. Make a quadrilateral with drinking straws. It is easy to push opposite angles and make the shape deform into a different quadrilateral.

Because the lengths of three sides define a unique triangle, for two triangles to be similar only one of the tests on the previous page needs to be made.

If you can establish that the angles are the same, you can conclude that the triangles are similar and carry out calculations to find lengths of sides.

Calculations of lengths of similar shapes

Examiner's tip

If you are asked to explain why two triangles are similar, look for reasons why the angles are equal. Usual reasons include 'opposite angles are equal', 'alternate angles are equal' or 'angles are in both triangles (common angles)'.

Example 6

The rectangles ABCD and PQRS are similar. Find the length of PQ.

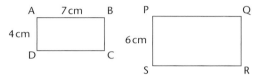

Since the widths of the rectangles are 6 cm and 4 cm, the scale factor is $6 \div 4 = 1.5$.

The length of $PQ = 7 \times 1.5 = 10.5$ cm.

Examiner's tip

It is always easier to spot the corresponding sides if the shapes are the same way round. It is worth spending time redrawing the shapes separately and the same way round and marking the lengths on the new diagram.

Example 7

In the triangle, angle ABC = angle BDC = 90°, AB = 6 cm, BC = 8 cm and BD = 4.8 cm.

(a) Explain why triangles ABC and BDC are similar.

(b) Calculate the length of DC.

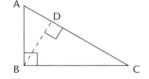

(a) In the triangles ABC and BDC, angle ABC = angle BDC = 90°.

The angle C is in both triangles.

Since the angle sum of a triangle is 180°, the third angles must be equal.

So, since all the corresponding angles are equal, the triangles are similar.

(b) First redraw the triangles so they are the same way round as each other.

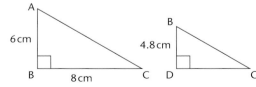

Since AB = 6 cm and BD = 4.8 cm the scale factor = $4.8 \div 6 = 0.8$.

$CD = 8 \times 0.8 = 6.4$ cm

Chapter 26 Enlargements

1. A scale model of a jumbo jet is made to a scale of $\frac{1}{500}$.
   (a) The length of the real aircraft is 70 m. What is the length of the model, in cm?
   (b) The wingspan of the model is 12 cm. What is the wingspan of the real aircraft, in m?

2. The diagram shows a plan of a garden. It is not drawn to scale.

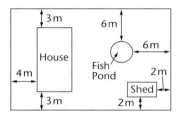

   The garden is 30 m × 20 m.
   The house is 14 m × 8 m.
   The shed is 2.4 m × 1.8 m.
   The fish pond has a diameter of 4 m.
   Draw a scale drawing of the garden, using a scale of $\frac{1}{200}$ or 1 cm to 2 m.

3. The scale of a map is 1 : 50 000.
   (a) A road is 2.5 km long. How long is the road on the map?
   (b) The distance between two villages on the map is 7 cm. What is the real distance between the villages?

4. A road map is drawn to a scale of 1 : 200 000.
   (a) How far, in kilometres, is 200 000 cm?
   (b) The distance from London to Cambridge is 87 km. How far is it on the map, in cm?
   (c) The distance from Exeter to Newcastle on the map is 293 cm. How far is the actual distance, in kilometres?

5. You need an A4 sheet of paper.
   (a) Measure the length and width of the paper.
   (b) Fold the paper in two (along the dotted line).

This size of paper is called A5.
Measure the length and width of this paper.
Divide the length of the A4 sheet by the length of the A5 piece. Do the same for the widths.
Is A4 paper similar to A5 paper? If so, what is the scale factor?
   (c) Fold the paper in two again. This size of paper is called A6.
   Repeat part (b) for A5 and A6 paper. What do you notice?
   (d) All the A-series of paper sizes work in the same way. What is the size of A3 paper?

6. The two rectangles in the diagram are similar. Find the length of the larger rectangle.

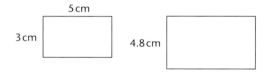

7. The line PQ is parallel to the line CD. The line QR is parallel to the line DE. Explain why the pentagons ABCDE and ABPQR are not similar.

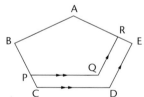

8. The triangles ABC and PQR are similar. Calculate the lengths of PQ and PR.

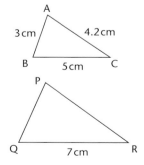

Exercise 26.2a continued

9. The triangles DEF and UVW are similar. Calculate the lengths UV and UW.

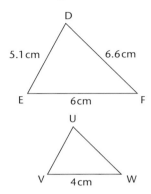

10. The lines PQ and BC are parallel.

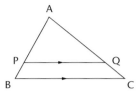

(a) Explain why triangles ABC and APQ are similar.

(b) If AB = 6 cm, BC = 8 cm and AP = 4.7 cm, calculate the length of PQ. Give your answer correct to three significant figures.

Exercise 26.2b

1. A toy car is a scale model of a real car, on a scale of $\frac{1}{50}$.
   (a) The length of the real car is 3.8 m. What is the length of the toy car, in cm?
   (b) The height of the toy car is 2.8 cm. What is the height of the real car, in m?

2. The diagram is a plan of Aftab's bedroom. It is not drawn to scale.

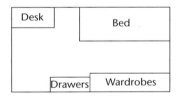

The room is 3.4 m × 2.5 m.
The bed is 2 m × 0.9 m.
The wardrobes are 1.8 m × 0.6 m.
The desk is 1.2 m × 0.6 m.
The drawers are 0.8 m × 0.45 m.
Draw a scale drawing of Aftab's bedroom using a scale of $\frac{1}{25}$.

3. A map is drawn to a scale of 1 : 25 000.
   (a) Two farms are 650 m apart. How far apart, in cm, are they on the map?
   (b) Lesley measures the distance she has walked in a day on the map. It measures 64 cm. How far has she actually walked, in kilometres?

4. A map of Europe is drawn to a scale of 1 : 15 000 000.
   On this map Britain measures approximately 6 cm, north to south. Estimate the actual distance, in kilometres, from the north of Britain to the south of Britain.

5. These two parallelograms are similar. Find the length of PQ.

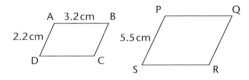

6. Explain why these quadrilaterals are **not** similar.

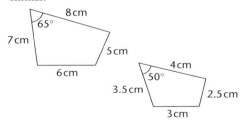

7. The triangles ABC and PQR are similar. Calculate the lengths of PQ and QR.

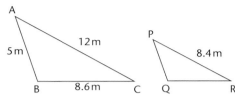

8. In the diagram, DE is parallel to BC, AB = 4.5 cm, BC = 6 cm and DE = 10 cm. Calculate the length BD.

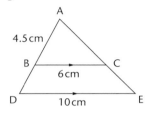

9. In the diagram, angle BAC = angle ADC = 90°, AD = 3 cm and DC = 5 cm.

(a) Explain why triangles ADC and BDA are similar.

(b) Calculate the length BD.

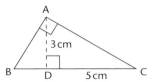

10. ABCD is a trapezium. The diagonals meet at O, and AB = 5 cm, DC = 8 cm and OD = 6 cm.

(a) Explain why triangles OAB and OCD are similar.

(b) Calculate the length of OB.

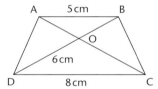

Key points

- Shapes are congruent if they are identical in size and shape.
- If one shape can be mapped onto another by a reflection, rotation or translation then the shapes are congruent.
- If shapes can fit together to fill the plane around a point, with no gaps, then the shapes are said to tessellate.

- Scales for maps and drawings can be given as, for example, $\frac{1}{200}$ or $1:200$ or 1 cm to 2 m.
- For shapes to be similar:
  - all corresponding sides must have proportional lengths
  - all corresponding angles must be equal.
- For similar triangles if one of the above conditions is true, the other must be.

1. Draw a set of axes with *x*-values from −4 to +4 and *y*-values from −4 to +4.
   (a) Draw triangle PAB with vertices at P(1, 1), A(1, 3) and B(3, 1).
   (b) Draw triangle QCD, the image of triangle PAB after it has been rotated through 90° clockwise about the origin.
   (c) Carry out the same transformation on triangle QCD. Label the new triangle REF.
   (d) Carry out the same transformation on triangle REF. Label the new triangle SGH.
   (e) How many lines of symmetry has the octagon ABCDEFGH?
   (f) Is the octagon regular? Give your reasons.

2. Look again at your diagram for question 1. Explain why trapezium ABCD is congruent to trapezium GHAB.

3. Using a parallelogram of your choice, draw a diagram to see whether it will tessellate. Will all parallelograms tessellate?

4. Sasha's bedroom is a rectangle which measures 3.1 m by 3.6 m. Her bed measures 1.9 m by 0.9 m. Sasha makes a scale drawing of the plan of her bedroom. She uses a scale of $\frac{1}{40}$.
   (a) What are the length and width of the bedroom on Sasha's scale drawing?
   (b) What are the length and width of the bed on Sasha's scale drawing?

5. A map is drawn to a scale of 1 : 25 000.
   (a) What length on the map represents 1 km on the ground?
   (b) The distance between two churches on the map is 18 cm. How far apart are they on the ground?

6. The two rectangles are similar. Calculate the height of the smaller rectangle.

7. Are these rectangles similar? Show a calculation to explain your answer.

8. (a) Explain why triangles ABC and PQR are similar.

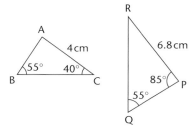

   (b) What is the scale factor?

9. Triangle PQR is similar to triangle ABC. Calculate the lengths PR and QR.

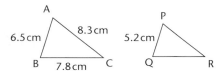

10. BC is parallel to DE, AB = 6.3 m, BD = 2.7 m and DE = 10.2 m. Calculate the length of BC.

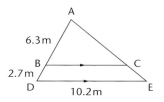

Comparing

Comparing data

Comparing sizes of sets of data

If there are two (or more) sets of data it is often necessary to make comparisons between them. For instance, if the information below gives the marks obtained by John and Aisha in their last five Maths tests, the question may arise as to who is better at Maths.

John	7	8	10	4	6
Aisha	8	9	7	8	6

One way to compare the two sets of figures is to calculate the mean of their scores.

John's mean $= 35 \div 5 = 7$
Aisha's mean $= 38 \div 5 = 7.6$

This would suggest that Aisha is better at Maths than John.

Whilst the mean is often a reliable way of comparing sets of data it is unwise to draw too many conclusions from such a small amount of data. It may be that the topics tested were just more suited to Aisha and, in any case, the difference is not large.

The three measurements of 'average', the mean, median and mode, can all be used to compare the sets of data.

Usually the mean, which takes into account all the data, is the most reliable but there are circumstances where this is not so. One or two very large or very small figures can distort a mean and give a false impression.

Example 1

Here are John and Aisha's score in the last ten Maths tests.

John	7	8	10	4	6	3	8	5	9	8
Aisha	8	9	7	8	6	9	8	7	6	9

Who is better at Maths?

John's mean = $68 \div 10 = 6.8$

Aisha's mean = $77 \div 10 = 7.7$

These figures suggest that Aisha is better at Maths and, since we now have more evidence, the conclusion is likely to be more reliable than before.

Example 2

The tables below give the sale prices of houses in two areas. Which area has the higher house prices?

Area A		Area B	
Price in pounds (£)	Number of houses	Price in pounds (£)	Number of houses
40 000–59 999	5	40 000–59 999	0
60 000–79 999	17	60 000–79 999	16
80 000–99 999	64	80 000–99 999	27
100 000–119 999	11	100 000–119 999	47
120 000–139 999	3	120 000–139 999	10

The modal class of Area A is £80 000–£99 999.

The modal class of Area B is £100 000–£119 999.

This suggests that the houses are more expensive in Area B.

Comparing spread of sets of data

When comparing sets of data, it is generally not sufficient to know that values in one set are, on average, 'bigger' than those in the other. It is also helpful to know whether one set of data is more spread out than the other.

The two measurements of spread that have been covered so far are the **range** and the **interquartile range**.

Look again at John and Aisha's scores in the last ten Maths tests.

John's range is $10 - 3 = 7$ Aisha's range is $9 - 6 = 3$

These figures show that John's spread of scores is greater than Aisha's.

Another way of stating the conclusion is to say that Aisha is more **consistent** than John.

Chapter 27 Comparing data

Example E

The cumulative frequency graphs below are for the house prices in Example 2.

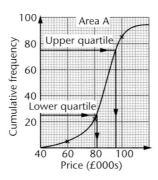

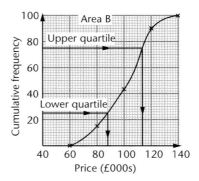

Use the interquartile ranges to compare the spread of the house prices in the two areas.

The interquartile range for area A = £95 000 – £82 000 = £13 000

The interquartile range for area B = £114 000 – £88 000 = £26 000

So the spread of house prices is greater in area B than in Area A.

The median, range and interquartile range of distributions can be compared visually, by means of a box plot.

For the houses in the example above, the the median in Area A is £89 000 and the median in Area B is £104 000. The box plots for the two areas look like this.

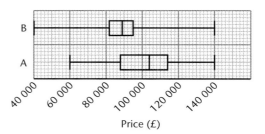

These box plots assume minimum prices of £40 000 and £60 000 and a maximum price of £139 999 for both areas. In fact, there is a strong likelihood that, for example, the maximum price in Area A is below £139 999.

Usually the interquartile range is the more reliable of the two measures of spread, since one or two very low or very high values can greatly distort the range. The interquartile range disregards extreme values.

It is sometimes possible to compare spreads of distributions if there is a marked difference in the shape of the frequency diagrams.

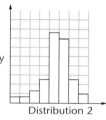

Distribution 1   Distribution 2

Looking at these frequency diagrams, you can see that the data in Distribution 1 are much more spread out than the data in Distribution 2.

Exercise 27.1a

1. In golf the lowest score is the best. Colin's mean score in this season's golf rounds is 71. His lowest score is 63 and his highest is 88. Vijay's mean score is 73. His lowest score is 66 and his highest score is 82. Make two comparisons of the two players' scores.

2. The table below shows the results of an investigation into costs of dental treatment in two towns.

	Median cost	Interquartile range
Town A	£19.25	£4.20
Town B	£16.50	£5.30

Make two comparisons of the cost of dental treatment in the two towns.

3. Here is a set of nine numbers.

   8   6   7   3   12   6   11   5   8

   Write down another set of nine numbers with the same median but a larger range.

4. The table below shows the amounts of rainfall, in millimetres, in twelve months in Moralia and Sivarium.

	J	F	M	A	M	J	J	A	S	O	N	D
Moralia	25	23	21	18	18	16	15	14	18	17	22	27
Sivarium	5	6	8	12	18	78	70	21	7	4	3	2

Find the mean and range for each of the places and state your conclusions.

5. Here are Tara's and Justin's marks in their last five English homeworks.

Tara	14	15	17	13	15
Justin	10	18	11	19	20

Calculate the mean and range of each of the two pupils' scores and state your conclusions.

Why might these conclusions be somewhat unreliable?

6. The numbers of letters delivered to the houses in two roads are shown in this table.

| Number of letters | Number of houses | |
	Jubilee Road	Riverside Road
0	2	0
1	27	5
2	18	16
3	11	29
4	5	18
5	3	5
6	1	4
7	0	3
8	0	2
9	0	1

Find the mode and range of the numbers of letters delivered in the two roads and state your conclusions.

7. The table shows how much pocket money (to the nearest pound) is received by pupils in Class 9a.

Amount of pocket money in pounds (£)	Number of pupils
2	1
3	5
4	10
5	7
6	4

(a) Calculate an estimate for the mean and range of the amounts of pocket money received.

(b) In class 9b the mean amount of pocket money is £3.80 and the range is £8. Compare the amounts of pocket money in the two classes.

8. The cumulative frequency diagrams below show the times of response to 100 alarm calls for two fire brigades.

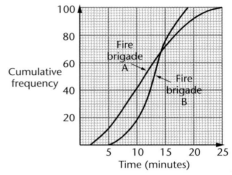

(a) Use the graphs to find the median and interquartile range of the response times for each fire brigade.

(b) Draw box plots to compare the results for each fire brigade.

(c) Comment on your results.

9. Panesh is buying light bulbs. Britelite have a mean life of 300 hours with a range of 200 hours. Lightglo have a mean life of 280 hours and a range of 20 hours. Which would you advise Panesh to buy? Explain why.

Exercise 27.1b

1. The median age of the Ribchester hockey team is 24 years 9 months and the range is 8 years 2 months. The median age of Sillington hockey team is 22 years 5 months and the range is 5 years 4 months. Make two statements to compare the ages of the two teams.

2. The lengths of time in minutes spent on homework by Gareth and Salima on five days in a week are listed below.

	M	Tu	W	Th	F
Gareth	50	60	45	80	70
Salima	20	80	100	30	55

Find the mean and range of the two pupils' times. State your conclusions.

3. The table shows the mean and interquartile range of the price of a 'standard basket of shopping' in two regions of the country.

	Mean	Interquartile range
Region A	£43.52	£3.54
Region B	£46.54	£1.68

Compare the prices in the two regions.

4. The table shows the amounts ~~~~t on Christmas presc~~~~ ~~~~ year 10.

Amount of money in pounds (£)	N~~~~ ~pupils
0.00–4.99	3
5.00–9.99	14
10.00–14.99	
15.00–19.99	3C
20.00–24.99	13
25.00–29.99	4

(a) Draw a cumulative frequency diagram and use it to find the median and interquartile range for the money spent.

(b) The median amount spent by Year 11 pupils was £19, the lower quartile was £17 and the upper quartile was £21.50. The minimum was £5 and the maximum £31. Draw a box plot for each of the years.

(c) Compare the distributions of money spent by Year 10 and Year 11 pupils.

5. The frequency diagrams show the number of children per family, in two classes.

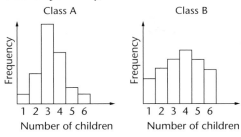

Use the modes and the shapes of the diagrams to compare the distributions.

6. These are the weekly earnings, in pounds, of the employees at a small firm.

96	120	120	125	137	145
157	190	200	220	590	

State, with reasons, which measurement of spread you would use to compare this firm with another, similar, small firm.

7. The lengths of 100 leaves from an ash tree in a park in a city centre are shown in the table. The lengths are measured to the nearest centimetre.

Length of leaf (cm)	Frequency
9	12
10	15
11	33
12	19
13	13
14	8

(a) Calculate an estimate for the mean length of leaf and estimate the range.

(b) Leaves from an ash tree from a country area have a mean length of 12.7 cm and a range of 4.2 cm. Compare the distributions of the leaves from the two different areas.

8. The table shows the means and interquartile ranges of two batsmen's scores in their last 20 innings.

	Mean	Interquartile range
Mike	43.4	6.4
Alec	47.8	15.2

Which batsman would you select? Explain why.

Correlation

The table below shows the amount of ice-cream sold by an ice-cream seller in ten days last summer.

Number of hours of sunshine	3	6	11	2	0	7	2	12	7	5
Number of ice-creams sold	120	200	360	100	50	250	150	470	330	230

The graph below shows this information plotted on a scatter diagram or scatter graph.

The number of hours of sunshine is plotted as the *x*-coordinate and the number of ice-creams sold is plotted as the *y*-coordinate.

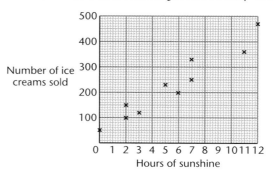

From the graph, it can be seen that, in general, the **more** hours of sunshine there were, the **more** ice-creams were sold.

This is an example of **positive correlation**.

Although the points are not exactly in a straight line, nevertheless there is a trend that the further to the right on the graph the higher the point is.

In graphs such as these the nearer the graph is to a straight line, the better the correlation is.

Examples of graphs showing positive correlation

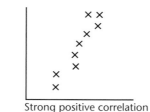

Perfect positive correlation   Strong positive correlation   Weak positive correlation

If there is no correlation the scatter diagram looks like this.

A shopkeeper in the same town as the ice-cream seller noted how many umbrellas were sold in the same ten days. This is the table.

Number of hours of sunshine	3	6	11	2	0	7	2	12	7	5
Number of umbrellas sold	6	5	2	9	11	4	8	0	5	7

The scatter diagram for this information looks like this.

The number of hours of sunshine is plotted as the x-coordinate and the number of umbrellas sold is plotted as the y-coordinate.

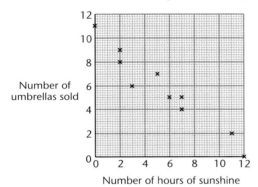

Here the trend is the other way round. In general, although the points are not exactly in a straight line, the **more** hours of sunshine there are the **fewer** umbrellas are sold.

This is an example of **negative correlation**.

Examples of graphs showing negative correlation

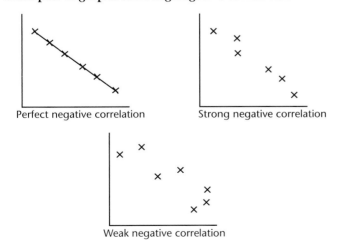

Perfect negative correlation

Strong negative correlation

Weak negative correlation

Examiner's tip

When commenting on a scatter diagram it is better (and quicker) to use the correct terms such as 'strong positive correlation' rather than using phrases like 'the more hours of sunshine, the more ice-creams are sold'.

Chapter 27 Correlation

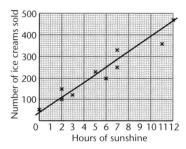

The line of best fit should reflect the slope of the points and have approximately the same number of points on either side.

Examiner's tip

When drawing a line of best fit:
1. put your ruler on the graph at the right slope
2. slide the ruler, keeping it at the same slope, until you have approximately the same number of points on either side.

Remember, lines of best fit do not necessarily go through the origin.

It may be easier to hold your ruler on its edge, vertical to the page, so you can see both sides of the line.

Lines of best fit

Look again at the graph for ice-cream and hours of sunshine.

A straight line has been drawn on it, passing through the cluster of points. There are as many points above the line as there are below it. This is the 'best' straight line that can be drawn to show the trend of the points. It is called the **line of best fit**.

It should ignore any points that obviously do not fit the trend. These are called **outliers**. A line of best fit should **not** be attempted if there is little or no correlation.

The line of best fit can be used to estimate values that are not in the original table. For example, you could estimate that for 5 hours of sunshine 210 ice-creams would be sold.

If the line of best fit is used to estimate values it must be recognised that:

- if the correlation is not good the estimate will probably not be a very good one
- estimates should **not** be made too far beyond the range of the given points. For example, in the above case, estimates should not be made for 15 hours of sunshine.

Examples of bad 'lines of best fit'

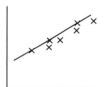 

Fault: Slope about right but does not have the same number of points on either side.

Fault: Same number of points either side but slope wrong.

Exercise 27.2a

1. The scatter diagram below shows the number of sunbeds hired out and the hours of sunshine at Brightsea.

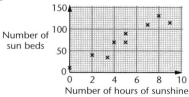

Comment on the results shown by the scatter graph.

2. A firm noted the number of days 'sick-leave' taken by its employees in a year, and their ages. The results are shown in the graph.

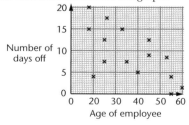

Comment on the results shown by the scatter graph.

3. The table below shows the Maths and Science marks of eight pupils in their last examination.

Pupil	A	B	C	D	E	F	G	H
Maths mark	10	20	96	50	80	70	26	58
Science mark	30	28	80	55	62	70	38	48

(a) Draw a scatter graph to show this information, with the Maths score on the x-axis.
(b) Comment on the graph.
(c) Draw a line of best fit.
(d) Use your line of best fit to estimate:
   (i) the mark in Science of a pupil who scored 40 in Maths
   (ii) the mark in Maths of a pupil who scored 75 in Science.

4. The table below shows the amount of petrol left in the fuel tank after the number of miles travelled.

Number of miles	50	100	150	200	250	300
Number of gallons	7	5.2	4.2	2.6	1.2	0.4

(a) Draw a scatter graph to show this information, with the number of miles on the x-axis.
(b) Comment on the graph.
(c) Draw a line of best fit.
(d) Use your line of best fit to estimate the number of gallons left after 170 miles.

5. In Kim's game 20 objects are placed on a table and you are given a certain time to look at them. They are removed or covered up and you then have to recall as many as possible. The table shows the amount of time given to nine people and the number of items they remembered.

Time in seconds	20	25	30	35	40	45	50	55	60
Number of items	9	8	12	10	12	15	13	16	18

(a) Draw a scatter graph to show this information, with the amount of time on the x-axis.
(b) Comment on the graph.
(c) Draw a line of best fit.
(d) Use your line of best fit to estimate the number of items remembered if 32 seconds are allowed.
(e) Why should the graph not be used to estimate the number of items remembered in three seconds?

6. Sanjay thinks that the more time he spends on his school work, the less money he will spend. Sketch a scatter graph that shows this.

Exercise 27.2b

1. A teacher thinks that there is a correlation between how far back in class a pupil sits and how well they do at Maths. To test this she plotted their last Maths grade against the row they sit in. Here is the graph she drew.

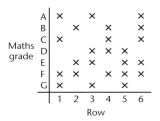

Was the teacher right? Give your reasons.

2. The scatter graph below shows the positions of football teams in the league and their mean crowd numbers, in thousands.

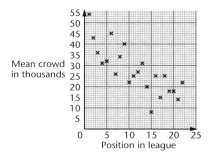

Comment on the graph.

3. The scatter graph shows the ages of people and the numbers of lessons they took before they passed their driving tests.

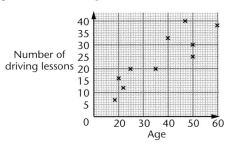

   Comment on the graph.

4. In Jane's class a number of pupils have part-time jobs. Jane thinks that the more time they spend on their jobs, the worse they will do at school. She asked ten of them how many hours a week they spend on their jobs, and found their mean marks in the last examinations. Her results are shown in the table.

Pupil	A	B	C	D	E	F	G	H	I	J
Time on part-time job (hours)	9	19	13	3	15	20	5	17	6	22
Mean mark in examination	50	92	52	70	26	10	80	36	74	24

(a) Plot a scatter graph to show Jane's results, with time in hours on the *x*-axis.

(b) Do the results confirm Jane's views? Are there any exceptions?

(c) Draw a line of best fit for the relevant points.

(d) Estimate the mean score of a pupil who spent 12 hours on their part-time job.

5. In an experiment, Tom's reaction times are tested after he has undergone vigorous exercise. The table shows Tom's reaction times and the amounts of time spent in exercise.

Amount of exercise (minutes)	0	10	20	30	40	50
Reaction time	0.34	0.46	0.52	0.67	0.82	0.91

(a) Draw a scatter graph to show this information, with the number of minutes of exercise on the *x*-axis.

(b) Comment on the graph.

(c) Draw a line of best fit.

(d) Use your line of best fit to estimate Tom's reaction time after 35 minutes' exercise.

6. Fiona thinks that the more she practises, the more goals she will score at hockey. Sketch a scatter graph to show this.

Time series

The table below shows the value, in thousands of pounds, of an ice-cream company's quarterly sales for 1995 to 1998.

	First quarter	Second quarter	Third quarter	Fourth quarter
1995	145	256	328	258
1996	189	244	365	262
1997	190	266	359	250
1998	201	259	401	265

The graph illustrates these figures.

Note that the points have been joined, in order, by dotted lines. In general, points on graphs should only be joined up when it makes sense to read information between the points. Here, since the figures are total sales it does not really make sense to do so. However it is often useful to join the points with dotted lines to show the **trend**.

The figures and the graph are an example of a **time series**. You can see there are **peaks** at each third quarter and **troughs** at each first quarter. With a repeating pattern or **cyclical** effect like this it is sometimes difficult to see trends.

Other examples of figures that may be cyclical are monthly or seasonal rainfall, or monthly or seasonal unemployment figures in certain areas.

Moving averages

Moving averages can help you see trends in figures that are cyclical. They are calculated as follows.

Look at the figures for the first four quarters above.

The mean = $(145 + 256 + 328 + 258) \div 4 = 246.75$

Then find the mean for the second group of consecutive quarters.

$(256 + 328 + 258 + 189) \div 4 = 257.75$

Notice that 1995's first quarter is omitted and 1996's first quarter is included.

Now find the next mean by omitting the 256 and including the next quarter, 244, that is:

$(328 + 258 + 189 + 244) \div 4 = 254.75$

The next mean is $(258 + 189 + 244 + 365) \div 4 = 264$

Continue until the last quarter is included, each time omitting the first figure and picking up the next one in the table.

If all the quarters' figures are put in order and numbered as below, the lines underneath move along one each time and indicate the numbers that should be used.

1	2	3	4	5	6	7	8	9	10	11	12	13	14	15	16
145	256	328	258	189	244	365	262	190	266	359	250	201	259	401	265

and so on.

Check that you agree with this complete list.

Quarters	1–4	2–5	3–6	4–7	5–8	6–9	7–10
Moving average	246.75	257.75	254.75	264	265	265.25	270.75

Quarters	8–11	9–12	10–13	11–14	12–15	13–16
Moving average	269.25	266.25	269	267.25	277.75	281.5

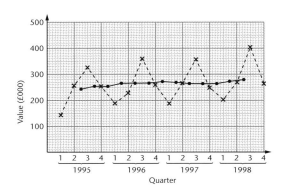

These points are now plotted on the graph, at the middle of each interval of points, for example at 2.5 for quarters 1 to 4, at 3.5 for the next four quarters and so on.

You can see that plotting the moving averages flattens out the peaks and troughs and gives a fairly flat graph, with possibly a slight overall increase.

Using four figures to find a moving average, as in the case in quarterly figures, gives a **four-quarter moving average**. If the figures varied monthly, 12-month moving averages may be found using the means of 12 consecutive months.

Exercise 27.3a

1. The table below shows the gross Accident and Health Insurance premiums (in millions of euros) paid in the Netherlands for the four quarters of 1997 to 1999.

	1st quarter	2nd quarter	3rd quarter	4th quarter
1997	43	17	15	15
1998	47	19	18	18
1999	57	26	22	13

(a) Plot these figures in a graph. Use a scale of 1 cm to each quarter on the horizontal axis and 2 cm to 10 million euros on the vertical axis.
(b) Calculate the four-quarter moving averages.
(c) Plot the moving averages on your graph.
(d) Comment on the general trend and the quarterly variation.

2. The table below shows the total sales (in megawatts) of Danish wind turbines in the years 1995 to 1998.

	1st quarter	2nd quarter	3rd quarter	4th quarter
1995	96.6	125.8	122.9	229.1
1996	74.1	143.1	173.0	335.9
1997	216.2	234.2	234.5	282.6
1998	168.8	239.7	282.1	525.4

(a) Plot these figures in a graph. Use a scale of 1 cm to each quarter on the horizontal axis and 2 cm to 100 megawatts on the vertical axis.
(b) Calculate the four-quarter moving averages.
(c) Plot the moving averages on your graph.
(d) Comment on the general trend and the quarterly variation.

3. The table below shows a company's quarterly sales (in £000s) of raincoats in the years 1996–99.

	1st quarter	2nd quarter	3rd quarter	4th quarter
1996	154	121	63	134
1997	132	106	72	108
1998	115	111	58	97
1999	110	93	47	82

(a) Plot these figures in a graph. Use a scale of 1 cm to each quarter on the horizontal axis and 2 cm to £20 000 on the vertical axis.

(b) Calculate the four-quarter moving averages.

(c) Plot the moving averages on your graph.

(d) Comment on the general trend and the quarterly variation.

4. The table below shows the daily audiences for a four-week Christmas pantomime season.

	Mon	Tues	Wed	Thurs	Fri	Sat
Week 1	256	312	324	452	600	580
Week 2	297	367	382	538	600	600
Week 3	248	327	325	495	570	583
Week 4	192	219	287	306	490	572

(a) Plot these figures in a graph. Use a scale of 1 cm to each day on the horizontal axis and 2 cm to 100 people on the vertical axis. Turn your graph paper to make the graph fit well.

(b) Calculate the six-day moving averages.

(c) Plot the moving averages on your graph.

(d) Comment on the general trend and the daily variation.

Exercise 27.3b

1. The table below shows a household's quarterly expenditure (in pounds) on fuel and light in the years 1995–98.

	1st quarter	2nd quarter	3rd quarter	4th quarter
1995	380	272	264	371
1996	432	285	207	272
1997	298	192	158	285
1998	310	208	182	291

(a) Plot these figures in a graph. Use a scale of 1 cm to each quarter on the horizontal axis and 2 cm to £100 on the vertical axis.

(b) Calculate the four-quarter moving averages.

(c) Plot the moving averages on your graph.

(d) Comment on the general trend and the quarterly variation.

(e) During this period, major insulation work was carried out on the house. When do you think that was?

2. The table below shows the number of bankruptcies in Auckland by quarters from 1995 to 1998.

	1st quarter	2nd quarter	3rd quarter	4th quarter
1995	60	61	72	57
1996	83	75	90	66
1997	62	96	99	79
1998	72	63	79	65

(a) Plot these figures in a graph. Use a scale of 1 cm to each quarter on the horizontal axis and 2 cm to 20 bankruptcies on the vertical axis.

(b) Calculate the four-quarter moving averages.

(c) Plot the moving averages on your graph.

(d) Comment on the general trend and the quarterly variation.

3. The table below shows the daily sales (in £000) of a shop over a three-week period.

	Mon	Tues	Wed	Thurs	Fri	Sat	Sun
Week 1	7.3	8.8	9.2	10.3	15.5	16.2	12.8
Week 2	6.7	7.8	10.1	11.8	14.7	17.9	11.3
Week 3	7.1	6.3	8.2	10.9	12.9	16.6	11.6

(a) Plot these figures in a graph. Use a scale of 1 cm to each day on the horizontal axis and 2 cm to £2000 on the vertical axis. Turn your graph paper to make the graph fit well.

(b) Calculate the seven-day moving averages.

(c) Plot the moving averages on your graph.

(d) Comment on the general trend and the quarterly variation.

4. The table below shows the monthly number (in hundred thousands) of US Citizens flying to Europe from 1997 to 1998.

	Jan	Feb	Mar	Apr	May	Jun
1997	5.8	5.4	7.6	7.5	10.3	11.8
1998	6.3	5.9	8.9	8.5	11.0	12.8
1999	6.4	6.2	10.3	9.3	11.5	13.2

	Jul	Aug	Sep	Oct	Nov	Dec
1997	10.9	9.9	10.2	8.0	6.8	7.0
1998	12.0	10.3	10.8	8.8	7.1	7.4
1999	12.5	11.0	11.1	9.4	8.2	7.5

(a) Plot these figures in a graph. Use a scale of 1 cm to two months on the horizontal axis and 1 cm to 100 000 citizens on the vertical axis.

(b) Calculate the 12-month moving averages.

(c) Plot the moving averages on your graph.

(d) Comment on the general trend and the monthly variation.

Key points

- The mean, median and mode can be used to compare distributions.
- The range and interquartile range can be used to compare the spread of distributions.
- The interquartile range disregards extreme values.
- Scatter graphs show the correlation between two variables.
- If there is reasonable correlation, a line of best fit can be drawn.
- The line of best fit should reflect the slope of the points and should have approximately the same number of points on either side.
- The line of best fit can be used to estimate values of one variable if the other is known.
- The line of best fit can only be used to estimate values within the range of the given data.
- A time series shows the variation of sets of figures over periods of time. These periods can be quarterly, daily, monthly, ... These are usually displayed on a graph.
- To calculate a moving average, for example for quarterly figures, first calculate the mean for the first four quarters. Then omit the first quarter and include the fifth quarter and find the new mean. Then omit the second quarter and include the sixth, and so on.
- The moving averages are plotted at the middle of the interval.

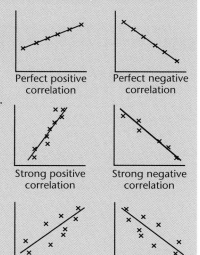

Perfect positive correlation | Perfect negative correlation

Strong positive correlation | Strong negative correlation

Weak positive correlation | Weak negative correlation

No correlation

Revision exercise 27a

1. Over the last month, David's mean journey time to work has been 43 minutes, with an interquartile range of 7 minutes. Angie's mean time is 32 minutes with an interquartile range of 12 minutes. Make two comparisons of David and Angie's journey times.

2. Eleven members of 10g and eleven members of 10f are given a Maths problem to solve. The times they took to solve the problem are shown in the table.

10f	17	15	11	9	6	27	18	21	6	19	8
10g	4	13	15	11	32	7	9	12	6	10	14

Find the median and range for each class and comment on the results. Why might the inter-quartile range be a better measurement to use?

3. The table shows the prices of a sample of 100 houses in the north-west of England.

Price (£000)	Number of houses
$20 < x \leqslant 40$	4
$40 < x \leqslant 60$	15
$60 < x \leqslant 80$	27
$80 < x \leqslant 100$	41
$100 < x \leqslant 120$	10
$120 < x \leqslant 140$	3

Use mid-interval values of £30 000, £50 000, £70 000, £90 000, £110 000 and £130 000 to estimate the mean house price in the sample. A similar sample in the south-east gave a mean of £107 000 and a range of £150 000. Compare the two areas.

4. A survey on 50 adults in each of England and France studied the amount of wine consumed in a year. The table shows the mean and interquartile range of the number of bottles consumed in each country. Compare the two countries.

	Mean	Interquartile range
England	21	9
France	46	8

5. These cumulative frequency diagrams show the marks obtained in examinations in French and English by 200 pupils in year 8.

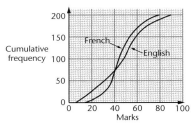

(a) Draw box plots for each of the languages.
(b) Use the median and interquartile range for each subject to compare the results.

6. The scatter graphs show heights and weights of ten boys and ten girls. Compare the two graphs, noting any differences and any similarities.

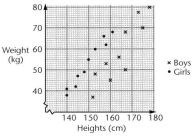

7. Brian thinks that the more he practises, the lower his golf score will be. State the type of correlation of which this is an example.

8. The table shows marks given, out of 30, by two judges for eight cats for quality of their coats.

Cat	A	B	C	D	E	F	G	H
Judge 1	17	23	15	28	22	18	27	14
Judge 2	7	23	9	27	13	15	25	4

(a) Draw a scatter diagram to show the judges' scores, with Judge 1 on the *x*-axis.

(b) Comment on the relationship between the two judges' scores.
(c) Draw a line of best fit.
(d) Judge 2 gave a ninth cat 18 marks. Estimate the marks that Judge 1 would give the same cat.

9. Market research predicts that the possible prices for replica shirts would lead to sales as in the table below.

Price (£)	20	25	30	35	40	45	50
Number of shirts	7600	7400	6800	5600	5400	4500	3600

(a) Draw a scatter graph for this information.
(b) Comment on the relationship between price and predicted sales.
(c) Draw a line of best fit.
(d) Estimate the sales of shirts if the price were fixed at £33.
(e) Why would it be wrong to predict the sales if the price were fixed at £65?

10. A survey is carried out at the checkout of a supermarket. It investigates the total cost of bills and the number of items bought. Sketch what you think the scatter diagram would look like, with number of items on the *x*-axis and total cost of the bill on the *y*-axis.

11. The table shows the number of unemployed people at the end of each quarter in Devon, to the nearest 100. The months indicate the end of the quarter for which the figures are given.

	January	April	July	October
1996	41 700	38 300	35 600	33 100
1997	33 800	28 500	24 600	23 500
1998	26 600	24 000	22 200	21 100
1999	23 800	20 900	18 900	17 700

(a) Plot these figures in a graph. Use a scale of 1 cm to each quarter on the horizontal axis and 2 cm to 10 000 people on the vertical axis.
(b) Calculate the four-quarter moving averages.
(c) Plot the moving averages on your graph.
(d) Comment on the general trend and the quarterly variation.

Identifying a locus

The locus of a point is the path or the region that the point covers as it moves according to a particular rule.

The plural of locus is loci.

The locus of a point 3 cm from A is a circle, centre A, radius 3 cm.

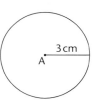

The locus of a point less than ($<$) 3 cm from A is the region inside a circle centre A, radius 3 cm.

The locus of a point greater than ($>$) 3 cm from A is the region outside a circle centre A, radius 3 cm.

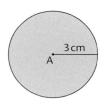

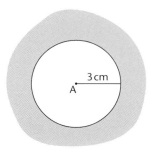

You need to know some other basic loci.

The locus of a point 2 cm from a straight line is a pair of lines parallel to that line, 2 cm away from it on either side.

	locus
2 cm $\updownarrow$	given line
2 cm $\updownarrow$	locus

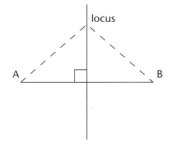

The locus of a point that stays an equal distance from two points is the perpendicular bisector of the line joining the two points.

The locus of a point that stays an equal distance from two intersecting lines is the pair of lines that bisect the angles between the lines. Can you see why this is so?

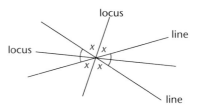

Drawing the perpendiculars to the lines from a point on the locus creates two congruent triangles.

Sketching loci

When you sketch a locus, draw it as accurately as you can but not to size or with accurately constructed bisectors. You must mark all the distances and angles that are equal.

Example 1

A line is 6 cm long. Sketch the locus of all points that are 2 cm from the line.

The locus is two parallel lines with a semicircle joining them at each end.

Example 2

Two towns A and B are 6 miles apart. Make a sketch and shade the region that is nearer to B than A.

Draw the perpendicular bisector of the line AB and shade the region on B's side of the line. The shading could go past B and up or down the page further.

The two parts of the line AB need to be shown as 3 miles each, and the 90° angle must be indicated.

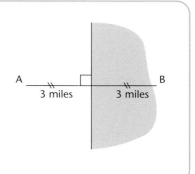

Constructing loci

You need to know two constructions.

1. The perpendicular bisector of a line

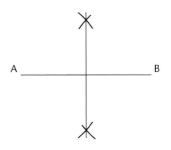

> This is the locus of a point that moves so that it is equidistant from two points.

Draw the line AB.
Open the compasses to a radius that is more than half the length of AB.
Put the compass point at A and draw an arc above and below the line.
Keep the compasses set to the same radius.
Put the compass point at B and draw an arc above and below the line.
Join the two points where the arcs meet.

2. The bisector of an angle

> This is the locus of a point that moves so that it is equidistant from two lines.

Draw an angle and mark the vertex (corner) A.
Put the point of the compasses at A and draw an arc to cut the lines forming the angle at B and C.
Put the point at B and draw an arc in the angle.
Keep the compasses set to the same radius.
Put the point at C and draw an arc in the angle to cut the arc just drawn.
Draw a straight line through A and the point where the arcs cut.
The bisector could be continued to the left of A. If the lines are extended, another bisector could be drawn, perpendicular to the first one.

Use these constructions, and what you already know, to draw various loci.

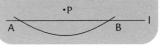

Examiner's tip

A similar method can be used to draw the perpendicular from a point P to a line l. Start with an arc, centre P. Use the points A and B as before.

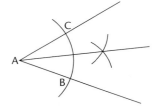

Examiner's tip

These methods are more accurate that just using measurement.

Example 3 | Draw a triangle ABC with sides AB = 5 cm, AC = 4 cm and A = 50°.

Use compasses to bisect angle A. Shade in the locus of the points inside the triangle that are nearer to AB than AC.

This diagram is half-size.

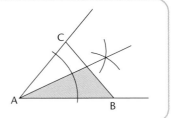

Exercise 28.1a

1. Draw a circle, centre A, radius 5 cm. Shade the locus of the points that are less than 5 cm from A.
2. Draw a rectangle 4 cm by 5 cm. Sketch the locus of the points that are 1 cm from the perimeter of the rectangle, outside the rectangle.
3. Draw a rectangle ABCD with AB = 6 cm and BC = 4 cm. Sketch the locus of the points inside the rectangle that are nearer to A than B.
4. Draw two parallel lines across the page, 4 cm apart. Draw the locus of the points that are 1 cm from the top line and 3 cm from the bottom line.
5. A fox never travels more than 5 miles from its den or earth. Draw a sketch to show the region where it travels.
6. Draw a line 7 cm long. Construct the perpendicular bisector of the line.
7. Draw an angle of 70°. Construct the bisector of the angle.
8. Construct a triangle ABC with AB = 8 cm, AC = 7 cm and BC = 5 cm. Use compasses and a ruler to bisect angle A. Shade the locus of the points inside the triangle that are nearer to AB than AC.
9. Draw a square ABCD with side 6 cm. Construct the locus of the points that are equidistant from A and C. What do you notice about the locus?
10. Draw triangle ABC with AB = 8 cm, A = 90° and B = 40°. Do a construction to find the locus of the points inside the triangle that are nearer to AC than BC.

Exercise 28.1b

1. Show, by shading, the locus of the points that are more than 4 cm from a fixed point A.
2. Draw a line 6 cm long. Show, by shading in a sketch, the locus of the points that are less than 2 cm from the line.
3. Draw an angle of 80°. Construct the bisector of the angle.
4. Draw a line AB 6 cm long. Construct the perpendicular bisector of AB.
5. Draw a square with side 4 cm. Label one corner A. Show the locus of the points inside the square that are less than 3 cm from A.
6. Draw a rectangle ABCD with sides AB = 7 cm and BC = 5 cm. Use compasses to construct the line equidistant from AB and AC.
7. Construct the triangle ABC with A = 30°, B = 50° and AB = 10 cm. Construct the locus of the points equidistant from A and B.
8. Two towns Bimouth and Tritown are 10 miles apart. Phoebe wants to live nearer to Bimouth than Tritown. Using a scale of 1 cm : 2 miles, make a scale drawing and show, by shading, the region where she can live.
9. Draw a triangle ABC with AB = 7 cm, A = 50° and B = 40°. Show, by shading, the locus of the points within the triangle that are nearer to AC than BC.
10. Sonia has a 20 metre flex on her lawnmower and the socket is in the middle of the back wall of her house. The back of the house is 12 m wide and her garden is a rectangle the same width as the house, stretching 24 m from the house. Using a scale of 1 cm : 4 m, make a drawing of her garden and show, by shading, the region she can reach with the mower.

Problems involving intersection of loci

Combining all you know about loci, you can answer more complicated questions involving more than one locus.

Example 4

Construct triangle ABC with AB = 7 cm, AC = 6 cm and BC = 4 cm. By using constructions, find the point that is equidistant from all three vertices. Mark this point D.

This diagram is not to scale

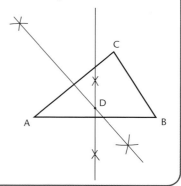

First you need the line equidistant from two vertices. If you choose AB you need to construct the perpendicular bisector of AB. Then you need to construct the perpendicular bisector of another side. Where they cross is the required point.

You could also bisect the third side and that line would also pass through the same point.

Example 5

Two points A and B are 4 cm apart. Show, by shading, the locus of the points that are less than 2.5 cm from A, and nearer to B than A.

You need to draw a circle, radius 2.5 cm and centre A. You also need to draw the bisector of the line AB. The region you require is inside the circle and on the B side of the bisector. Here the region required is shaded.

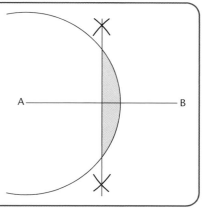

Example 6

Erica wants to put a rocking chair in her room. She wants the chair more than 0.5 m from a wall and less than 2 m from corner A. This is a sketch of her room. Using a scale of 1 cm : 1 m, make a scale drawing of the room and show, by shading, the region where the chair can be placed.

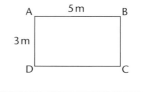

Draw the rectangle and then add lines 0.5 cm from each side. Draw a circle, centre A radius 2 cm. In this diagram the regions not required are shaded, leaving the blank region where the chair can be placed.

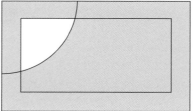

Example 7

Find the centre of the rotation that maps triangle ABC onto triangle A'B'C'.

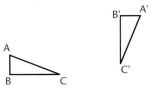

The centre of rotation must be equidistant from A and A'. It will be on the perpendicular bisector of AA'. Arcs have been omitted to make the diagram clearer.

The centre must also be equidistant from C and C'. The centre of rotation will be the point where the two perpendicular bisectors meet.
The centre must also be equidistant from B and B'. Construct the perpendicular bisector of BB' to check.

centre of rotation

Examiner's tip

You can either shade the region required or shade the regions not required. It is often easier to do the latter if the regions are at all complicated.

Exercise 28.2a

Draw all of these accurately.

1. Two points A and B are 5 cm apart. Show, by shading, the region that is less than 3 cm from A and more than 4 cm from B.
2. A rectangle ABCD has sides AB = 5 cm and BC = 4 cm. Draw the rectangle and show, by shading, the region inside the rectangle that is nearer to AB than CD, and less than 3.5 cm from B.
3. Draw a triangle ABC with AB = 6 cm, A = 60° and B = 55°. Use constructions to find the point D that is equidistant from all three sides.
4. Draw a rectangle ABCD with sides AB = 4 cm and BC = 3 cm. Show the points that are equidistant from AB and BC and 3.5 cm from A.

5. Draw a triangle ABC with sides AB = 9 cm, BC = 6 cm and AC = 5 cm. Show, by shading, the region inside the triangle that is nearer to AB than BC and more than 3 cm from C.

6. Two towns, Hilldon and Baton are 20 miles apart. It is proposed to build a new shopping centre within 15 miles of Hilldon but nearer to Baton than Hilldon. Using a scale of 1 cm : 5 miles, make a drawing and show the region where the shopping centre can be built.

7. Richard's bedroom is rectangular with sides 4 m and 6.5 m. He wants to put a desk within 1 metre of a longer wall and within 2.5 m of the centre of the window in the middle of one of the shorter walls. Using a scale of 1 cm : 1 m, make a scale drawing and show, by shading, the region where the desk can be placed.

8. Kirsty has a triangular patio with sides 6 m, 4 m and 5 m. She wants to put a plant pot on the patio more than 2 m from any corner. Using a scale of 1 cm : 1 m, make a drawing and show, by shading, where she can put the plant pot.

9. This is a sketch of a plot of land that Arun wants to use for camping.

He wants to put a tap in the field within 35 m of the gate, G, which is at the middle of one of the shorter sides. He also wants to it to be within 25 m of his farm which is at corner F. Using a scale of 1 cm : 10 m, make a scale drawing of the land. Show, by shading, the position where the tap can be placed.

10. A field is in the shape of a quadrilateral ABCD with AB = 25 m, BC = 30 m, A = 90°, B = 106° and C = 65°. The farmer wants to put the scarecrow within 15 m of corner A and nearer to CD than CB. Using a scale of 1 cm to 5 m, draw the field and show, by shading, the region where the scarecrow can be placed.

Exercise 28.2b

1. Show, by shading, the locus of the points that are more than 2 cm from a point A and less than 3 cm from point A.

2. Two points A and B are 4 cm apart. Show, by shading, the locus of all the points that are less than 2.5 cm from A and more than 3 cm from B.

3. Draw a triangle ABC with AB = 6 cm, A = 40° and B = 35°. Use constructions to find the point D that is equidistant from A and B and 4 cm from C.

4. Draw a square with side 4 cm. Show, by shading, the region within the square that is more than 2 cm from ever vertex.

5. Draw a triangle ABC with AB = 6 cm, AC = 5 cm and A = 55°. Bisect the angle A. Draw the perpendicular bisector of AB. Show, by shading, the region that is inside the triangle, nearer to AB than AC and nearer to B than A.

6. Dave and Clare live 7 miles apart. They set out on bikes to meet. They ride directly towards each other. When they meet, Dave has ridden less than 5 miles and Clare less than 4 miles. Using a scale of 1 cm : 1 mile, make a scale drawing showing where they could have met.

7. Tariq's garden is a rectangle ABCD with AB = 10 m and BC = 4 m. He wants to put a rotary washing line in the garden. It must be more than 4 m from corner C and more than 1 m from side AB. Using a scale of 1 cm : 1 m, make a scale drawing of the garden and show where he can put the rotary washing line.

8. The distances between three towns Arbridge, Beaton and Ceborough are AB = 25 miles, AC = 40 miles and BC = 30 miles. A new garage is to be built as near as possible to all three towns. Use a scale of 1 cm : 5 miles and make constructions to find the point D where the garage should be placed.

9. Sasha has a rectangular garage 2 m by 5 m. It has a door at one end. She wants to put a hook in the ceiling. It must be midway between the two longer sides, less than 3.5 m from the door end and less than 2.5 m from the other end. Make a scale drawing of the ceiling using a scale of 1 cm : 1 m. Show by shading the region where the hook can be fixed.

10. This is a sketch of the playing field in Towbridge.

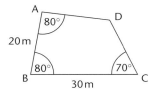

A new swing is to be placed in the field. It must be within 15 m of A and nearer to AB than AD. Use a scale of 1 cm : 5 m to make a drawing and show the region where the swing can be placed.

Key points

- A locus is the path or region where a point can move according to a rule.
- The locus of a point x cm from point A is a circle centre A, radius x cm.
- The locus of a point equidistant from two points A and B is the perpendicular bisector of the line AB.

- The locus of a point equidistant from two parallel lines is a line parallel to the two lines and midway between the lines.
- The locus of a point equidistant from two non-parallel lines is the bisector(s) of the angle(s) between the lines.

Revision exercise 28a

1. Draw an angle of 65° and construct the bisector of the angle.
2. Draw a line AB, 5 cm long. Construct the perpendicular bisector of AB.
3. Draw a triangle ABC with AB = 6 cm, A = 40° and B = 60°. Find the point D that is equidistant from A and B and also equidistant from AB and AC. Show your construction lines.

4. Two points A and B are 6 cm apart. Show, by shading, the locus of the points that are less than 5 cm from A and more than 5 cm from B.
5. Draw a rectangle ABCD with sides AB = 4 cm and BC = 3 cm. Show the locus of the points outside the rectangle that are within 2 cm of the sides of the rectangle.
6. Draw a triangle ABC with AB = 8 cm, A = 47° and AC = 5 cm. Show the locus of the points inside the triangle that are nearer to AB than BC.

7. This diagram shows the position of three schools.

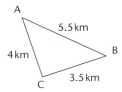

It is decided to build a swimming pool for the three schools. It must not be more than 3.5 km from any of the schools. Using a scale of 2 cm : 1 km, make a scale drawing and show the region where the pool can be located.

8. Carterknowle Church hall is rectangular with sides AB = 12 m and BC = 5 m. The main door is at corner C. A spotlight is to be fixed on the ceiling, more than 6 m from the main door, more than 5 m from the opposite corner and nearer to AB than AD. Using a scale of 1 cm : 1 m, make a scale drawing of the hall and show the region where the light can be fitted.

9. This is a plan of the floor area of a shop.

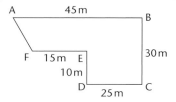

All the corners are 90° except A and F. A heat detector is placed at A and another at D. They both have a range of 20 m and do not work round corners. Using a scale of 1 cm : 5 m, make a scale drawing of the plan and show, by shading, the region that is not covered by heat detectors.

10. This is a sketch of Sanjay's patio.

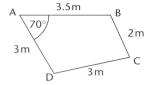

He wants to place a plant pot on the patio, within 1 metre of AB, nearer AB than AD, and no more than 2 metres from A. Using a scale of 2 cm : 1 m, make a scale drawing of the patio and show the region where the pot can be placed.

Answers

Chapter 1

1. (a) 0.7   70%
    (b) 0.4   40%
    (c) 0.75   75%
    (d) 0.333   33.3%
    (e) 0.667   66.7%
2. (a) 37%   (d) 34.5%
    (b) 83%   (e) 125%
    (c) 8%
3. (a) 0.01   (c) 0.04
    (b) 0.17   (d) 1.6
4. 23%
5. 80%
6. 30%

1. (a) 16%   (f) 8%
    (b) 24%   (g) 90%
    (c) 25%   (h) 20%
    (d) 30%   (i) 5%
    (e) 73%   (j) 130%
2. 20%
3. 24%
4. 15%
5. 28%

1. 30%
2. 20%
3. 16%
4. 60%
5. £252
6. 68 kg
7. £3.24
8. £10.40

1. (a) 9 litres   (b) 4 litres
2. (a) 10 bags   (b) 4 bags
3. (a) 300 g   (b) 160 g
4. (a) 12 kg   (b) 10 kg
5. (a) 200 ml   (b) 100 ml
6. (a) 500 ml   (b) 40 ml

1. £8   £12
2. 20 ml
3. 4 litres of red, 12 litres of white
4. 25 litres black, 10 litres white
5. 30 kg
6   0.5 litres
7. £8, £4, £6

1. (a) 0.313   (c) 0.175
    (b) 0.571   (d) 0.267
2. (a) 31.3%   (c) 17.5%
    (b) 57.1%   (d) 26.7%
3. 7%
4. 13.3%
5. 12.5%
6.   25%
7. 6%
8. £55.25
9. £540, £777.60
10. 300 g
11. £180, £240

Chapter 2

1. (a) (i)   (ii) 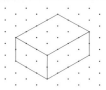 (iii)   (b) 7

2. (a) 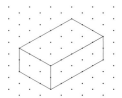 (b) 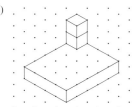 (c)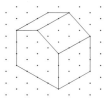

1. (a)–(d) Check students' drawings
2. (a) 3.0 cm, 94°, 56°   (c) 6.7 cm, 27°, 63°
    (b) 6.7cm, 72°, 43°   (d) 9.1 cm, 30°, 20°
3. (a) 5.0 cm   (c) 2.7 cm
    (b) 4.8 cm   (d) 2.0 cm

1. (a)–(d) Check students' drawings
2. (a) 3.0 cm, 5.7 cm   (c) 5.6 cm, 4.6 cm
    (b) 5.2 cm, 2.7 cm   (d) 1.3 cm, 3.5 cm
3. (a)–(c) Check students' drawings
4. (a) 117°, 26°, 36° each to the nearest degree
    (b) 49°, 65°, 65° each to the nearest degree
    (c) 81°, 36°, 63°
5. (a) and (b) Check students' drawings

1.

2.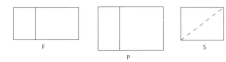

3.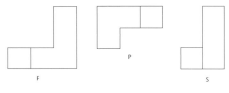

4.

5.

6.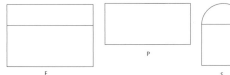

7.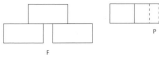

8.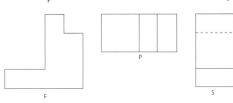

9.

10.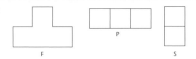

Revision exercise 2a (page 29)

1. (a) BC = 7.5 cm
   (b) angle SRT = 51°
   (c) PR = 3.5 cm
   (d) XY = 6.4 cm, angle XYZ = 68°
   (e) AC = 7.2 cm
   (f) LM = 7.2 or 4.4 cm, angle MLN = 71°
       or 109°
2. (a)–(d) Check students' drawings
3. 68°, 68°, 44°
4. (a) Check students' drawings
   (b) Check students' drawings
       (i) Rhombus
       (ii) Use set square and ruler
   (c) Check students' drawing
       (square-based pyramid)
   (d) Check students' drawings
5. (a) Check students' drawings
   (b) 4.3 cm

6. (a)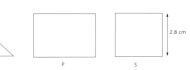
   (b)

7.

8.

Chapter 3

Exercise 3.1a (page 32)

1. $C = x \times y$ or xy
2. $A = m \times n$ or mn
3. $h = n \times t$ or nt
4. $F = 32 + 1.8 \times C$ or $32 + 1.8C$
5. $B = s + n \times u$ or $s + nu$
6. $R = m \div p$ or $R = \dfrac{m}{p}$
7. $T = 30 + 40 \times k$ or $30 + 40k$
8. $A = \frac{1}{2} \times b \times h$ or $\frac{1}{2}bh$
9. $d = 1.65 \times p$ or $1.65p$
10. $i = e \div r$ or $\dfrac{e}{r}$

Exercise 3.2a (page 32)

1. $C = 180$
2. $A = 42$
3. $h = 340$
4. $F = 104$
5. $B = 25.01$
6. $R = 6.015$
7. $T = 390$
8. $A = 15$
9. $d = 330$
10. $i = 5.44$

Answers

Exercise 3.3a (page 33)

1. $5x$
2. $3y + 2z$
3. a^3
4. $a^2 + b^2$
5. $4p + 3q$
6. $3a + 3b$
7. $4m + 3n$
8. $2x + y$
9. $3p - 3q$
10. $a^2 - a - 6$
11. $2a + a^2$
12. $5b$
13. $6b^2$
14. $6ab$
15. $5ab$
16. $3b - 3a$
17. p^3q^2
18. $2a^3 b^3$
19. $4xy^2 + 2xy$
20. $6a^5$

Exercise 3.4a (page 34)

1. Add 1, 5
2. Add 2, 10
3. Add 2, 49
4. Add 10, 50
5. Double, 32
6. Repeat, 5
7. Multiply by 3, 243
8. Subtract 2, 19
9. Subtract 3, -2
10. Divide by 2, $4\frac{1}{2}$

Exercise 3.5a (page 35)

1. 2, 3, 4, 5
2. 2, 4, 6, 8
3. 1, 3, 5, 7
4. 6, 7, 8, 9
5. 3, 6, 9, 12
6. 4, 7, 10, 13
7. 2, 7, 12, 17
8. 10, 20, 30, 40
9. 0, 7, 14, 21
10. 1, 0, -1, -2

Exercise 3.6a (page 36)

1. n
2. $2n + 2$
3. $4n$
4. $2n - 2$
5. $4n + 3$
6. $6n - 5$
7. $10n + 1$
8. $3n + 2$
9. $100n + 1$
10. $27 - 2n$

Exercise 3.7a (page 37)

1. $2a + 2b$
2. $3x + 6$
3. $8x + 4$
4. $a^2 + 2a$
5. $y^2 - y$
6. $2 - 2x$
7. $5p - 5q$
8. $9x - 3$
9. $6x + 4$
10. $4x - 6$
11. $x^2 + xy$
12. $3x^2 + 3x$
13. $3y - xy$
14. $2xz + 3yz$
15. $12p^2 - 20pq$
16. $-3x - 3y$
17. $-3x + 3y$
18. $-a^2 - ab$
19. $-a^2 + ab$
20. $-6p^2 + 12pq$

Exercise 3.8a (page 38)

1. (a) $3m$ (b) $3m = 6$
2. (a) $5b$ (b) $5b = 240$
3. (a) $y + 6$ (b) $y + 6 = 21$
4. $x - 5 = 61$
5. $2x + 2(x + 3) = 14$
6. $5w = 20$
7. $a + 3 = 2a$
8. $2x - 3 = 14$
9. $2(y - 3) = 14$
10. $3z + z - 7 = 29$

Exercise 3.9a (page 39)

1. $a = 4$
2. $b = 6$
3. $c = 8$
4. $d = 2$
5. $e = 11$
6. $f = -1$
7. $g = 9$
8. $h = 6$
9. $j = -4$
10. $k = 8\frac{1}{2}$

Exercise 3.10a (page 40)

1. $x = 1$
2. $x = 2$
3. $x = -1$
4. $x = 2$
5. $x = 2$
6. $x = 1$
7. $x = 3$
8. $x = -1$
9. $x = 3\frac{1}{2}$
10. $x = 2\frac{1}{3}$

Exercise 3.11a (page 41)

1. $x = 4$
2. $x = 1$
3. $x = 4$
4. $x = -2$
5. $x = 6\frac{1}{2}$
6. $x = 3$
7. $x = 3$
8. $x = 3.4$
9. $x = 3\frac{3}{4}$
10. $x = 1\frac{3}{4}$

Exercise 3.12a (page 42)

1. $x = 4$
2. $x = 3$
3. $x = 3$
4. $x = -3$
5. $x = 1$
6. $x = 2$
7. $x = -1.4$
8. $x = 5$
9. $x = 0$
10. $x = 4$

Exercise 3.13a (page 42)

1. $m = 2$
2. $b = 48$
3. $y = 15$
4. $x = 66$
5. $x = 2$
6. $w = 4$
7. $a = 3$
8. $x = 8\frac{1}{2}$
9. $y = 10$
10. $z = 9$

Answers

Exercise 3.14a (page 42)

1. $x = 7$
2. $x = 6$
3. $x = -1$
4. $x = 1\frac{1}{3}$
5. $x = 14$
6. $x = 5$
7. $x = -10$
8. $x = 1.4$
9. $y = -4$
10. $z = 9$
11. $x = -4\frac{3}{4}$
12. $x = 7\frac{2}{3}$
13. $x = -2$
14. $y = 6$
15. $z = 14$
16. $x = -7$
17. $x = -3$
18. $x = 4$
19. $z = \frac{5}{9}$
20. $x = -7$

Exercise 3.15a (page 43)

1.
2.
3.
4.
5.

Revision exercise 3a (page 43)

1. (a) $x = 14$
   (b) $x = 3$
   (c) $x = 1\frac{1}{3}$
   (d) $x = 5$
   (e) $x = \frac{3}{4}$
2. (a) $2x + y$
   (b) $4a^2b + ab$
   (c) $24yz$
   (d) p^3q^2
   (e) $6x^3y$
3. (a) $6a - 2b$
   (b) $14c - 28$
   (c) $6x^2 - 10x$
   (d) $10xy + 15xz$
   (e) $24x - 9x^2$
4. (a) $2n + 1$
   (b) $5n$
   (c) $n + 10$
   (d) $-2n$
   (e) $20 - 3n$
5. (a) $P = 65$
   (b) $P = 25$
   (c) $P = 25$
6. (a) $3(x - 7) + x = -5$
   (b) $x = 4$
7. (a) $x = -4$
   (b) $x = 1$
   (c) $x = 5$
   (d) $x = 3$
   (e) $x = 0$
8. (a)
   (b)
9. (a) $x = 6$
   (b) $x = 2$
   (c) $x = -16$
   (d) $x = -3.4$
   (e) $x = 2\frac{1}{4}$

Chapter 4

Note:

In many questions the students' responses may differ from those given below according to the size of groups chosen, their accuracy of drawing and measuring angles, and so on.

Exercise 4.1a (page 46)

1.

Mark	Frequency
0–9	2
10–19	5
20–29	9
30–39	10

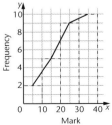

2.

Length (cm)	Frequency
50–59	4
60–69	6
70–79	12
80–89	10
90–99	4

3.

Time	Frequency
20–29	1
30–39	3
40–49	8
50–59	10
60–69	5
70–79	2
80–89	1

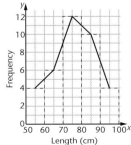

1.

Length	Frequency	Angle
50–59	4	40°
60–69	6	60°
70–79	12	120°
80–89	10	100°
90–99	4	40°

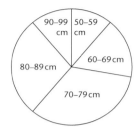

2.

Height	Frequency	Angle
130–149	2	24°
150–169	7	84°
170–189	6	72°
190–209	9	108°
210–229	6	72°

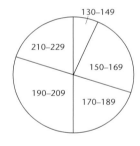

3.

Savings	60°
Clothes	144°
CDs etc.	720°
Going out	84°

4. Note that the angles are rounded.

Protein	22°
Sugar	36°
Starch	270°
Fat	5°
Sodium	4°
Fibre	5°
Other	18°

1.

Country	Number
USA	221
France	250
Majorca	61
Germany	180
Italy	119
Spain	119
Other	50

2.

Plant	Area (km²)
heather	2.5
fern	1.4
grass	3.6
gorse	1.5
other	1

3.

Gas	Percentage
nitrogen	78
oxygen	21
argon and others	1

1. Likely score between 40 and 50 marks

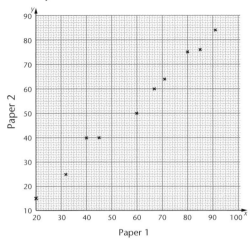

Paper 1

2.

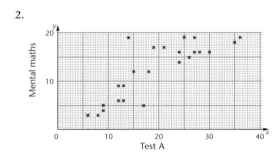

Test A

3.

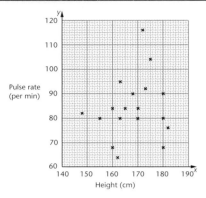

Height (cm)

Answers should mention:
(i) data is too scattered
(ii) both 185 and 140 are outside the range of heights plotted.

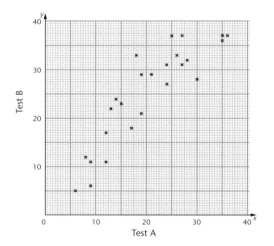

Test A

Exercise 4.1a:

    Question 1 modal class is 20−29

    Question 2 modal class is 70−79

    Question 3 modal class is 50−59

1. mean: 2.3(84),
    median: 2,
    mode: 3,
    range: 6

2. mean: 158(.106),
    median: 152.5,
    mode: none,
    range: 49.6

Answers

1. Group A: Maths: mean 53.6, median 56.5, mode 61, range 62
   Group A: English: mean 53.7, median 53, mode 53, range 17
   Group B: Maths: mean 56.8, median 47, mode 43, range 59
   Group B: English: mean 53, median, 53, mode 53, range 13
   Comments could include some or all of the following:
   Group A have the same median for maths and English and the means are virtually the same but the range for maths is greater; Group B – the mean for maths is greater than for English, the range is greater, the median for English is higher.

   Group B seem better at English than Group A – the same mean, median and mode but the range is smaller.

2. Town A: mean 14.75, median 15, mode 15, range 7
   Town B: mean 14.75, median 15, mode 16, range 11
   Town B has the higher mode but the range of temperature is smaller for A.

3. Town A: mean 7.55, median 8, mode 8, range 7
   Town B: mean 7.25, median 7, mode 8, range 7
   Both towns have the same mode and range but the mean and median are smaller for town B.

1. (a)
1.	7 9
2.	4 5 6 6 6 7 8 8 9 9
3.	1 1 2 3 5 6 8 8 9
4.	0 1 1 2

   (b) Median weight is 3.1 kg

2. (a) Test 2 because fewer scored high marks
   (b) Test 1: median = 66, Test B: median = 52

1. There seems to be a relationship: the taller a person is, the heavier they are.
   A height of 160 cm could give a weight of around 55−60 kg.
   A weight of 50 kg could give a height of around 145−155 cm.

   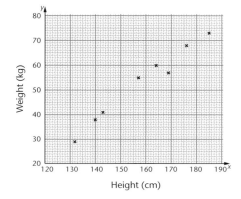

2.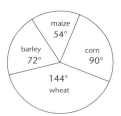

   Angles are: barley 72°, maize 54°, corn 90°, wheat 144°

3. Details are:

Hot meal	90°	97 pupils
Cold meal	68°	73 pupils
Packed lunch	79°	85 pupils
Chip shop	102°	110 pupils
Home	22°	24 pupils

   97 pupils is one quarter of the sample.
   Therefore year group is 388 pupils.

4. Machine A: mean: 200, mode: 199,
    median: 200, range: 6
    Machine B: mean: 200, mode: 201,
    median: 201, range: 7
    The range for B is higher so machine A may be
    more consistent; the mean is the same for
    both but the median for B is higher than that
    for A as is the mode so B may be better for
    customers and worse for manufacturers?

5. (a)

4.	4 9
5.	2 3 8 9
6.	2 3 4 4
7.	2 6 5
8.	1
9.	2 3

    (b) Median is 6.3

Chapter 5

Exercise 5.1a (page 63)

1. 9.5
2. 0.38 to 2 d.p.
3. 2.98
4. 0.07
5. 0.073p
6. 5 litres – e.g. 2×2 litres
    costs more!
7. 1 litre
8. 12×330 ml
9. £1240, £930, £620
10. £45.30

Exercise 5.2a (page 65)

1. negative ÷ negative = positive
2. $12.4 \times 1 = 12.4$ so the answer should be less than 12.4
3. $30 \times 4 = 120$; $30 \times 40 = 1200$
4. $8 = \sqrt{64}$, $9 = \sqrt{81}$ so the answer lies between 8 and 9
5. The square of a number between 0 and 1 is smaller than the number.
6. $7 \times 20p = £1.40$
7. $10 \times £13 = £130$
8. $60 \div 4 = 15$
9. $210 \div 30 = 7$
10. $\frac{50}{4 \times 8} \approx \frac{50}{30} \approx 1.6$

Exercise 5.3a (page 68)

1. £8065
2. £16 808
3. £3074.68
4. £6978.60
5. £2149.16
6. £1.32
7. 19 715
8. £41.84
9. $0.78
10. (a) $3225
    (b) £266.27

Exercise 5.4a (page 71)

1. (a) 207.42
    (b) 485.35
    (c) 27.90
2. (a) 47.9
    (b) 6.8
    (c) 90.6
3. 8.7
4. 625
5. 4.3
6. 115 miles
7. 235 grams
8. 38.7%
9. £10.92 or £10.93
10. £6.33 or £6.34

Revision exercise 5a (page 72)

1. 11.6 to 1 d.p.
2. 33p
3. £636 and £424
4. 16 miles
5. £15
6. $6 \times £4 = £24$, so
    he has enough.
7. £2117.50
8. £2.49
9. 4.6
10. £9.94

Chapter 6

Exercise 6.1a (page 76)

1. 4
2. 9
3. Check students' drawings
4. circle
5. 4
6. (a) 4   (b) 4
7. (a) 10   (b) 8
8. Rotational symmetry order 4
9. Check students' drawings
10. Check students' drawings

Exercise 6.2a (page 79)

1. 78°
2. $a = 55°$, $d = 57°$
3. 35°, 145°, 145°
4. 93°
5. 76°
6. $x = 112°$, $y = 68°$, $p = 126°$, $q = 54°$, $r = 61°$
7. 23, 180, 139, 360, 50, equal 130, add to 180°
8. 60°
9. 125°
10. Check students' drawings

Exercise 6.3a (page 83)

1. 74°, alternate angles
2. 67°, allied angles
3. Check accurate drawing
4. Check accurate drawing
5. $a = 110°$, $b = 70°$, $c = 70°$
6. $a = 82°$, $b = 67°$, $c = 31°$
7. 75°, 105°, 105°
8. 64°, 116°, 116°
9. Check sketch and description
10. $a = 40°$, $b = 72°$, $c = 68°$

Exercise 6.4a (page 87)

1. 131°
2. 58°
3. 78°
4. 110°, 121°, 97°, 90°, 122°
5. 126°, 132°, 115°, 145°, 100°, 102°
6. 40°, 140°
7. 150°
8. 15
9. (a) 720°   (b) 1440°
10. 107°

Revision exercise 6a (page 88)

1. 8
2. 4
3. Answer should be similar to
4. 70° and 70°, or 40° and 100°

5. 73°
6. $a = 120°$, alternate angles
   $b = 50°$, alternate angles
   $c = 72°$, corresponding angles
   $d = 97°$, allied angles

7. 58°
8. 900°
9. (a) 36°   (b) 144°
10. 140°

Chapter 7

Exercise 7.1a (page 91)

1.
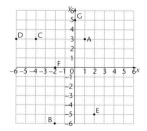

2.
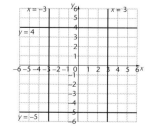

3. Points $(-3, -12)$, $(0, 0)$, $(3, 12)$
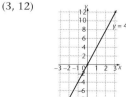

Answers

4. Points $(-3, 0)$, $(0, 3)$, $(3, 6)$

5. Points are $(-4, -10)$, $(0, 2)$, $(2, 8)$

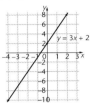

6. Points are $(-2, -5)$, $(0, -3)$, $(4, 1)$

7. Points are $(-2, -10)$, $(0, -4)$, $(4, 8)$

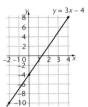

8. Points are $(-3, -14)$, $(0, -2)$, $(3, 10)$

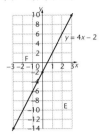

9. Points are $(-2, 9)$, $(0, 5)$, $(4, -3)$

10. Points are $(-4, 8)$, $(0, -4)$, $(2, -10)$

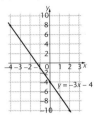

1. Points are $(0, 3)$, $(5, 0)$

2. Points are $(0, 7)$, $(2, 0)$

3. Points are $(-3, -6)$, $(0, 1.5)$, $(3, 9)$

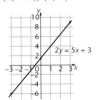

4. Points are $(-2, -5.5)$, $(0, -2.5)$, $(4, 3.5)$

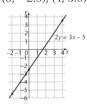

5. Points are $(-6, 0)$, $(0, 3)$, $(4, 5)$

6. Points are $(-3, -5)$, $(0, -4)$, $(6, -2)$

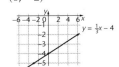

7. (a) Points are $(-3, -10)$, $(0, 2)$, $(3, 14)$

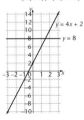

(b) $(1.5, 8)$

8. (a) $y = 2x + 3$, points are $(0, 3)$, $(2, 7)$, $(5, 13)$
$2x + y = 7$ points are $(0, 7)$, $(3.5, 0)$

(b) $(1, 5)$

1.

x	-3	-2	-1	0	1	2	3
x^2	9	4	1	0	1	4	9
5	5	5	5	5	5	5	5
$y = x^2 + 5$	14	9	6	5	6	9	14

2.

x	-4	-3	-2	-1	0	1	2	-1.5
x^2	16	9	4	1	0	1	4	2.25
$3x$	-12	-9	-6	-3	0	3	6	-4.5
-7	-7	-7	-7	-7	-7	-7	-7	-7
$y = x^2 + 3x - 7$	-3	-7	-9	-9	-7	-3	3	-9.25

3.

x	-6	-5	-4	-3	-2	-1	0	1	2	-2.5
$-x^2$	-36	-25	-16	-9	-4	-1	0	-1	-4	-6.25
$-5x$	30	25	20	15	10	5	0	-5	-10	12.5
6	6	6	6	6	6	6	6	6	6	6
$y = -x^2 - 5x + 6$	0	6	10	12	12	10	6	0	-8	12.25

4. Because $x = 1$ and $x = 2$ give the same value, an extra column for $x = 1.5$ must be included.

x	-2	-1	0	1	2	3	4	1.5
x^2	4	1	0	1	4	9	16	2.25
$-3x$	6	3	0	-3	-6	-9	-12	-4.5
1	1	1	1	1	1	1	1	1
$y = x^2 - 3x + 1$	11	5	1	-1	-1	1	5	-1.25

5. (a)

x	-3	-2	-1	0	1	2	3
x^2	9	4	1	0	1	4	9
-2	-2	-2	-2	-2	-2	-2	-2
$y = x^2 - 2$	7	2	-1	-2	-1	2	7

(b)

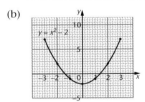

6.

x	-3	-2	-1	0	1	2	3
$-x^2$	-9	-4	-1	0	-1	-4	-9
4	4	4	4	4	4	4	4
$y = -x^2 + 4$	-5	0	3	4	3	0	-5

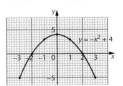

Answers

7. (a)

x	-1	0	1	2	3	4	5
x^2	1	0	1	4	9	16	25
$-3x$	3	0	-3	-6	-9	-12	-15
$y = x^2 - 3x$	4	0	-2	-2	0	4	10

(b)

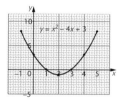

(c) $x^2 - 3x = 0$ when the curve cuts the x-axis. Solutions are $x = 0$ and $x = +3$.

8. (a)

x	-1	0	1	2	3	4	5
x^2	1	0	1	4	9	16	25
$-4x$	4	0	-4	-8	-12	-16	-20
3	3	3	3	3	3	3	3
$y = x^2 - 4x + 3$	8	3	0	-1	0	3	8

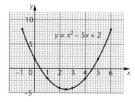

(b) $x^2 - 4x + 3 = 0$ when the curve crosses $y = 0$, so $x = 1$ or 3.

9. (a)

x	-1	0	1	2	3	4	5	6	2.5
x^2	1	0	1	4	9	16	25	36	6.25
$-5x$	5	0	-5	-10	-15	-20	-25	-30	-12.5
2	2	2	2	2	2	2	2	2	2
$y = x^2 - 5x + 2$	8	2	-2	-4	-4	-2	2	8	-4.25

(b) $x^2 - 5x + 2 = 0$ when the curve crosses the x-axis, so $x = 0.4$ or 0.5 and 4.5 or 4.6.

10. (a)

t	0	1	2	3	4	5
t^2	0	1	4	9	16	25
$d = 5t^2$	0	5	20	45	80	125

(b)

(c) Time = 3.6 seconds

1. $y = x^3 - 12x + 2$

x	-3	-2	-1	0	1	2	3	4
x^3	-27	-8	-1	0	1	8	27	64
$-12x$	36	24	12	0	-12	-24	-36	-48
2	2	2	2	2	2	2	2	2
$y = x^3 - 12x + 2$	11	18	13	2	-9	-14	-7	18

2. $y = x^3 - x^2 + 5$

x	-2	-1	0	1	2	3	4
x^3	-8	-1	0	1	8	27	64
$-x^2$	-4	-1	0	-1	-4	-9	-16
5	5	5	5	5	5	5	5
$y = x^3 - x^2 + 5$	-7	3	5	5	9	23	53

3. $y = \dfrac{8}{x}$

x	-8	-4	-2	-1	1	2	4	8
$y = \dfrac{8}{x}$	-1	-2	-4	-8	8	4	2	1

4. (a) $y = x^2 - 2x - 3$

x	-2	-1	0	1	2	3	4
x^2	4	1	0	1	4	9	16
$-2x$	4	2	0	-2	-4	-6	-8
-3	-3	-3	-3	-3	-3	-3	-3
$y = x^2 - 2x - 3$	5	0	-3	-4	-3	0	5

$y = x + 1$ has points $(-2, -1)$, $(0, 1)$, $(4, 5)$.

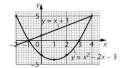

(b) $(-1, 0)$, $(4, 5)$

5. (a) $y = x^3 - 3x$

x	-3	-2	-1	0	1	2	3
x^3	-27	-8	-1	0	1	8	27
$-3x$	9	6	3	0	-3	-6	-9
$y = x^3 - 3x$	-18	-2	2	0	-2	2	18

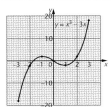

(b) $x^3 - 3x = 0$ when it cuts $y = 0$.
$x = -1.8, 0, 1.8$

6. (a) The points are worked out in question 1.

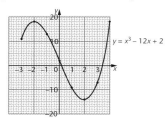

(b) The only two values in the range of the graph are $x = 0.2$ and $x = 3.2$.

7. The points are worked out in question 3.

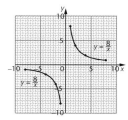

8. (a) $y = \dfrac{5}{x}$

x	-5	-4	-2.5	-2	-1	1	2	2.5	4	5
$y = \dfrac{5}{x}$	-1	-1.25	-2	-2.5	-5	5	2.5	2	1.25	1

(b), (c)

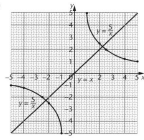

(d) When the line and the curve cross $x^2 = 5$. $x = \sqrt{5} = -2.2$ or 2.2.

9. (a) $y = x^3 - 4x$

x	-3	-2	-1	0	1	2	3
x^3	-27	-8	-1	0	1	8	27
$-4x$	12	8	4	0	-4	-8	-12
$y = x^3 - 4x$	-15	0	3	0	-3	0	15

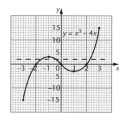

(b) $x^3 - 4x - 2 = 0$ can be written as $x^3 - 4x = 2$.
So the solution is where $y = x^3 - 4x$ and $y = 2$ intersect.
Therefore the solution is $x = -1.8$, -0.4 or $+2.1$.

10. (a) $y = 4x^2 - 5$

x	-3	-2	-1	0	1	2	3
x^2	9	4	1	0	1	4	9
$4x^2$	36	16	4	0	4	16	36
-5	-5	-5	-5	-5	-5	-5	-5
$y = 4x^2 - 5$	31	11	-1	-5	-1	11	31

Points for $y = 3x + 2$ are $(-3, -7)$, $(0, 2)$, $(3, 11)$.

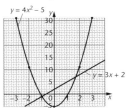

(b) (i) $4x^2 - 5 = 20$ has solutions where $y = 4x^2 - 5$ and $y = 20$ cross.
The solutions are $x = -2.4$ and $x = 2.4$.

(ii) $4x^2 - 3x - 7 = 0$ can be rearranged as $4x^2 - 5 = 2 + 3x$.
This has solutions where $y = 4x^2 - 5$ meets $y = 3x + 2$. The solutions are $x = -1$ and $x = 1.8$.

Exercise 7.5a (page 105)

1. $A = 078°$, $B = 112°$, $C = 207°$, $D = 290°$
   It is difficult to measure an angle accurately. Be as careful as you can, but in exams you are normally allowed to be ±1° out on your angle measurement.

2. $270°$

3. $304°$

4. $045°$

5. (a) $138°$
   (b) $252°$

6.

7.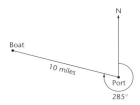

8. (a)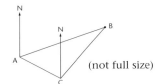
   (not full size)

   (b) $280°$

9. (a) sketch

   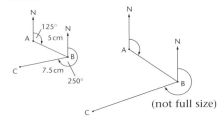
   (not full size)

   (b) (i) $210°$
       (ii) CA measures 6 cm which represents 1.2 km.

10. (a) R from Q is $87° + 49° = 136°$
    (b) R from P is $87° + 180° - 37° = 230°$
    (c) P from R is $230° - 180° = 50°$

1.

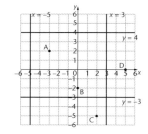

2. (a) Points are $(-3, -9)$, $(0, 0)$, $(3, 9)$

(b) Points are $(-3, 11)$, $(0, 5)$, $(3, -1)$

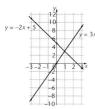

(c) Point of intersection $(1, 3)$

3. (a) Points are $(-2, -8)$, $(0, -4)$, $(5, 6)$

(b) Points are $(0, 2)$, $(5, 0)$

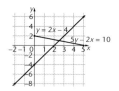

(c) Point of intersection $(2.5, 1)$

4. (a) $y = x^2 - 6x + 3$

x	-1	0	1	2	3	4	5	6	7
x^2	1	0	1	4	9	16	25	36	49
$-6x$	6	0	-6	-12	-18	-24	-30	-36	-42
3	3	3	3	3	3	3	3	3	3
$y = x^2 - 6x + 3$	10	3	-2	-5	-6	-5	-2	3	10

(b) $x^2 - 6x + 3 = 0$ is where $y = x^2 - 6x + 3$ and $y = 0$ intersect. The solution is $x = 0.6$ or $x = 5.4$.

5. (a) $y = -x^2 - x + 12$

x	-5	-4	-3	-2	-1	0	1	2	3	4	-0.5
$-x^2$	-25	-16	-9	-4	-1	0	-1	-4	-9	-16	-0.25
$-x$	5	4	3	2	1	0	-1	-2	-3	-4	0.5
12	12	12	12	12	12	12	12	12	12	12	12
$y = -x^2 - x + 12$	-8	0	6	10	12	12	10	6	0	-8	12.25

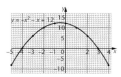

(b) The solution is $x = -4$ or $x = 3$.

6. (a) $h = 40t - 5t^2$

t	0	1	2	3	4	5	6	7	8
t^2	0	1	4	9	16	25	36	49	64
$40t$	0	40	80	120	160	200	240	280	320
$-5t^2$	0	-5	-20	-45	-80	-125	-180	-245	-320
$h = 40t - 5t^2$	0	35	60	75	80	75	60	35	0

(b)

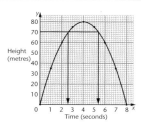

(c) Time is 2.6 or 5.4 seconds.

7. (a) $C = \dfrac{n}{4} + 20$

n	0	40	80	120	160	200
$C = \dfrac{n}{4} + 20$	20	30	40	50	60	70

(b) Points are (0, 25), (100, 45), (200, 65)

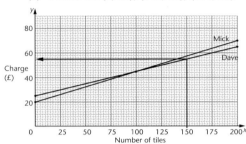

(c) Dave is cheaper. He charges £55.

8. (a) A from B is 67° + 180° = 247°

(b) C from B is 247° − 93° = 154°

(c) A from D is 197° − 180° = 017°

9. Sketch   Scale drawing

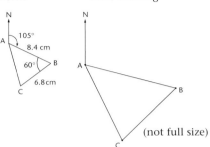

Distance of A from C is 7.8 cm which represents 390 metres. The bearing of A from C is 335°.

10. Sketch   Scale drawing

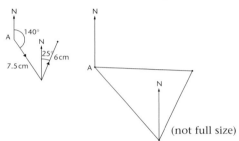

The distance from the start is 7.3 cm which represents 14.6 miles. The bearing from the start is 092°.

Chapter 8

Exercise 8.1a (page 114)

1.
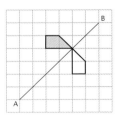

2.
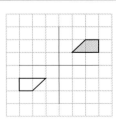

3.
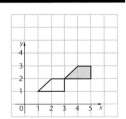

4.

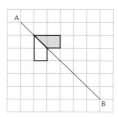

5.

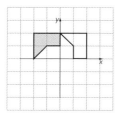

6.

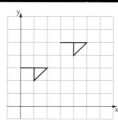

7.

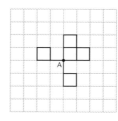

8.

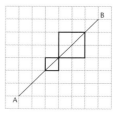

9. (a) Reflection in the line $x = 3$
   (b) Translation of 2 units to the right and 1 unit down

10.

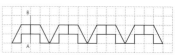

Exercise 8.2a (page 120)

1.

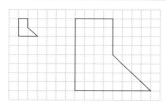

2.

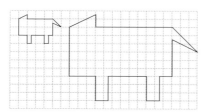

3.

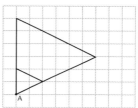

4.

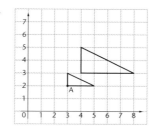

5.

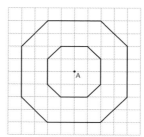

6.

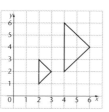

7.

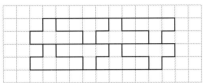

(a) one horizontal line of symmetry, no rotational symmetry

(b) S: no line of symmetry, rotational symmetry of order 2

8.

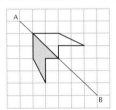

It is soon clear that the shape will tessellate.

9. The exterior angle of a hexagon is 360° ÷ 6 = 60°.
The interior angle is 180° − 60° = 120°.
360 ÷ 120 = 3. So the angles fit at a point.
Therefore a regular hexagon will tessellate.

10. The transformation is an enlargement of scale factor 2, centre O.

Revision exercise 8a (page 123)

1.

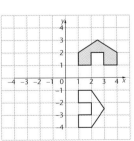

2.

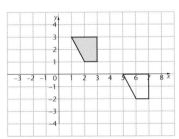

3.

4.

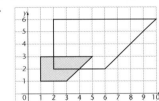

5. (a)

This letter has two lines of symmetry and rotational symmetry of order 2.

(b)

R

This letter has no lines of symmetry and no rotational symmetry.

6. The exterior angle of a regular octagon is 360° ÷ 8 = 45°.
The interior angle is 180° − 45° = 135°.
135 will not divide exactly into 360. Therefore a regular octagon will not tessellate.

7.

It is soon obvious this shape will tessellate.

8.

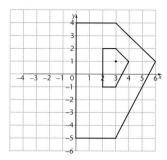

9. (a)

Four lines of symmetry and rotational summetry order 4

(b)

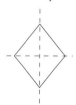

Two lines of symmetry and rotational symmetry order 2

(c)

Infinitely many lines of symmetry and infinite rotational symmetry

10. (a) A onto B is a reflection in the y-axis (or $x = 0$).
    (b) B onto D is a reflection in the line $y = 2$.
    (c) B onto C is a translation 8 units to the right and 1 unit down.
    (d) D onto E is an enlargement scale factor 2, centre $(0, 0)$.
    (e) F onto B is a translation 4 units to the left and 7 units up.
    (f) A onto G is a reflection in the line $y = x$.
    (g) G on to B is a rotation of 90° anticlockwise centred on $(0, 0)$.

Chapter 9

Activity 2 (page 129)

(a) multiples of 4
(b) the 'odd' multiples of 3 i.e. 9, 15, 21, ...
(c) multiples of 8

Activity 3 (page 129)

(a) e.g. $4^2 - 3^2 = 16 - 9 = 7$
$= 4 + 3$ or $2 \times 3 + 1$
$5^2 - 4^2 = 25 - 16 = 9$
$= 5 + 4$ or $2 \times 4 + 1$
Proof: if the two numbers are n and $n + 1$:
$(n + 1)^2 - n^2 = n^2 + 2n + 1 - n^2 = 2n + 1$

(b) e.g. $3, 4, 5 \Rightarrow 3 \times 5 = 15$
$4^2 = 16$
$16 - 15 = 1$
$6, 7, 8 \Rightarrow 6 \times 8 = 48$
$7^2 = 49$
$49 - 48 = 1$
Proof: if the three numbers are n and $n + 1$ and $n + 2$:
$n, n + 1, n + 2 \Rightarrow n(n + 2)$
$= n^2 + 2n$
$(n + 1)^2 = n^2 + 2n + 1$
Difference $= 1$ because
$(n^2 + 2n + 1) - (n^2 + 2n)$
$= 1$

Activity 4 (page 129)

(a) Square A has 1 large square and 4 small squares, i.e. $1^2 + 2^2$ squares
Square B has 1 large square, 4 (two by two) squares and 9 (one by one) squares, i.e. $1^2 + 2^2 + 3^2$
Square C: $1^2 + 2^2 + 3^2 + 4^2$

(b) In the sixth diagram:
$1^2 + 2^2 + 3^2 + 4^2 + 5^2 + 6^2$
$= 91$
In the tenth diagram: 385

Activity 5 (page 130)

A good opportunity to make sure that students understand the task and can work systematically, and also to check to see if they have thought of any extension question, for example, investigating 2s and 5s or 3s and 7s or ... table of solutions up to 14:

1	X	8	1(3) + 1(5)
2	X	9	2(3) + 0(5)
3	1(3) + 0(5)	10	0(3) + 2(5)
4	X	11	2(3) + 1(5)
5	0(3) + 1(5)	12	4(3) + 0(5)
6	2(3) + 0(5)	13	1(3) + 2(5)
7	X	14	3(3) + 1(5)

All numbers are possible when you have a group of 'successes' as large as the value of the smaller number – here 3. Thus as soon as three numbers can be made, all other numbers can be made by adding 3 to each of them thus: $8 + 3 = 11$, $9 + 3 = 12$, $10 + 3 = 13$ and these numbers can be used to continue so that $11 + 3 = 14$ and so on.

In order for there to be no gaps, after a certain point the numbers must be co-prime or relatively prime (have no common factors other than 1).

The smallest number after which all totals are possible is, for n and m: $nm - (n + m)$

So for 2 and 5 all numbers are possible after 3, for 3 and 7 all are possible after 11.

Activity 6 (page 130)

(a) yes
(b) prime
(c) square numbers

Activity 7 (page 131)

Only those prisoners in cells whose numbers are square numbers will escape because square numbers have an odd number of factors.

Activity 8 (page 131)

Subtracting the squares results either in a loop or the final answer of zero, as in the example.
Adding the squares will either result in a loop or the final answer of 1.

Activity 9 (page 131)

Student's own work.

Activity 10 (page 132)

1. (a) 9   (c) 17
   (b) 5   (d) 6

2. (a) 5   (c) 13
   (b) 5

Revision exercise 9a (page 132)

1. (a) 49   (e) 12
   (b) 121   (f) 13
   (c) 61   (g) 3
   (d) 14   (h) 3

2. (a) $2 \times 2 \times 2 \times 3$
   (b) $2 \times 3 \times 5$
   (c) $3 \times 3 \times 5$
   (d) $2 \times 2 \times 3 \times 5$
   (e) $2 \times 2 \times 2 \times 2 \times 2 \times 2$

3. 17 178, 17 172 and 17 177 are definitely not square numbers.

Chapter 10

1. (a) 12 cm²
   (b) 20 m²
   (c) 35 cm²
   (d) 31.5 m²
   (e) 30.38 cm²
   (f) 12.6 m²
   (g) 7.5 m²
   (h) 16.38 cm²
   (i) 24 m²

2. 9 square units
3. 10 square units
4. 17.5 square units
5. approximately 10.8 cm²
6. approximately 15.4 cm²

1. (a) 37.7 cm
   (b) 28.3 cm
   (c) 62.8 m
   (d) 51.2 cm
   (e) 47.8 m

2. (a) 31.4 cm
   (b) 44.0 cm
   (c) 100.5 m
   (d) 113.7 m
   (e) 33.3 m

3. 57.5 m
4. 94.2 cm
5. 15.9 cm

1. (a) 50.3 cm²
   (b) 804 m²
   (c) 401 m²
   (d) 581 m²
   (e) 249 cm²
2. 7.07 m²
3. 0.503 m²

4. 124 cm²
5. 4.52 m²
6. 145 cm²
   $3 \times 7^2 = 3 \times 49$
   $= 3 \times 50$
   $= 150$ cm²

1. 100.9 m³
2. Check dimensions of Student Book
3. (a) 60 cm³
   (b) 94 cm²
4. 1 400 000 cm³, 1400 litres

5. 4050 cm³
6. (a) 1.326 m³
   (b) 0.0624 m³ or 62 400 cm³
7. 3 m
8. 2880 cm³

1. 7.5 m²
2. (a) 8 square units
   (b) 6 square units
   (c) 14 square units
3. 240 cm²
4. 1195 cm²,
   $3 \times 20^2 = 1200$ cm²

5. 7.54 m
6. (a) 466 cm²
   (b) 81.7 cm
7. length = 27.1 cm, area = 298 cm²
8. 275 cm²

9. (a) 22.3 cm³
   (b) 7.9 cm
10. (a) 34. 9 m³
    (b) 51.92 m²
    (c) 3.6 (or 4) litres

Chapter 11

1. 0.2
2. $\frac{5}{7}$
3. (a) $\frac{2}{3}$
   (b) $\frac{3}{5}$
4. (a) $\frac{13}{52} \left(\frac{1}{4}\right)$
   (b) $\frac{39}{52} \left(\frac{3}{4}\right)$

5. (a) $\frac{1}{5}$
   (b) $\frac{4}{5}$
6. $\frac{40}{49}$
7. 0.005

1. 0.3
2. $\frac{4}{12} \left(\frac{1}{3}\right)$
3. $\frac{2}{20} \left(\frac{1}{10}\right)$
4. 0.1

5. 0.65
6. $\frac{5}{11}$
7. It may do both, it may do neither.

1. (a) 0.154
   (b) 0.255
2. The coin was not thrown enough times.
3. (a) (i) $\frac{103}{500}$
       (ii) $\frac{96}{500}$
   (b) Yes as all are close to expected value of 100
4. (a) 0.41
   (b) 0.59
5. (a) It is a large survey.
   (b) (i) $\frac{3}{16}$ (0.188)
       (ii) $\frac{3}{8}$ (0.375)
6. 0.35(5)
7. Choose a bead, note its colour and replace it. Repeat this a large number of times (suggest at least 200)
   Probability $= \frac{\text{number of reds}}{\text{number of trials}}$

1. 0.35
2. $\frac{2}{9}$
3. $\frac{7}{8}$
4. 0.3
5. 0.29
6. $\frac{8}{20}$ (0.4)
7. May not choose any, may choose more than one, not mutually exclusive, …
8. 0.87
9. (a) (i) $\frac{105}{250}$ (0.42)
       (ii) $\frac{25}{250}$ (0.1)
   (b) Any reasonable reason e.g. May be fewer drivers over 65.
10. Spin the spinner a large number of times (e.g. 200). Record number of 1s, 2s, 3s, 4s. If they all come up approximately equally often, conclude it is fair.

Chapter 12

1. (a) 5.48; 5.5
   (b) 12.08; 12.1
   (c) 0.21; 0.2
   (d) 0.57; 0.6
   (e) 9.02; 9.0
   (f) 78.04; 78.0
   (g) 7.01; 7.0
   (h) 0.07; 0.1
2. (a) 7.4
   (b) 7.42
3. 9.43
4. (a) 0.333
   (b) 0.286
   (c) 0.273
   (d) 0.308
5. 7.3
6. 9.34

1. (a) 460
   (b) 250
   (c) 120
   (d) 1000
   (e) 5700
   (f) 9900
   (g) 8800
   (h) 200
2. (a) 130; 100
   (b) 450; 500
   (c) 550; 500
   (d) 4560; 4600
   (e) 1410; 1400

1. (a) 67 900
   (b) 54.1
   (c) 1800
   (d) 1 564 400
   (e) 0.0068
   (f) 1.5
   (g) 0.09
   (h) 45.3
2. (a) 4200
   (b) 710
3. $(2 \times 3.65 \times 2.2) + (2 \times 2.44 \times 2.2)$
   $= 26.796 = 27 \text{ m}^2$
4. (a) £252.72 i.e. £253
   (b) £12 888 i.e. £13 000
5. (a) 77 000
   (b) 680
   (c) 0.81
   (d) 3.1
   (e) 0.71
6. (a) 1.19
   (b) 5.48
   (c) 2.53
   (d) 2.20
   (e) 0.47
7. (a) $6 \times 0.2 = 1.2$, 1.0999, 1.1
   (b) $20 \div 4 = 5$, 4.9519, 5.0
   (c) $40 \times 20 = 800$, 585.1269, 590
   (d) $10 \div 0.05 = 200$, 190.7949, 190
   (e) $2 \times 6 \times 1 = 12$, 12.417 44, 12

1. $\frac{3}{6}, \frac{3}{5}, \frac{4}{6}, \frac{3}{4}, \frac{4}{5}, \frac{5}{6}, \frac{6}{5}, \frac{5}{4}, \frac{4}{3}, \frac{6}{4}, \frac{5}{3}, \frac{6}{3}$
2. 0.000280, 0.0014, 0.0042, 0.0098, 0.0126, 0.5
3. 1 560 005, 156 005, 15 605, 15 565, 15 065
4. 0.62, 0.6, 0.0624, 0.006 004

1. (a) 6 : 24   (b) 12 : 30   (c) 32 : 40   (d) 12 : 15 : 18
2. (a) 1 : 10   (b) 4 : 1   (c) 4 : 5   (d) 1 : 18
3. (a) 1 : 5   (b) 1 : 3   (c) 1 : 4   (d) 1 : 300
4. 2.28 km
5. 2.8 cm
6. (a) 40   (b) 14 800 − 15 000
7. (a) 1 : 50   (b) 1 : 5   (c) 1 : 60
8. (a) 1 : 3   (b) £120, £360
9. (a) 13, 21, 34, 55, 89   (b) Tends to 1.618…

1. (a) 54   (d) 400   (g) −3
   (b) −72   (e) 420   (h) −22
   (c) −42   (f) 108   (i) −16

2.

×	8	−6	4	−3
10	80	−60	40	−30
−7	−56	42	−28	21
3	24	−18	12	−9
−5	−40	30	−20	−15

3. (a) $(-3 \div -6) \times 2 = 1$
   (b) $(-6 \div -3) - 2 = 0$
   (c) $(2 - -3) \times -6 = -30$
4. (a) −9   (d) 27   (g) 39
   (b) −1.33   (e) −27   (h) 144
   (c) $-\frac{3}{4}$ (f) −9

1. (a) 3^5
   (b) 7^3
   (c) 8^5
   (d) $3^2 \times 5^3$
   (e) $2^3 \times 3^2 \times 4^5$
2. (a) 5^5 (d) 3^{11}
   (b) 6^9 (e) 8^5
   (c) 10^7 (f) 5^{-6}
3. (a) 3^{-3} (c) 5^{-4}
   (b) 2^7

1. (a) 5×10^3
   (b) 5×10^1
   (c) 7×10^4
   (d) 4.6×10^1
   (e) 2×10^{-2}
   (f) 5.46×10^5
   (g) 4.5×10^{-4}
   (h) 1.6×10^7
2. (a) 500
   (b) 400 000
   (c) 0.006
   (d) 4500
   (e) 0.0084
   (f) 0.0000287
   (g) 9700
   (h) 0.000 055

(a) $5000 + 70\,000 \Rightarrow 75\,000 \Rightarrow 7.5 \times 10^4$
(b) $7\,000\,000 - 3000 \Rightarrow 6\,997\,000 \Rightarrow 6.997 \times 10^6$
(c) $3000 + 300 \Rightarrow 3300 \Rightarrow 3.3 \times 10^3$
(d) $6000 - 500 \Rightarrow 5500 \Rightarrow 5.5 \times 10^3$

1. (a) 49 (b) 144 (c) 625 (d) 1600 (e) 121
2. (a) 7 (b) 11 (c) 13 (d) 17 (e) 5
3. (a) 64 (b) 125 (c) 216 (d) 1000 (e) 512
4. (a) 7 (b) 9 (c) 11 (d) 100
5. (a) 7.48 (b) 5.20 (c) 7.75 (d) 16.73 (e) 26.04

1. (a) $\frac{1}{3}$ (d) $\frac{1}{100}$
   (b) $\frac{1}{6}$ (e) $\frac{1}{640}$
   (c) $\frac{1}{49}$
2. (a) 16   (d) 67
   (b) 9   (e) 1000
   (c) 52

Answers

1. (a) 7.90   (c) 0.24
   (b) 13.12   (d) 0.68
2. 12.69 (2 a.p.)
3. (a) 0.429   (c) 0.154
   (b) 0.182   (d) 0.538
4. £29.36
5. (a) 130   (e) 7900
   (b) 540   (f) 9800
   (c) 1000   (g) 8900
   (d) 1240   (h) 100
6. (a) 6790   (e) 0.0059
   (b) 57.1   (f) 1.8
   (c) 1900   (g) 0.08
   (d) 1 576 400   (h) 40.3
7. 0.000 98, 0.0098, 0.0926, 0.9, 0.9042, 0.914
8. (a) 1 : 4   (c) 1 : 3
   (b) 1 : 2   (d) 1 : 30

9. 2.15 km or 215 000 cm
10. (a) 1 : 200
    (b) 1 : 500 000
    (c) 1 : 10
11. (a) 72   (f) 96
    (b) −150   (g) −9
    (c) −18   (h) −3
    (d) 36   (i) −8
    (e) 30
12. (a) −20   (e) −125
    (b) −2   (f) −17
    (c) −2.5   (g) 119
    (d) 75   (h) 400
13. (a) 7^5 (e) 8^6
    (b) 6^9 (f) 6^{-2}
    (c) 10^{12} (g) 4^{-1}
    (d) 3^{12} (h) 9^4

14. (a) 7.6×10^3
    (b) 8.99×10
    (c) 6×10^4
    (d) 4.66×10^2
    (e) 5.6×10^{-2}
    (f) 5.646×10^5
    (g) 5.5×10^{-3}
    (h) 2.4×10^7
15. (a) 6000
    (b) 500
    (c) 0.007
    (d) 450
    (e) 0.084
    (f) 0.002 87
    (g) 4700
    (h) 0.0555
16. (a) 1.4×10^4
    (b) 6.8×10^4

Chapter 13

Activity 2 (page 173)

Exterior angle of a triangle

Proof: Either $b = p$ (because they are corresponding angles)
        $a = s$ (because they are alternate angles)
∴ $a + b = p + s$
    $a + b = r$
∴ $r = p + s$
Or $p + s + q = 180°$ (because angles in a triangle add up to 180°)
    $a + b + q = 180°$ (because angles on a straight line add up to 180°)
∴ $a + b = p + s$
∴ $r = p + s$

Exercise 13.1a (page 173)

1. (a) $x = 116°$ (c) $a = 120°$
   (b) $b = 83°$ (d) $c = 12°$
2. $x = 48°$ (angles in a triangle), $d = 132°$, $e = 120°$
   (angles on a straight line) or $d = 132°$
   (exterior angle of a triangle), $x = 48°$
   ∴ $e = 360° − (48° + 132° + 60°)$
   (angles round a point) = 120°
   or $y = 42°$ (angle sum of triangle)
   ∴ $e = 78° + 42° = 120°$
   (exterior angle of triangle)
   or $e = 180° − 60° = 120°$
   (angles on a straight line)

3. (a) $x = 94°$
   (b) $a = 70°$
   (c) $a = 120°$
   (d) $y = 80°$
4. (a) $x = 129°$
   (b) $x = 30°$
   (c) $x = 135°$
   (d) $a = 28.75°$
   (e) $5x = 180°$ ∴ $x = 36°$ ∴ $y = 3 \times 36° = 108°$

Activity 3 (page 175)

1. The lines OA, OP and OB are equal because they are radii.
2. The angles marked *a* are equal to each other because they are opposite angles in an isosceles triangle.
3. The angles marked *b* are equal to each other because they are opposite angles in an isosceles triangle.
4. In triangle APB angles $a + a + b + b = 180°$ because the sum of the angles of the triangle $= 180°$.
   So $2(a + b) = 180°$ and angle APB $= a + b = 90°$.

Finding the centre of the circle:

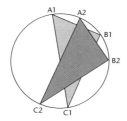

Exercise 13.2a (page 176)

1. (a), (b) *a* and *b* are equal, (isosceles triangle)
   c and *d* are equal (isosceles triangle)
   $\therefore c = d = 50°$
   $b = 90° - 50° = 40°$ (angle in a semi-circle)
   $\therefore a = 40°$
   (c) $a = b = 40°$ as angle at O $= 80°$
   (exterior angle of triangle)
2. $a = 44°$ (angle sum of triangle)
3. (a) $x = y = 90°$ (angles in semi-circle);
   $a = 360° - (90° + 90° + 99°) = 81°$
   (b) Because the angles at A and B would both have to be 90° which they are not.
4. The interior angle of a hexagon $= 120°$.
   The base angles of the isosceles triangle are 30° and 30° therefore angle APB $= 120° - 30° = 90°$.
   Therefore AB is a diameter.

Activity 4 (page 177)

	Odd number of sides			Even number of sides		
Number of sides	3	5	7	4	6	8
right angles	1	3	5	4	5	6

For an odd number of sides the relation is 'number of sides − 2' or $n - 2$.

For an even number of sides it is 'half the number of sides + 2' or $^n/_2 + 2$.

Some 'What if … ?' questions could be:
• if sides are not vertical or horizontal, or
• if the polygons are concave …

Exercise 13.3a (page 177)

1. (a) 12
   (b) 8
   (c) 9
   (d) 3
   (e) 15
2. Yes
   (a) $360° \div 6 = 60°$
   (b) Exterior angle $= 120°$ therefore it has three sides and is therefore an equilateral triangle and it is possible.
3. (a) 176.4°
   (b) 179.64°

Exercise 13.4a (page 179)

1. $a = 80°, b = 50°, c = 30°,$
   $d = 80°, e = 30°, f = 80°,$
   $g = 70°$
2. $p = 90°, q = 40°, r = 115°,$
   $s = 115°, t = 65°, x = 140°,$
   $y = 135°$

Revision exercise 13a (page 180)

1. (a) $x = 58°, z = 58° + 71° = 129°$ (b) $x = 140°$
   (c) The fourth angle of the quadrilateral
   $= 360° - (50° + 70° + 130°) = 110°, x = 70°, y = 60°$
2. (a) $x = 47°$ (b) $x = 40°$ (c) $x = 135°$
3. The exterior angle $= 360° \div 20 = 18°$ so the interior angle $= 162°$.
4. The interior angle $= 168°$ so the exterior angle $= 12°$.
   The number of sides $= 360 \div 12 = 30$.
5. $a = 90°, b = 100°, c = 30°, d = 70°, e = 140°, f = 80°,$
   $g = 35°, h = 145°, i = 80°$

Chapter 14

Exercise 14.1a (page 183)

1. (a) Frequencies:
   2, 3, 4, 1
   (b) 12
   (c) 11.4

2. (a) 40
   (b) 1078
   (c) 26.95
3. (a) 27
   (b) 5

4. 10.15
5. (a) 26
   (b) 2
   (c) 2.3

Exercise 14.2a (page 187)

1. (a) 17.5 cm ≤ length < 18.5 cm
   (b) 34.5 m ≤ length < 35.5 m
   (c) 4.5 g ≤ mass< 9.5 g, 9.5 g ≤ mass < 14.5 g
   (d) 1.5 s ≤ time < 3.5 s, 3.5 s ≤ time < 5.5 s

2. (a) 15 cm
   (b) 2.25 m
   (c) 82.5 kg
   (d) 83 kg, 88 kg
   (e) 35.5 s, 45.5 s

3. (a) 5.2 s to 1 d.p.
   (b)
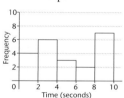

4. (a) 72.8 cm
   (b)
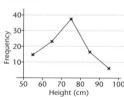

5. 1.5 m
6.
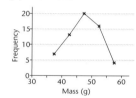

7. (a) 30 ≤ y < 40
   (b) 35.4 cm
8.
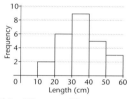

9. (a) 20 g
   (b) 47.25 g to 2 d.p.

10. (a)

Mass (g)	40−50	50−60	60−70	70−80	80−90
Frequency	5	13	21	8	3

   (b) 63.2 g

Revision exercise 14a (page 191)

1. 4.4 to 1 d.p.
2. (a) 30
   (b) 93
   (c) 4
   (d) 3
3. (a) 5
   (b) 6
   (c)

Number of videos	0	1	2	3	4	5	6
Frequency	5	7	2	1	6	8	3

   (d) 2.9 to 1 d.p.

4. (a) 3
   (b) 25
   (c) 4
   (d) 4.44
   (e) Very few people ate more than 5 chocolates but they have affected the mean.

5. (a)

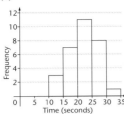

(b) 20 cm

6. 22.0 cm

7. (a)

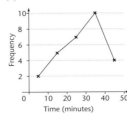

(b) 28.2 minutes to 1 d.p.

8. (a) $8 < x \leqslant 12$
   (b) 16 m
   (c) 10.6 m
9. (a) 28
   (b) (i) 3.5
       (ii) 7.5
   (c) 3.3 hours to 1 d.p.
10. (a) 5.45 hours
    (b) by the manufacturer advertising how long they last.

Chapter 15

1., 2.

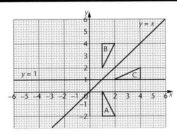

3., 4.

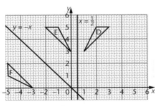

5., 6.

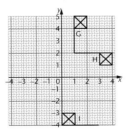

7., 8.

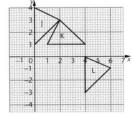

9. (a) Rotation through 90° clockwise about (2, 4).
   (b) Reflection in the line $x = -1$.
   (c) Reflection in the line $y = -x$.
   (d) Rotation through 180° about (4, 2).
   (e) Rotation through 90° clockwise about (3, −1).
   (f) Reflection in the line $y = -2\frac{1}{2}$.

1., 2.

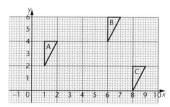

3., 4.

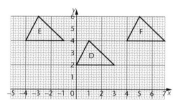

5., 6.

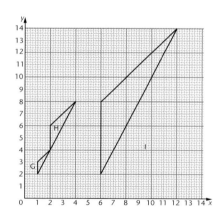

7., 8.

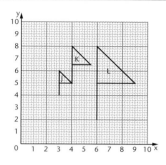

9. (a) Translation through $\begin{pmatrix} 3 \\ -5 \end{pmatrix}$.

   (b) Enlargement, scale factor 2, centre (0, 4).

   (c) Translation through $\begin{pmatrix} -8 \\ -3 \end{pmatrix}$.

   (d) Enlargement, scale factor $2\frac{1}{2}$, centre (0, 0).

   (e) Translation through $\begin{pmatrix} -6 \\ 4 \end{pmatrix}$.

   (f) Enlargement, scale factor $\frac{1}{3}$, centre (5, 3).

1. Translation through $\begin{pmatrix} 7 \\ -2 \end{pmatrix}$.

2. Translation through $\begin{pmatrix} 4 \\ 2 \end{pmatrix}$.

3. Translation through $\begin{pmatrix} a+b \\ c+d \end{pmatrix}$.

4. Enlargement, scale factor 6. centre (0, 2).
5. Enlargement, scale factor 3, centre ($1\frac{1}{2}$, 5).
6. Enlargement, scale factor pq.

7. Rotation through 90° anticlockwise about (1, 1).
8. Rotation through 90° anticlockwise about ($\frac{1}{2}$, $-\frac{1}{2}$).
9. Rotation about the point where the mirror lines cross. (The angle is twice the angle between the mirror lines, but students are unlikely to spot this.)
10. Rotation through 90° anticlockwise about (5, 0).
11. Translation through $\begin{pmatrix} 4 \\ -4 \end{pmatrix}$.

1., 2.

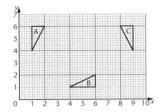

3. Reflection in the line $x = 5$.

4., 5.

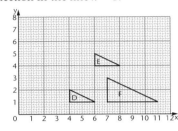

6. Enlargement, scale factor 2, centre (1, 1).

7. (a) Translation through
$$\begin{pmatrix} -6 \\ 0 \end{pmatrix}.$$

   (b) Reflection in the line $x = -4\frac{1}{2}$.
   (c) Rotation through 90° clockwise about (0, 2).
   (d) Rotation through 90° anticlockwise about $(-3, -1)$.
   (e) Enlargement, scale factor $\frac{1}{2}$, centre (0, 2).
   (f) Reflection in the line $y = -x$.

Chapter 16

Exercise 16.1a (page 214)

1. (a) $\frac{1}{7} = \frac{2}{14} = \frac{5}{35}$
   (b) $\frac{4}{9} = \frac{16}{36} = \frac{32}{72}$

2. (a) $\frac{3}{4}$
   (b) $\frac{4}{5}$
   (c) $\frac{1}{2}$
   (d) $\frac{2}{9}$

3. (a) 1
   (b) $\frac{5}{6}$
   (c) $\frac{17}{20}$
   (d) $\frac{5}{6}$
   (e) $\frac{31}{40}$
   (f) $\frac{11}{12}$

4. (a) $\frac{1}{7}$
   (b) $\frac{1}{2}$
   (c) $\frac{5}{12}$
   (d) $\frac{1}{4}$
   (e) $\frac{7}{24}$
   (f) $\frac{13}{36}$

5. (a) $\frac{9}{10}$
   (b) $\frac{23}{30}$
   (c) $\frac{11}{12}$
   (d) $\frac{11}{40}$
   (e) 0
   (f) $\frac{2}{7}$

Exercise 16.2a (page 216)

1. (a) $1\frac{3}{4}$
   (d) $3\frac{3}{4}$
   (b) $2\frac{2}{5}$
   (e) $12\frac{1}{2}$
   (c) $5\frac{2}{3}$

2. (a) $\frac{3}{2}$
   (d) $\frac{18}{7}$
   (b) $\frac{13}{5}$
   (e) $\frac{37}{4}$
   (c) $\frac{43}{8}$

3. (a) $4\frac{7}{12}$
   (d) $10\frac{5}{18}$
   (b) $3\frac{9}{10}$
   (e) $1\frac{1}{7}$
   (c) $6\frac{3}{20}$
   (f) $4\frac{5}{8}$

4. (a) $1\frac{1}{5}$
   (d) $1\frac{13}{20}$
   (b) $3\frac{1}{8}$
   (e) $1\frac{1}{2}$
   (c) $3\frac{1}{6}$
   (f) $4\frac{8}{15}$

5. (a) $1\frac{5}{12}$
   (d) $5\frac{1}{8}$
   (b) $2\frac{3}{4}$
   (e) $4\frac{7}{10}$
   (c) $3\frac{17}{20}$
   (f) $\frac{9}{14}$

6. $1\frac{3}{8}$

Exercise 16.3a (page 218)

1. $\frac{1}{6}$
2. $\frac{2}{5}$
3. $\frac{2}{9}$
4. $\frac{2}{9}$
5. $2\frac{2}{3}$
6. $\frac{2}{3}$

7. $\frac{9}{10}$
8. $9\frac{3}{4}$
9. $5\frac{1}{2}$
10. $1\frac{3}{4}$
11. $1\frac{3}{5}$
12. $2\frac{2}{3}$

Exercise 16.4a (page 220)

1. 14
2. 6
3. 16.5
4. −11
5. −2
6. −12
7. 0.14
8. −46.5 to 3 s.f.
9. 0.368 to 3 s.f.
10. −22.6 to 3 s.f.

Exercise 16.5a (page 222)

All answers are to 3 s.f.

1. (a) 209 (b) 2.37
2. (a) 0.722 (c) 0.852
   (b) 1.26 (d) −0.649
3. (a) 33.7° (c) 65.5°
   (b) 74.5° (d) 21.8°
4. 83 700
5. 7.57
6. 3.71×10^2
7. 600
8. 3.77
9. 0.0678
10. 7.46

Exercise 16.6a (page 223)

1. £20, £30, £50
2. 30°, 60°, 90°
3. £8
4. 3.2 litres
5. £150
6. £220
7. £146
8. 330 g
9. Michael £18.90; Iain £22.50
10. 1 : 4.43 : 0.732

Exercise 16.7a (page 226)

1. (a) 1.06.
   (b) 1.09
   (c) 1.175
   (d) 1.0125
   (e) $\frac{6}{5}$
   (f) $\frac{11}{9}$
2. (a) 0.94
   (b) 0.91
   (c) 0.825
   (d) 0.9875
   (e) $\frac{4}{5}$
   (f) $\frac{7}{9}$
3. £1340
4. 3525
5. £167
6. £38774
7. £19.26
8. £18 575
9. $(1.07)^{10} = 1.97$. So it does not quite double.
10. (a) £4051.69
    (b) £5033.40

Exercise 16.8a (page 229)

1. £53.75
2. 40
3. £75.05
4. £12 500
5. 1540
6. £2.25
7. £27 000
8. £24 000
9. £480
10. (a) 79p (b) 1.24

Revision exercise 16a (page 231)

1. (a) $1\frac{1}{6}$ (f) $\frac{2}{5}$
   (b) $4\frac{13}{20}$ (g) $1\frac{1}{2}$
   (c) $\frac{7}{18}$ (h) 8
   (d) $\frac{7}{12}$ (i) $1\frac{1}{5}$
   (e) $2\frac{1}{3}$ (j) $2\frac{1}{2}$
2. $2\frac{1}{6}$ inches
3. (a) −8 (c) 4
   (b) −12
4. (a) 11.16 (i) 1.44
   (b) −63.6 (j) 9.43
   (c) 0.147 (k) 1.74
   (d) 0.969 (l) −50.7
   (e) 21.4° (m) 2.52
   (f) 2.94 (n) 56.0°
   (g) 19 683 (o) 1.45
   (h) 1840
5. £375
6. 19
7. 11 865
8. 1679
9. 331
10. £520
11. 918
12. 234

Chapter 17

1. $V = -21$
2. $P = 55$
3. $T = 2$
4. $M = 10$
5. $R = 24$
6. $L = 2\frac{1}{6}$
7. $D = \frac{8}{25}$
8. $M = 0.563$
9. (a) $S = 720$ m
   (b) $S = 30.625$ m
10. $A = 111.5$

1. $5a + b$
2. $6ab - 4ac$
3. $-a^2 + 2b^2$
4. $2x^2 - 4xy + y^2$
5. $2b^2 - a^2$
6. Will not simplify.
7. $3a^3 + 7a^2$
8. $14abc$
9. $2x^2$
10. 0
11. $5a + 8$
12. $19x - 9$
13. $2b + 16$
14. $3x - 1$
15. $2x^2 + 5xy - 12y^2$

1. $x^2 + 5x + 6$
2. $a^2 + 7a + 12$
3. $a^2 + 3a + 1$
4. $2x^2 - 3x - 2$
5. $2x^2 + x - 6$
6. $6a^2 - 4ab - 2b^2$
7. $4a^2 + 7ab - 2b^2$
8. $6a^2 - 13ab + 6b^2$
9. $8a^2 - 18ab + 9b^2$
10. $20 - 7b - 6b^2$
11. $6a^2 - 5ab + b^2$
12. $14a^2 + 13ab + 3b^2$
13. $a^2 + 4a + 4$
14. $16x^2 - 24xy + 9y^2$
15. $9x^2 - 6xy + y^2$
16. $a^2 - 4$
17. $9a^2 - b^2$
18. $25x^2 - 4y^2$
19. $4a^2 - ab - 3b^2$
20. $10a^2 + 3ab - 4b^2$

1. $12a^5$
2. $2a^2$
3. $9a^6$
4. $6a^5b^3$
5. Cannot simplify
6. $5ab^2$
7. $3p^3$
8. $2abc^2$
9. $3t$
10. $2a^3b^3$

1. $n^2 + 1$
2. $2n^2 - 1$
3. $n^2 + 4$
4. $3n^2 + 3$
5. $n^2 - 2$
6. $3n^2 + 4$
7. $4n^2 - 3$
8. $n^2 + 2n + 1$
9. $n^2 + 3n - 2$
10. $n^2 + n$

1. $2(a + 4)$
2. $a(3 + 5a)$
3. $2a(b - 3c)$
4. $5ab(a + 2b)$
5. $x^2y(2y - 3x)$
6. $3ab(a - 2b)$
7. $2(6x - 3y + 4z)$
8. $3b(3a + 2b)$
9. $2ac(2a - c)$
10. $5y(3x - 1)$
11. $2a(3a^2 - 2a + 1)$
12. $3a^2b(1 - 3ab)$
13. $5abc(abc - 2)$
14. $a^2b(2 - 3b^2 + 7a^2)$
15. $a(4bc - 3c^2 + 2ab)$

1. $(x + 3)(x + 2)$
2. $(x + 5)(x + 1)$
3. $(x + 4)(x + 2)$
4. $(x + 4)(x + 1)$
5. $(x + 1)(x + 1)$
6. $(x - 6)(x - 1)$
7. $(x - 2)(x - 5)$
8. $(x - 3)(x - 1)$
9. $(x - 7)(x - 2)$
10. $(x - 4)(x - 2)$
11. $(a + 6)(a + 2)$
12. $(a - 3)(a - 3)$
13. $(b - 8)(b - 4)$
14. $(x + 3)(x + 8)$
15. $(x - 5)(x - 4)$

1. $(x - 4)(x + 2)$
2. $(x - 1)(x + 5)$
3. $(x + 2)(x - 3)$
4. $(x + 6)(x - 1)$
5. $(x + 3)(x - 1)$
6. $(x - 6)(x + 3)$
7. $(x - 10)(x + 1)$
8. $(x + 7)(x + 2)$
9. $(y + 11)(y - 2)$
10. $(x + 4)(x - 3)$
11. $(a - 2)(a + 10)$
12. $(a - 9)(a + 3)$
13. $(b + 2)(b + 10)$
14. $(x - 5)(x + 5)$
15. $(x - 7)(x + 7)$

Exercise 17.9a (page 247)

1. $b = a + c$
2. $x = \dfrac{3a}{y + w}$
3. $t = \dfrac{v - u}{a}$
4. $T = HA$
5. $T = \dfrac{P - C}{3}$
6. $u = 2P - v$
7. $r = \dfrac{C}{2\pi}$
8. $q = \sqrt{\dfrac{A - pr}{p}}$
9. $x = \frac{1}{2}P - y$
10. (a) $n = \dfrac{C - A}{32}$
    (b) $n = 56$

Exercise 17.10a (page 250)

1. (a) $x = -4, -3, -2, -1$
   (b) $x = 2, 3, 4, 5$
2. $x \leqslant 7$
3. $x > 2$
4. $x < 4$
5. $x \leqslant 1$
6. $x \geqslant 5$
7. $x > \frac{1}{2}$
8. $x < 3$
9. $x > 2.5$
10. $x \geqslant 4$
11. $x < -1$
12. $x \geqslant 2$
13. $x < 4$
14. $x > -3$
15. $x < -8.5$

Exercise 17.11a (page 251)

1. $2x + 3 = 23$, $x = 10$, their ages are 10 and 13.
2. $3a + 15 = 180$, $a = 55$, the angles are 55°, 55°, 70°.
3. $30 + 20x \leqslant 240$, $x \leqslant 10.5$, the most lengths he can hang is 10.
4. $2x + 28 = 616$, $x = 294$, there are 294 boys and 322 girls.
5. (a) $2x + 60 \geqslant 225$, $x \geqslant 82.5$
   (b) The smallest distance is $82\frac{1}{2}$ miles (accept 83 miles).
6. (a) $5x - 6$
   (b) (i) $5x - 6 = 19$, $x = 5$
       (ii) Adult bikes £5, Child bikes £3.
7. (a) $6x \leqslant 40$, $x \leqslant 6\frac{2}{3}$ (b) 6 m by 12 m
8. Patrick has x cars. $3x + 11 = 41$, $x = 10$
   Mark 14, Patrick 10, Iain 17

Revision exercise 17a (page 254)

1. (a) $h = 20$
   (b) $h = 1$
   (c) $h = -11.8$
2. $A = 222$
3. (a) $5a$
   (b) $ab^2 + a^2b$
   (c) $4ab + 2ac$
   (d) $4x^2 + xy$
4. (a) $7x + 5$
   (b) $4x - 6$
   (c) $a + 16$
   (d) $4x + 4$
   (e) $4x - 1$
   (f) $3x^2 - 18xy + 4y^2$
5. (a) $x^2 + 8x + 7$
   (b) $a^2 + 2a - 15$
   (c) $2y^2 - 2y - 4$
   (d) $2x^2 - 9x - 5$
   (e) $4a^2 - 3ab - b^2$
   (f) $6 - 7c + 2c^2$
   (g) $x^2 - 25$
   (h) $x^2 + 4xy + 4y^2$
   (i) $14x^2 + 11xy - 15y^2$
6. (a) $2a^5$
   (b) $5a$
   (c) a^5
   (d) $24a^4b^4$
   (e) $3xz$
   (f) $4a^2c$
7. (a) $n^2 + 3$
   (b) $2n^2 + 1$
   (c) $n^2 + n + 2$
8. (a) $3(a + 2b - 4c)$
   (b) $a(2 + 3b)$
   (c) $ab(a - 3b)$
   (d) $2xy(x - 3)$
   (e) $7ab(c + 2a)$
   (f) $3(3a^2 + b^2 - 2c^2)$
   (g) $5(pq - 2)$
   (h) $2a(1 - 2a + 3a^2)$
   (i) $50ac(2b - 1)$
9. (a) $(x + 4)(x + 1)$
   (b) $(x - 2)(x - 4)$
   (c) $(x - 8)(x - 2)$
   (d) $(x + 5)(x + 3)$
   (e) $(x - 7)(x + 1)$
   (f) $(x + 2)(x - 5)$
   (g) $(x - 6)(x - 2)$
   (h) $(x - 5)(x + 3)$
   (i) $(x - 10)(x + 7)$
   (j) $(x + 12)(x + 4)$
   (k) $(x - 9)(x + 2)$
   (l) $(x + 10)(x - 2)$
10. (a) $y = x + 3b$
    (b) $u = 2t - v$
    (c) $a = 2b - P$
    (d) $q = \dfrac{p - m}{x}$
    (e) $P = \dfrac{100I}{TR}$
    (f) $S = \dfrac{v^2 - u^2}{2a}$
11. (a) $x > 2\frac{1}{2}$
    (b) $x \leqslant 2$
    (c) $x \geqslant 6$
    (d) $x < 2$
    (e) $x \leqslant 8$
    (f) $x \leqslant 2\frac{1}{3}$
    (g) $x > -2$
    (h) $x < \frac{1}{2}$
    (i) $x > -\frac{1}{4}$
12. (a) $3x + 3$
    (b) $3x + 3 = 39$, $x = 12$
    (c) Their ages are 10, 12, 17.
13. (a) $24x + 320 \leqslant 500$, $x \leqslant 7.5$
    (b) She buys 7 packets of crisps.
14. (a) $5x + 110 = 360$, $x = 50$
    (b) The angles are 50°, 90°, 150°, 70°.

Answers

Chapter 18

Compost B

Height (h cm)	Frequency	Cumulative frequency
$0 \leqslant h < 10$	1	1
$10 \leqslant h < 20$	1	2
$20 \leqslant h < 30$	5	7
$30 \leqslant h < 40$	10	17
$40 \leqslant h < 50$	14	31
$50 \leqslant h < 60$	13	44
$60 \leqslant h < 70$	10	54
$70 \leqslant h < 80$	5	59
$80 \leqslant h < 90$	1	60

median = 49 cm
interquartile range = 61 cm − 38 cm = 23 cm

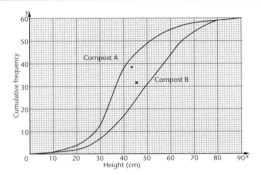

The median is higher for compost B but the interquartile range is wider. This implies a greater variation in plant heights.

Exercise 18.1a (page 262)

1. (a)

Speed (v km/h)	Cumulative frequency Policeman A	Cumulative frequency Policeman B
$10 \leqslant v < 30$	6	2
$30 \leqslant v < 50$	18	6
$50 \leqslant v < 70$	30	10
$70 \leqslant v < 90$	43	29
$90 \leqslant v < 110$	47	48
$110 \leqslant v < 130$	50	50

   (b) medians: A = 62 km/h, B = 88 km/h
   lower quartiles: A = 43 km/h, B = 76 km/h
   upper quartiles: A = 80 km/h, B = 96 km/h

   (c) For A the median speed is lower than for B and the interquartile range is higher than for B. A might be in a built-up area, B on a motorway

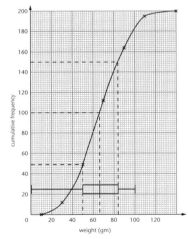

2. (a) lower quartile at 4, median at 8, upper quartile at 15.5
   (b) lower quartile at 2.7, median at 5.4, upper quartile at 8.1

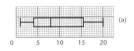

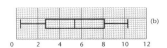

1. (a)

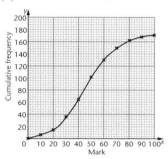

(b) (i) £780   (iii) 60
    (ii) £940 − £600 = £340

2. (a)

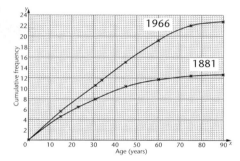

(b) (i) 46   (iii) 170 − 118 = 52
    (ii) 58 − 32 = 26   (iv) 60% = 102

3. (a)

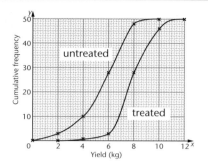

Treated: median = 7.8 kg,
interquartile range 8.9 − 7.1 = 18 kg
Untreated: median = 5.8 kg,
interquartile range 6.7 − 4.4 = 2.3 kg
The average yield is higher for the treated
trees as there is less variation.

(b) Yes, the chemical was effective.

4.

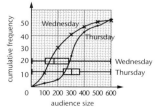

1881: median 22 years,
interquartile range 40.5 − 10 = 30.5
1966: median 33 years,
interquartile range 53 − 15 = 38
The average age was much higher in 1966
and the ages were more spread out.

5. (a)

Audience size	50–99	100–199	200–299	300–399	400–499	500–599
CF Wednesdays	11	31	41	47	51	52
CF Thursdays	3	6	24	43	48	53

(b) medians: Wednesday = 170, Thursday = 305
lower quartiles: Wednesday = 105, Thursday = 245
upper quartiles: Wednesday = 280, Thursday = 365

Answers

Chapter 19

1. (a) $30 \div 5 = 6$, 6.4536
   (b) $100 \times 3 = 300$, 339.365
   (c) $40 \div 10 = 4$, 5.0933
   (d) $4 \times 10 \times 20 = 800$, 1153.26
2. (a) $(10 \times 20) \div 60 = 3$, 5.4557
   (b) $(5^2 \times 50) \div 100 = 12.5$, 12.2409
   (c) $\sqrt{(5 \times 5)} = 5$, 5.047 77
3. (a) $0.4 \times 90 \div 8 = 4.5$
   (b) $10 \div 6 = 1.667$
   (c) $10 \times 0.07 = 0.7$
   (d) $1 \div 5 = 0.2$
4. Students' own work

1. (a) £330.75 − £300 = £30.75
   (b) £1169.86 − £1000 = £169.86
   (c) £491.73 − £450 = £41.73
   (d) £5832 − £5000 = £832
   (e) £39 323.88 − £30 000 = £9323.88
2. Five years at 8% produces £1469.32, four years at 9% produces £1411.58

1. £168
2. £212
3. £25.99
4. £20.50
5. £33.24

1. 60.48 m/s
2. 10.5 miles
3. 7.5 hours
4. £2.475 per kg,
   £1.895 per kg, £1.96 per kg
5. 277.78 cm³
6. 2.45×10 people/km²

1. $3\frac{1}{2}$ minutes
2. 3 weeks
3. 5.7m
4. £16.10
5. 6650 km
6. 0.097 cm²

1. (a) 960   (b) 1400   (c) 132
2. £271.84
3. compound interest = £135.69
   simple interest = £126
   difference = £9.69
4. 45.1 m.p.h.
5. £65.27
6. 576 492
7. 5.19 litres/minute,
   3.85 minutes or 3 minutes 51 seconds
8. 500 g
9. 1.43×10^3 kg/m³
10. 11m
11. 1.44×10^8 km
12. 75.36 kg

Chapter 20

(a) 15 cm²   (d) 200 cm²
(b) 351 cm²   (e) 152 cm²
(c) 168 cm²

(a) 5 cm   (d) 11.31 cm
(b) 11.18 cm (e) 11.4 cm
(c) 5.39 cm   (f) 13 cm

Activity 4 (page 278)

- (a) 8 cm
  (b) 5.66 cm
  (c) 16 cm
  (d) 28.91 cm
  (e) 168.93 cm
  (f) 4 cm
  (g) 14.28 cm
  (h) 8.94 cm

- (a) 250.4 m
  (b) 28.62 m
  (c) 4.9 m
  (d) $30 + 21.2 + 37.08$
  $= 88.28$ cm
  (e) 3.23 m

Activity 5 (page 279)

Students' own work

- (a) 13.61 cm
  (b) 7.28 cm
  (c) 3.62 cm
  (d) 41.71 cm
  (e) 7 cm
  (f) 11.44 cm

- (a) 36.87°
  (b) 39.8°
  (c) 26.57°
  (d) 21.8°
  (e) 70.87°

- (a) 4.36 cm
  (b) 1.66 cm
  (c) 4.15 cm
  (d) 6.8 cm
  (e) 4.14 cm
  (f) 42.1°
  (g) 61.3°
  (h) 26.5°
  (i) 51°
  (j) 60°

- (a) 68°
  (b) 4.14 m
  (c) 18.8 m
  (d) 44.1 km
  (e) 36.4°
  (f) 64°
  (g) 582 m
  (h) east 53.6 km, south 45 km

Activity 6 (page 285)

(a) $\sqrt{2}$
(b) $\sqrt{3}$, $\sin 30° = \frac{1}{2}$, $\cos 30° = \frac{\sqrt{3}}{2}$
(c) 2, $\cos 45° = \frac{\sqrt{2}}{2}$, $\sin 45° = \frac{\sqrt{2}}{2}$

Revision exercise 20a (page 286)

1. (a) 7.21 cm
   (b) 12.57 cm
   (c) 6.81 cm
2. 25.06 cm
3. 36.06 km
4. (a) 35.1°   (d) 8.43 cm
   (b) 61.0°   (e) 7.55 cm
   (c) 59.7°   (f) 16.49 cm
5. 43.5 m

6. 45.6°, 45.6°, 88.8°
7. 4.6 m
8. (a) west 4.83 km,
      north 1.29 km
   (b) 131.6°, 6.46 km
9. $PQ^2 = (4 - 1)^2 + (3 - 1)^2$
   $= 9 + 4$
   $= 13$
   $PQ^2 = \sqrt{13} = 3.61$

Chapter 21

Exercise 21.1a (page 288)

1. $x = 10$
2. $x = 0$
3. $x = 3$
4. $x = 2\frac{2}{3}$
5. $x = -5$
6. $x = 2\frac{3}{4}$
7. $x = 4$
8. $x = 6$
9. $x = 2$
10. $x = \frac{6}{7}$
11. $x = 25$
12. $x = 20$
13. $x = 12\frac{1}{2}$
14. $x = 2.74$
15. $x = 4.81$
16. $x = 1.63$
17. $x = 3.64$
18. $x = 5.30$
19. $x = -1.44$
20. $x = 0.511$

Exercise 21.2a (page 289)

1. $x < 1$
2. $x > 2$
3. $x > 3$
4. $x \geqslant 5$
5. $x > 6$
6. $x \leqslant -2$ or $x \geqslant 2$
7. $-5 < x < 5$
8. $x \leqslant -1$ or $x \geqslant 1$
9. $-4 \leqslant x \leqslant 4$
10. $-4 < x < 4$

Exercise 21.3a (page 291)

1. $x = 2(2x - 30)$ $x = 20$
   Angles are 20° and 10°
2. $3(x + 4) = 27$ $x = 5$
3. $x = 3(32 - x)$ $x = 24$
   24 girls, 8 boys
4. $\frac{2}{3}x = x - 20$ $x = 60$
5. $2n - 5 = 3(n - 2)$, $n = 1$
6. $4(p + 12) = 28p$, $p = 2$,
   children pay £2, adults £14
7. $3x + 5(x + 2) = £38$,
   $x = 3.5$, a one course meals cost £3.50
8. $x^2 < 36$, $-6 < x < 6$, the number lies between
   -6 and $+6$

Exercise 21.4a (page 294)

1. $x = 4, y = 1$
2. $x = 2, y = 3$
3. $x = 1, y = 3$
4. $x = 2, y = 3$
5. $x = 5, y = 1$
6. $x = 4, y = 2$
7. $x = 2, y = 1$
8. $x = 3, y = 2$
9. $x = 2, y = 3$
10. $x = 1, y = 5$
11. $x = 2, y = 1$
12. $x = 2\frac{1}{2}, y = 1\frac{1}{2}$
13. $x = 2\frac{1}{2}, y = 1$
14. $x = 3, y = -1$
15. $x = 3\frac{1}{2}, y = 1$

Exercise 21.5a (page 296)

1. $x = 2, y = 3$
2. $x = 1, y = 1$
3. $x = 2, y = 3$
4. $x = 1, y = 2$
5. $x = 5, y = 6$
6. $x = 2, y = 1$
7. $x = -1, y = 2$
8. $x = -2, y = -3$
9. $x = -1, y = 4$
10. $x = 2\frac{9}{26}, y = 3\frac{5}{26}$

Exercise 21.6a (page 298)

1. $x = 2$ or 3
2. $x = 5$ or 1
3. $x = -2$ or -4
4. $x = -4$ or -1
5. $x = -1$ twice
6. $x = 6$ or 1
7. $x = 2$ or 5
8. $x = 1$ or 3
9. $x = 7$ or 2
10. $x = 2$ or 4
11. $x = 4$ or -2
12. $x = -5$ or 1
13. $x = 3$ or -2
14. $x = -6$ or 1
15. $x = -3$ or 1
16. $x = 6$ or -3
17. $x = 10$ or -1
18. $x = -7$ or -2
19. $x = -11$ or 2
20. $x = -4$ or 3

Exercise 21.7a (page 300)

1.
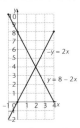

$x = 2, y = 4$

2.
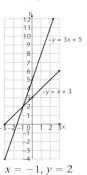

$x = -1, y = 2$

3.
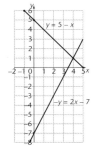

$x = 4, y = 1$

4.
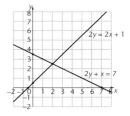

$x = 2, y = 2.5$

5. (a)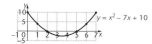
$y = x^2 - 7x + 10$

(b) $x = 2$ or 5.

6. (a)
$y = x^2 - x - 2$

(b) $x = -1$ or $+2$

7. (a)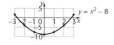
$y = x^2 - 8$

(b) $x = -2.8$ or $+2.8$

8. (a)
$y = x^2 + x - 3$

(b) $x = -2.3$ or $+1.3$

9. (a)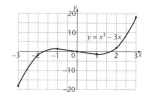
$y = x^3 - 3x$

(b) $x = -1.7$, 0 or 1.7

10. (a)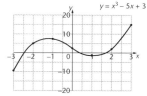
$y = x^3 - 5x + 3$

(b) $x = -2.5$, 0.7 or 1.8

Exercise 21.8a (page 302)

1. $x = 1.7$
2. $x = 2.8$
3. $x = 2.8$
4. $x = 4.6$
5. $x = 3.5$
6. $x = 2.29$
7. $x = -3.7$
8. $x = -2.5$
9. $x = 2.59$
10. $x = -2.62$

Exercise 21.9a (page 304)

1. (a) $x + y = 47$, $x - y = 9$
   (b) $x = 28$, $y = 19$
2. (a) $2c + 3d = 27.5$, $3c + d = 18.5$
   (b) A cassette costs £4, a disc costs £6.50
3. (a) $s + 2b = 13$, $2s + b = 11$
   (b) Small holds 3 litres, large holds 5 litres
4. $a + 2c = 31$, $2a + 3c = 54$, adult costs £15, child costs £8
5. $x(x + 2) = 63$ or $x^2 + 2x - 63 = 0$, the numbers are 7 and 9

Exercise 21.10a (page 306)

1. $x > 2$
2. $y < 2x$
3. $3x + 4y > 12$
4. $y < 2x + 1$
5.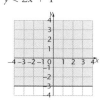

6.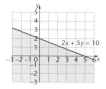
$2x + 5y = 10$

7.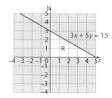
$3x + 5y = 15$

The region required is labelled R.

8.
$y = 2x$

The required region is labelled R.

1. $t = \dfrac{s}{a + 2b}$

2. $t = \dfrac{Pb}{b - a}$

3. $r = \sqrt{\dfrac{A}{\pi}}$

4. $a = \dfrac{cd}{b - c}$

5. $c = \dfrac{ab}{a + d}$

6. $s = \dfrac{bx + 2ax}{1 + b}$

7. $x = \dfrac{s + bs}{b + 2a}$

8. $t = \dfrac{ab}{1 - bs}$

9. $c = \sqrt{a - b}$

10. $P = \dfrac{100A}{100 + RT}$

1. (a) $x = 3$
   (b) $x = 4$
   (c) $x = 6$
   (d) $x = 6$
   (e) $x = 4$
   (f) $x = 1\frac{2}{7}$
   (g) $x = 25$
   (h) $x = 5$

2. (a) $x < 3$
   (b) $x \leqslant 4$
   (c) $x > 5$
   (d) $-7 \leqslant x \leqslant 7$
   (e) $x \leqslant -4$ or $x \geqslant 4$
   (f) $-6 < x < 6$

3. $\dfrac{x}{3} = 3x - 24$, $x = 9$

4. (a) $x + 20$
   (b) $3x + 2(x + 20) = 340$,
   $x = 60$, a lolly costs 60p,
   an ice-cream costs 80p

5. (a) $x - 25$
   (b) $x - 25 = \dfrac{4x}{5}$, $x = 125$,
   Marcia is 125 cm,
   Carole is 100 cm

6. (a) $x = 7, y = 8$
   (b) $x = 2, y = 3$
   (c) $x = 3, y = 1$
   (d) $x = 5, y = 2$
   (e) $x = -1, y = 2$
   (f) $x = -2, y = 3$
   (g) $x = 1, y = 2$
   (h) $x = \frac{1}{2}, y = 2\frac{1}{2}$
   (i) $x = \frac{1}{2}, y = -1$

7. (a) $x = 2$ or 4
   (b) $x = -2$ or -3
   (c) $x = 3$ or -1
   (d) $x = 5$ or -2
   (e) $x = 4$ or 1
   (f) $x = -5$ or -2
   (g) $x = 7$ or -2
   (h) $x = -15$ or -2
   (i) $x = 5$ or 4
   (j) $x = -3$ or -1
   (k) $x = 12$ or -3
   (l) $x = -9$ or 2

8. (a)

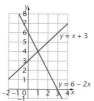

   $x = 1, y = 4$

   (b)

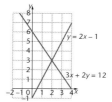

   $x = 2, y = 3$

9. (a)

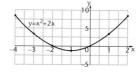

   (b) $x = -2$ or 0

10. (a)

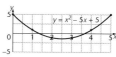

   (b) $x = 1.4$ or 3.6

11. (a)

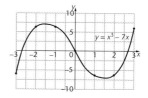

   (b) $x = -2.6$, 0 or 2.6

12. (a) $x = 2.6$
   (b) $x = 3.3$
   (c) $x = 2.1$
   (d) $x = -3.4$

13. $3c + 2a = 139$, $2c + a = 81$, crisps cost 23p per packet, apple juice costs 35p per can.

14. $y - x = 200$ or $-x + y = 200$, $3x + 2y = 2400$, medium hold 400 g, large hold 600 g

15. (a) $y < 3$
   (b) $3x + 4y > 12$

16. (a)

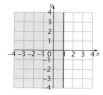

(b)

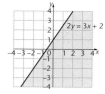

17.

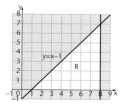

The region required is labelled R.

18.

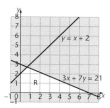

The region required is labelled R.

Chapter 22

Exercise 22.la (page 313)

1. 7 cm (accept 6−8 cm)
2. 110° (accept 100°−120°)
3. 8−10 m
4. 25 g
5. 4 kg

Exercise 22.2a (page 316)

1. discrete: 6-way, 2 pockets; continuous: 20 cm
2. discrete: 2 compartments, 3 pen holders;
   continuous: size (H)31.5 cm, (W)44.5 cm, (D)11.5 cm
3. 108 goals, 167 games, or first goal
4. 30 minutes, six minutes, 5 yards
5. Check students' own work
6. (c)
7. to nearest m or 10 m
8. (a) to nearest mile (b) to nearest $\frac{1}{4}$ or $\frac{1}{2}$ mile
9. to the age in years at the last birthday
10. to nearest ounce or nearest 25 g

Exercise 22.3a (page 318)

1. (a) LB 26.5 cm, UB 27.5 cm
   (b) LB 29.5 cm, UB 30.5 cm
   (c) LB 127.5 cm, UB 128.5 cm
2. (a) LB 5 cm, UB 15 cm
   (b) LB 25 cm, UB 35 cm
   (c) LB 145 cm, UB 155 cm
3. (a) LB 5.55 cm, UB 5.65 cm
   (b) LB 0.75 cm, UB 0.85 cm
   (c) LB 11.95 cm, UB 12.05 cm

4. (a) LB 1.225 m, UB 1.235 m
   (b) LB 0.445 m, UB 0.455 m
   (c) LB 9.075 m, UB 9.085 m
5. (a) 10.615 s, UB 10.625 s
   (b) LB 9.805 s, UB 9.815 s
   (c) LB 48.095 s, UB 48.105 s
6. 56.5 and 57.5
7. 4.65 and 4.75
8. 467.5 and 468.5
9. 34.905 and 34.915
10. 0.6335 and 0.6345

1. 50 mph
2. 4.5 m/s
3. 8.75 g/cm³
4. 7500 people/km²
5. 20 mph
6. 18 km/h
7. 6 g
8. 20 square miles
9. 11.4 s
10. 75 km/h

1. 85 cm–1 m
2. $\frac{1}{4}$ to $\frac{1}{3}$ mile or 0.4–0.5 km
3. (a) Check students' answers   (b) Check students' answers
4. (a) Jenny's
   (b) Jenny should have used cm. Her line was 10.2 cm long. Suni could not measure his line as accurately as 9.68 cm with a ruler.
5. nearest second
6. 16.5 cm by 27.5 cm
7. (a) 6.81 cm   (b) to the nearest $\frac{1}{100}$ cm
8. 7.36 m/s
9. 15 km/h
10. 8 cm³

Chapter 23

1. (a) 5 minutes
   (b) 30 litres
   (c) 10
2. (a) 9.30 a.m.
   (b) 11 a.m.–12 noon
3.

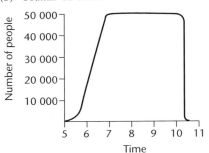

4. (a) £42
   (b) 225
   (c) (i) £30
       (ii) 8p
5.

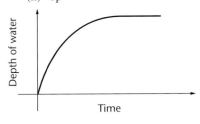

1. (a) 2
   (b) −3
   (c) 1
2. (a) 2.5
   (b) 0.75
   (c) $-\frac{2}{3}$
3. (a) 6
   (b) 0
   (c) −5
   (d) $\frac{1}{4}$
4. (a) $-\frac{1}{2}$
   (b) 2.5
   (c) 1
   (d) 0.75
5. speed = 2 m/s
6. AB: 0.4, AC: 2.5, BC: −1
7. 0.26; the cost per minute, in £
8. Check students' graphs
   (a) 3
   (b) 3
9. Check students' graphs
   (a) 2
   (b) 5
   (c) 4
10. Check students' graphs
    (a) −2
    (b) −3
    (c) −1

Exercise 23.3a (page 331)

1. (a) $y = 3x + 2$ (c) $y = 5x$
   (b) $y = -x + 4$
2. (a) $y = 4x + 2$ (c) $y = 2x$
   (b) $y = \frac{1}{3}x + 4$
3. (a) $y = -x + 5$ (c) $y = -\frac{1}{2}x - 3$
   (b) $y = -1.5x + 1$
4. (a) $3, (0,-2)$ (c) $-2, (0,7)$
   (b) $5, (0,2)$
5. (a) $-2, (0,5)$ (c) $-1.2, (0,2)$
   (b) $-2, (0,3.5)$
6. $C = 0.08m + 30$
7. $C = 0.26x$
8.

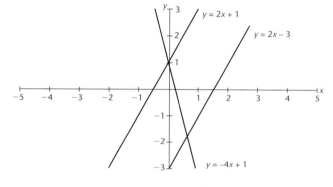

9. (a) (ii) (c) (iii)
   (b) (i)

10. (a)

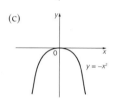

(b)

(c)

(d)

Revision exercise 23a (page 334)

1. (a) 5 miles
   (b) The train was not moving
   (c) The train was braking for the station
   (d) CD
2.

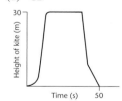

3. (a) 2.5
   (b) -1
   (c) 0
4. AB: -0.5, AC: 2.5, BC: $\frac{1}{4}$

5. (a) $2, (0, -3)$
   (b) $0.5, (0, 0)$
   (c) $\frac{1}{3}, (0, \frac{2}{3})$
   (d) $-0.4, (0, 2)$
6. (a) Check students' graphs
   (b) 1.2
   (c) 2.6 km
   (d) $d = 1.2t + 2.6$
7. (a) $y = 1.5x + 1$
   (b) $y = -0.5x + 2$
   (c) $y = -4x$
   (d) $y = 1.5x - 3$
8. Check students' sketches
9. (a)

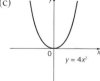

(b)

(c)

10. (a) (iii)
    (b) (iv)
    (c) (ii)
    (d) (i)

Answers

Chapter 24

1.

First sock	Second sock
brown	brown
brown	red
brown	green
red	brown
red	red
red	green
green	brown
green	red
green	green

2.

First game	Second game	Third game
Alex	Alex	Alex
Alex	Alex	Meiling
Alex	Meiling	Alex
Meiling	Alex	Alex
Alex	Meiling	Meiling
Meiling	Alex	Meiling
Meiling	Meiling	Alex
Meiling	Meiling	Meiling

3.

First	Second	Third
B	B	B
B	B	G
B	G	B
G	B	B
B	G	G
G	B	G
G	G	B
G	G	G

(a) $\frac{1}{8}$
(b) $\frac{3}{8}$

4.

5	6	7	8	9	10
4	5	6	7	8	9
2nd Spin 3	4	5	6	7	8
2	3	4	5	6	7
1	2	3	4	5	6
	1	2	3	4	5

1st Spin

(a) $\frac{1}{25}$
(b) $\frac{4}{25}$

5.

C	X	X	X	X
D	X	X	X	X
S	X	X	X	X
H	X	X	X	X
	H	S	D	C

(a) $\frac{1}{16}$
(b) $\frac{4}{16} = \frac{1}{4}$

6.

6	X	X	X	X
5	X	X	X	X
4	X	X	X	X
3	X	X	X	X
2	X	X	X	X
1	X	X	X	X
	H	S	D	C

(a) $\frac{1}{24}$
(b) $\frac{2}{24} = \frac{1}{12}$
(c) $\frac{3}{24} = \frac{1}{8}$

7.

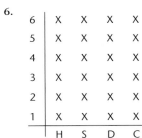

Card   Coin

(a) $\frac{1}{8}$
(b) $\frac{2}{8} = \frac{1}{4}$

8.

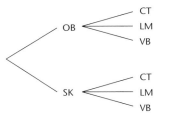

Starter   Main Course

Answer $= \frac{1}{6}$

1. $\frac{8}{10} = \frac{4}{5}$
2. 0.75
3. $\frac{8}{52} = \frac{2}{13}$
4. $\frac{9}{100} = 0.09$
5. 0.24
6. (a) $\frac{1}{3}$
   (b) $\frac{1}{9}$
7. $\frac{1}{169}$
8. 0.3, Likely to be influenced by one another, therefore not independent.

1.

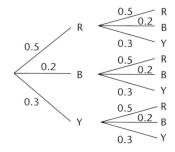

1st Ball   2nd Ball

(a) 0.49
(b) 0.21
(c) 0.21
(d) 0.42

2.

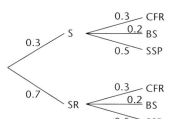

Starter   Main Course

(b) (i) 0.14
    (ii) 0.15

3.

Monday   Tuesday

(a) 0.64
(b) 0.32

4. (a)

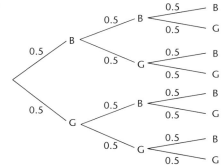

1st   2nd

(b) (i) 0.25
    (ii) 0.38

5.

B   G

(a) 0.125
(b) 0.375

1. 0.55

2. 0.06

3. $\frac{1}{100}$

4. 0.64

5.

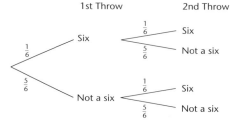

2nd die	6	6	12	18	24	30	36
	5	5	10	15	20	25	30
	4	4	8	12	16	20	24
	3	3	6	9	12	15	18
	2	2	4	6	8	10	12
	1	1	2	3	4	5	6
		1	2	3	4	5	6

1st die

(a) $\frac{1}{36}$

(b) $\frac{4}{36} = \frac{1}{9}$

(c) $\frac{3}{36} = \frac{1}{3}$

6.

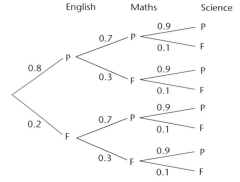

1st Throw   2nd Throw

$\frac{1}{6}$ Six $\frac{1}{6}$ Six

    $\frac{5}{6}$ Not a six

$\frac{5}{6}$ Not a six $\frac{1}{6}$ Six

    $\frac{5}{6}$ Not a six

(a) $\frac{1}{36}$

(b) $\frac{11}{36}$

7.

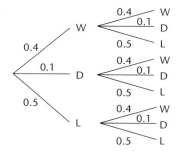

(a) 0.25

(b) 0.75

(c) 0.08

8. (a) 0.01

     (b) 0.18

9.

English	Maths	Science

0.8 P — 0.7 P — 0.9 P
       0.1 F
     0.3 F — 0.9 P
       0.1 F
0.2 F — 0.7 P — 0.9 P
       0.1 F
     0.3 F — 0.9 P
       0.1 F

(a) 0.504

(b) 0.398

Chapter 25

Exercise 25.1a (page 351)

1. (a) 40 cm²

     (b) 42 cm²

     (c) 30 cm²

2. (a) 46 cm²

     (b) 14 cm²

     (c) 42 cm²

3. (a) 24 cm²

     (b) 15 cm²

     (c) 21 cm²

4. $x = 8$, $y = 4$, $z = 8$

5. 3 cm

Exercise 25.2a (page 355)

1. (a) 30 cm²

     (b) 16 cm²

     (c) 15 cm²

2. (a) 20 cm²

     (b) 12.5 cm²

     (c) 21 cm²

3. (a) 9 cm²

     (b) 18 cm²

     (c) 12 cm²

4. $a = 4$, $b = 3$, $c = 1\frac{1}{2}$

5. 9 cm

Exercise 25.3a (page 358)

1. 837 cm³

2. 217 cm³

3. 402 cm³

4. (a) 525 cm³

     (b) 405 cm³

     (c) 67.5 cm³

5. 46.8 cm³

6. 1680 cm³

7. 16 cm

8. 9.07 cm

9. 5.1 cm

10. 6.44 m

1. (a) length
   (b) area
   (c) area
2. (b) and (c)
3. (a) and (c)
4. (a), (b) and (c)
5. (a) none
   (b) length
   (c) area

1. (a) 26.6 cm²
   (b) 27 cm²
   (c) 29.8 cm²
2. (a) 30.0 cm²
   (b) 30 cm²
   (c) 31.9 cm²
3. 550 cm³
4. 2.22
5. 15.3 cm
6. (a) area
   (b) length
   (c) volume

7. (a) 2
   (b) 2
   (c) 3
8. (a) 61.8 cm²
   (b) 23.4 cm³
9. (a) 23.5 cm²
   (b) 36.8 cm²
10. (a) 5.66 cm
    (b) 39.6 cm²

Chapter 26

1.

Translation through the vector $\begin{pmatrix} 12 \\ 0 \end{pmatrix}$.

Translation twice the distance between the mirror lines.

2. (a) Rotation through 180° about the midpoint of OC
   (b) Reflection in OD
   (c) Rotation through 180° about the midpoint of OE
   (d) Reflection in OF
   (e) Rotation through 180° about the midpoint of OG
   (f) Reflection in OH
   (g) Rotation through 180° about the midpoint of OA

3. (a) Rotation through 180° about O
   (b) Translation through vector $\overrightarrow{CO}$
   (c) Enlargement, centre C, scale factor 2
4. (a) Because triangle 1 can be reflected (in OD) onto triangle 4
   (b) Because triangle 1 can be translated (vector $\overrightarrow{OG}$) onto triangle 6
5. A can be mapped onto B by rotation through 90° clockwise about O; A can be mapped onto C by translation through vector $\begin{pmatrix} -4 \\ -2 \end{pmatrix}$

   (Also B can be mapped onto C by rotation (through 90° anticlockwise about (−1, −3))
6. (a) 144°
   (b)

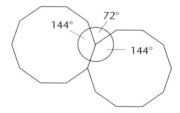

Since 2 × 144° = 288° there is a gap of 72° that cannot be filled by a decagon.

Exercise 26.2a (page 372)

1. (a) 14 cm
   (b) 60 m
2. Check students' drawings
3. (a) 5 cm
   (b) 3.5 km
4. (a) 2 km
   (b) 43.5 cm
   (c) 586 km
5. (a) 29.7 cm × 21 cm
   (b) 21 cm × 14.8 cm, yes, scale factor 1.4
   (c) size 14.8 cm × 10.5 cm, same scale factor
   (d) 42 cm × 29.7 cm

6. 8 cm
7. The sides are not proportional.
8. 4.2 cm, 5.88 cm
9. 3.4 cm, 4.4 cm
10. (a) Angle A is common, angle APQ = angle ABC (corresponding angles), angle AQP = angle ACB (corresponding angles); one of these can be substituted by 'angle sum of triangle'.
    (b) 6.27 cm

Revision exercise 26a (page 375)

1. (a)–(d)

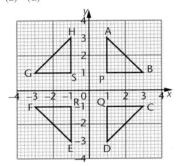

   (e) 4
   (f) No because, for example, AB ≠ BC
2. Because ABCD can be mapped onto GHAB by a rotation through 90° anticlockwise about O

3.

   Yes

4. (a) 9 cm, 7.75 cm
   (b) 4.75 cm, 2.25 cm
5. (a) 4 cm
   (b) 4.5 km
6. 2.4 m
7. 6 ÷ 8 = 0.75, 3 ÷ 5 = 0.6, the ratios are not equal so the triangles are not similar
8. (a) 180 − (40 + 55) = 85 so the angles are the same
   (b) 1.7
9. 6.64 cm, 6.24 cm
10. 7.14 m

Chapter 27

In comment questions, clearly students may use other words.
Answers read off graphs are approximate particularly from lines of best fit as different students will have different lines.

Exercise 27.1a (page 379)

1. Colin has, on average, lower scores. Vijay is more consistent. (ranges 25 and 16)
2. Town A's dental charges are higher but Town B has a greater spread of charges.
3. Nine numbers with a median of 7 and a greater range than 9.

4. Both means 19.5, ranges: Moralia 13, Sivarium 76, so on average they are the same but there is a much wider spread in Sivarium.
5. Tara: mean 14.8, range 4; Justin: mean 15.6, range 10; Justin, on average, is better but more inconsistent. Five homeworks is a very small sample on which to make judgement.

6. Jubilee Road: mode 1, range 6; Riverside Road: mode 3, range 8; Riverside Road has more letters and a wider spread.

7. Mean £4.30, range £4 to £5 (Cannot tell exactly, since the amounts are given to the nearest pound.) Class 9b on average get less pocket money but with a greater spread of amounts.

8. (a) A: median 12, IQR 8; B: median 13.5, IQR 4
   (b) Fire brigade A is quicker on average but has a greater spread (variation in times).

(c)

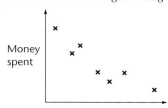

9. Probably LightGlo for greater guaranteed life. If buying large numbers, you may choose Britlite as the spread is not so important when buying in bulk.

Exercise 27.2a (page 384)

1. Fairly strong positive correlation
2. Weak negative correlation
3. (a)

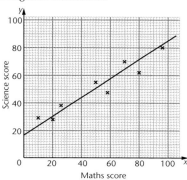

   (b) Reasonably strong positive correlation
   (d) (i) 45
       (ii) 84

4.

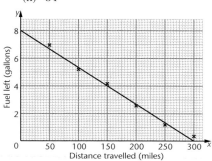

   (b) Strong negative correlation
   (d) 3.5 gallons

5.

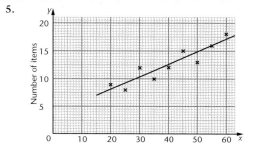

   (b) Fairly strong positive correlation
   (d) 11
   (e) Too far out of the range of the given data

6. e.g.

Money spent / Time on school work

1. (a), (c)

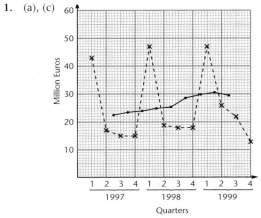

(b) moving averages 22.5, 23.5, 24, 24.75, 25.5, 28, 29.75, 30.75, 29.5

(d) general upward trend, though slight dip at end; always much higher in first quarter

2. (a), (c)

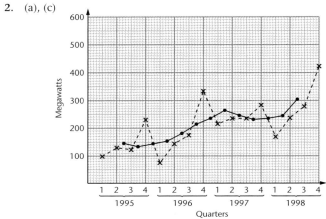

(b) moving averages 143.6, 138.0, 142.3, 154.8, 181.5, 217.1, 239.8, 255.2, 241.9, 230.0, 231.4, 243.3, 304

(d) general upward trend, with a few exceptions; first quarter always lowest, fourth quarter always highest

3. (a), (c)

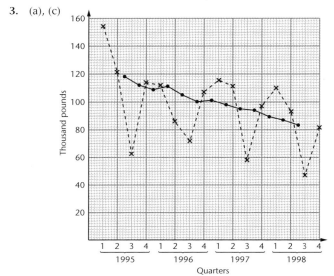

(b) moving averages 118, 112.5, 108.75, 111, 104.5, 100.25, 101.5, 98, 95.25, 94, 89.5, 86.75, 83

(d) general downward trend; third quarter always lowest

4. (a), (c)

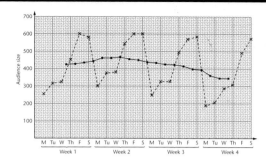

(b) moving averages 420.7, 427.5, 436.7, 446.3, 460.7, 460.7, 464, 455.8, 449.2, 439.7, 432.5, 427.5, 424.7, 415.3, 397.3, 391, 359.5, 346.2, 344.3

(d) general trend upwards and then downwards; Monday always least, Friday and Saturday always highest

Revision exercise 27a (page 391)

1. David takes longer but his times are more consistent.

2. 10f median 15, range 21; 10g median 11, range 28; 10g students do question more quickly on average but with a wider spread of times. The interquartile range would eliminate extremes.

3. Mean £79 400, range a maximum of £120 000. House prices are higher in the south-east, with a wider spread of prices.

4. French adults drink more wine than English adults. Spread of amounts is similar but slightly wider in England.

5. (a)

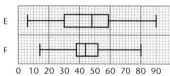

(b) French: median 44, IQR 14; English: median 48, IQR 29; The French marks are higher but with a wider spread.

6. Both have positive correlation. Girls have stronger correlation than boys. Boys are heavier and taller than girls.

7. Negative

8. (a)

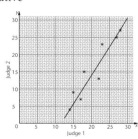

(b) A fair degree of positive correlation

(d) 22−23

9. (a)

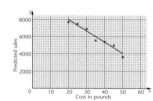

(b) Fairly strong negative correlation

(d) 6200

(e) Too far outside the range of the data.

10. e.g.

11. (a), (c) See graph

(b) moving averages 37 175, 25 200, 32 750, 30 000, 27 600, 25 800, 24 675, 24 075, 23 475, 22 775, 22 000, 21 175, 20 325

(d) general trend downwards; quarter up to January always highest, then steadily lower

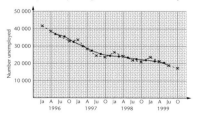

Answers

Chapter 28

Check the accuracy of students' drawings.
The diagrams in these answers are not accurate but
are given as a guide.

1.

The locus is the shaded area.

2.

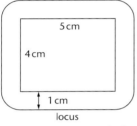

The locus is four straight lines and four
quarter-circles, all 1 cm outside the rectangle.

3.

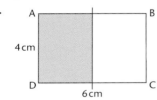

The line drawn is the bisector of AB. The
region shaded is the locus.

4.

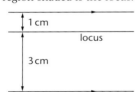

The locus is the line parallel to the two given
lines.

5.

The region is inside a circle radius 5 miles.

6.

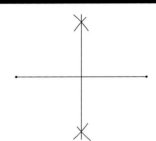

7.

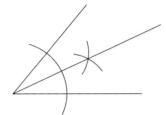

8.

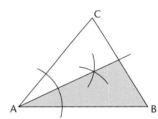

9.

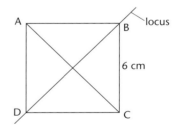

The locus passes through B and D.

10. Bisect angle C and the locus is the region on
the side nearer to A.

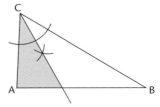

The regions **not** required are shaded in these answers.

1.

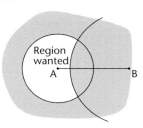

2.

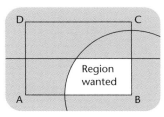

3. The point equidistant from all sides is the intersection of the bisectors of the angles.

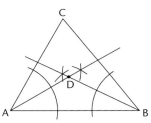

4.

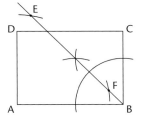

The points are marked E and F.

5. The region required is labelled R.

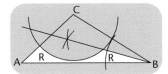

6. The region required is labelled R.

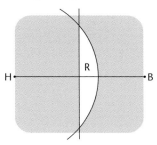

7. The regions required are labelled R.

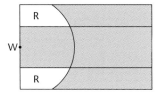

8. The region required is labelled R.

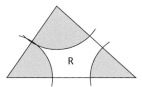

9. The region required is labelled R.

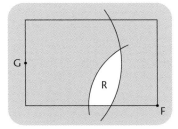

10. The region required is labelled R.

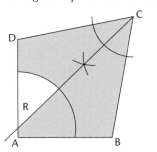

1.

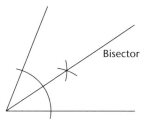

Bisector

2.

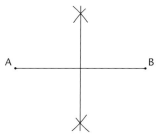

3.

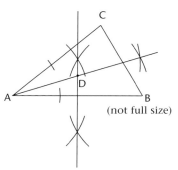

(not full size)

4. (Drawn half size)

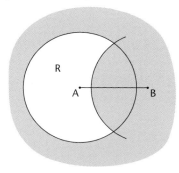

5. The region required is labelled R.

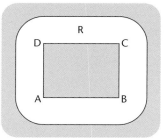

6. The region required is labelled R.

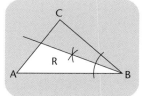

7. The region required is R. Note that it includes a region outside the triangle ABC.
(Drawn half size)

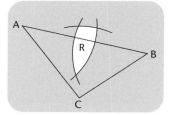

8.

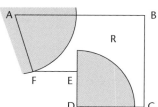

9. The region not covered by heat detectors is labelled R. (Drawn half size)

10. The region required is labelled R.

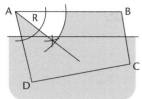

Index

3D shapes (polyhedra) 17
3D shapes, symmetry properties 75–76
3D shapes, volume 357–358

abundant numbers 134
accuracy in answers 272, 273, 315, 320
advice and tips 1, 4
algebra 31, 43
allied angles (co-interior angles) 82
alternate angles 81, 82
analysing data 259–261, 263
angles, allied 82
angles, alternate 81, 82
angles, bisectors 395
angles, co–interior 82
angles, corresponding 81, 82
angles, cosine (cos) 280, 283, 285, 286
angles, exterior 85–86, 119, 172–173, 180
angles, in circles 179, 180
angles, in polygons 177, 180
angles, in semicircles 175, 180
angles, included 20
angles, interior 85–86, 119, 172–173, 180
angles, non–included 22
angles, notation for 82–83
angles, sine (sin) 280, 282, 284, 286
angles, tangent (tan) 280, 281, 286
approximation 157–158, 159, 171
area, units of 136, 146
areas, circles 141, 146, 360
areas, triangles 135, 146
average value, selection 190, 376–377, 391
averages, moving 387–388, 391

axes 90
axes of rotational symmetry 75, 76

bands (intervals) 44–45
bar graphs (histograms) 184, 191
bearings 103–104, 107
best fit lines 384, 391
bird's–eye view 25
bisectors of angles and lines 394, 395, 397, 398, 400
bounds of measurement 317
box–and–whisker plots 261–262, 263, 378
brackets 36–37
brackets, expanding 237–238, 254

calculators, using 220–221
cancelling fractions 212
Cartesian plane 89
centre of enlargement 116
centre of rotation 195, 207, 209, 398
checking answers 64–65, 72, 265, 273
circles, angles in 179, 180
circles, area 141, 146, 360
circles, circumference 138–139, 146, 360
coin tossing 336–337
co–interior angles (allied angles) 82
collecting like terms 236, 254
column vectors 200, 201, 206–207, 209
compasses 28
compound interest 266, 273
compound measures 269–270, 273
compound units 318–319, 320
congruent shapes 109, 123, 365, 374
consecutive numbers 125–129

continuous measures 315, 320
conversion graphs 67
coordinates 89–90
co–primes 133
correlation 51, 381–384
corresponding angles 81, 82
cosine (cos) 280, 283, 285, 286
coursework 2–3
cube numbers 132
cube roots 132
cubes 75, 76
cubic equations, solving 301, 309
cuboids 76
cuboids, surface area 235
cuboids, volume 143–144, 146, 357
cumulative frequency tables and diagrams 259–262, 263, 378, 379
currency exchange rates 67–68
cyclic quadrilateral 179, 180
cyclical effects 387
cylinders, surface area 234
cylinders, volume 357, 358, 361

data, analysing 259–261, 263
data, comparing 56, 376–379
data, continuous 184–187
data, discrete 181–182
data, grouping 44–45, 47, 181–182
decimal places, rounding 70, 158, 171, 272, 273
decimal representations of fractions 6–7, 16
deficient numbers 134
density 270, 273, 319, 320
dependent events 342
dimensions 359–360, 361
discrete measures 315, 320
distance fallen 235

elevations 24–26
enlargements 116–117, 123, 202–203, 209, 368–369
equations, cubic 301, 309
equations, curved lines 330–331, 334
equations, forming 37, 250–251, 254, 290, 309
equations, quadratic 296–297, 299, 303, 309
equations, simultaneous 292–293, 295, 298, 303, 309
equations, solving 39–41, 43
equations, straight lines 329–330, 334
equivalent fractions 211
estimation 64–65, 312–313, 320
event 148
examinations 1
exchange rates (currency) 67–68
expanding brackets 36–37, 40, 41, 43
exterior angles 85–86, 119, 172–173, 180

factorising expressions 243–245, 254
factorising quadratic equations 296–297, 309
factors 132, 243–245, 254
finding the distance from a point to a line 19
foreign currency exchange rates 67–68
formulae 31, 233–234
formulae, rearranging 246–247, 308
fractions 6–7, 16, 211–213, 217–218, 231
fractions, cancelling 212
fractions, mixed numbers 214–215, 231
fractions, top–heavy 214–215, 231
frequency chart (frequency diagram) 45
frequency polygons 45–46, 59, 185, 191

gnomons 131
gradients 325–326, 329–330, 334
graphical methods of solving equations 298–299
graphs, curved lines 330–331, 334
graphs, gradients 325–326, 329–330, 334
graphs, harder functions 99–101
graphs, quadratic 93–96
graphs, showing regions 305–306, 309
graphs, story 321–322, 324, 334

graphs, straight–line 89–91, 92–93, 107, 329–330
grids 134
gross pay 66
gross taxable pay 66
grouping data 44–45, 47

hexagons 74, 366
histograms (bar graphs) 184, 191
hypotenuse 276, 279, 280
hypothesis testing 256, 257, 263

included angle 20
independent events 342, 343, 344, 348
indices (powers) 166–167, 169, 171, 239–240, 254
inequalities 42, 248–249, 254
inequalities, forming 250–251, 290
inequalities, graphical methods 305–306, 309
inequalities, solving 289
insurance 267–269, 273
interior angles 85–86, 119, 172–173, 180, 366
interquartile range 377, 378, 391, 260
intersecting loci 397–398
intervals (bands) 44–45, 317
inverse operations 65
isometric drawings 17
isosceles triangles 79

letters standing in for unknowns 31–32
linear equations, solving 287–288
lines, best fit 384, 391
lines, notation for 82–83
lines, straight 77, 329–330
lines, symmetry 73–74
loci, constructing 395, 396
loci, identifying 393–394, 400
loci, intersecting 397–398
loci, sketching 394
lower bound 317
lowest terms 212

maps 369, 374
mean 54–56, 59, 181–182, 186, 190, 191, 376–377, 391
measurements, bounded 317
measurements, estimating 312–313, 320
median 54–56, 58, 59, 190, 261, 262, 263, 376, 378, 391
methods for solving numerical problems 69–70, 72

minimum values of graphs 94–95, 99
mixed numbers 214–215, 231
mode 54–56, 58, 59, 181, 186, 187, 190, 191, 376, 377, 391
money problems, solving 66–68, 72
moving averages 387–388, 391
multiples 130, 132
mutually exclusive outcomes 150, 155, 342, 348

negative correlation 383, 391
negative numbers 37, 164, 171, 218–219, 231
net pay (take–home pay) 66
non–included angle 22
nth term in a sequence 35, 43, 241–242, 254
number of units per cost unit 62
number patterns 131
numbers, abundant 134
numbers, consecutive 125–129
numbers, cube 132
numbers, deficient 134
numbers, mixed 214–215, 231
numbers, negative 218–219, 231
numbers, perfect 134
numbers, prime 132
numbers, square 132, 133

octagons 74
octagons, sum of angles 86
opposite angles 78
order of rotational symmetry 74, 75
ordering numbers 162
outcomes 148
outcomes, grid 337, 339, 348
outcomes, listing 336–337, 348
outcomes, table 337, 348
outliers 384

parabolas 93, 107
parallel lines 19, 81, 87
parallelograms, area 350–351, 361
patterns 113, 364–365
peaks and troughs 387
pension contributions 66–67
pentagons 73, 74, 366
pentagons, sum of angles 84–85, 86, 119
percentage change 229, 231
percentages 6–7, 8, 9–11, 16
perfect numbers 134
perpendicular bisectors 394, 395, 397, 398, 400
perpendicular distance 19
perpendicular height 136

personal (tax) allowance 66
pi (π) 139
pie charts 47, 49, 59
pilot surveys 258, 263
planes of symmetry 76
plans 24–26
points, sum of angles 78
polygons 73–74, 87
polygons, frequency 185, 191
polygons, interior and exterior angles 177, 180
polygons, sum of angles 84–86, 88
polyhedra (3D shapes) 17, 75–76, 357–358
population density 270, 319, 320
positive correlation 382, 391
possibility space 337, 339, 348
possible outcomes, listing 336–337, 348
powers (indices) 166–167, 169, 171, 239–240, 254
price per unit 61
prime factors 130
prime numbers 132
prisms, volume 357–358, 361
probability, correct way of expressing 149
probability, mutually exclusive outcomes 150, 155
probability, of an outcome not happening 148–149, 155
probability, of two outcomes 342–344
probability, relative frequency 152–153, 155
profit and loss 9–11, 16
proportion 222–223
proportional change 225–226, 231
Pythagoras' theorem 274–278, 286

quadrants 89
quadratic equations, solving 296–297, 299, 303, 309
quadratic graphs 93–96
quadrilateral, cyclic 179, 180
quadrilaterals, sum of angles 78, 79
quartiles 260, 261, 263
questionnaires 256–258, 263
range 54–56, 59, 182, 186, 187, 191, 262, 263, 377, 378, 391
rate of change 325
ratio 12–13, 14–15, 16, 61–62, 72, 163, 171, 222–223
rearranging formulae 246–247
reciprocals 170, 171
rectangles, perimeter 33

reflections 109–110, 123, 193–194, 209
relative frequency 152–153, 155
roots 169, 171
rotational symmetry 74, 75, 117–118, 123
rotations 110–111, 123, 195–197, 207, 209
rounding numbers 47, 70, 157–158, 159, 171, 265, 273

scale drawings 368–369, 374
scale factor 116–117, 202, 209, 369
scatter diagrams (scatter graphs) 51–52, 59, 382–383, 391
semicircles, angles in 175, 180
sequences 34–36, 241–242, 254
set–square 19
sharing in a given ratio 14–15, 16, 62
significant figures 160, 171, 265, 273, 315, 320
similar shapes 116, 123, 370–371, 374
simplifying expressions 236, 239–240, 254
simultaneous equations, solving 292–293, 295, 298, 303, 309
sine (sin) 280, 282, 284, 286
skewed distribution 190
solving cubic equations 301, 309
solving equations using graphical methods 298–299
solving equations using trial and improvement 301, 309
solving inequalities 289
solving linear equations 287–288
solving numerical problems 69–70, 72
solving quadratic equations 296–297, 299, 303, 309
solving simultaneous equations 292–293, 295, 298, 303, 309
speed 269–270, 273, 318–319, 320, 326
spheres 76
spreadsheets 66, 70
square numbers 132, 133
square roots 132
squares 129
standard form of numbers 168–169, 171
steepness of a graph 325
stem and leaf tables 58, 59
story graphs 321–322, 324, 334
substitutions 233–234
subtended 179

sum of angles 77–79, 85–86, 87, 119
surds 285
surveys 257, 263
surveys, pilot 258, 263
symmetry 73–74, 87, 117–118, 123
symmetry, rotational 74, 75

table of possible outcomes 337, 348
tally charts 44–45
tangent (tan) 280, 281, 286
taxation 66–67
terms in a sequence 34–35, 241–242, 254
terms, nth 35, 43, 241–242, 254
tessellations 118–119, 123, 366, 374
tetrahedra 75, 76
three–figure bearings 103
time series 386–387, 391
transformations 109–113, 116–117, 193–197, 199–203, 206–207, 209, 364–369
translations 112, 123, 199–201, 206–207, 209
transversal line 81
trapezia, area 354, 361
tree diagrams 338, 339, 346, 348
trends 387
trial and improvement method for solving equations 301, 309
trial and improvement methods 69–70
triangle spotty paper 17
triangles, area 135, 146
triangles, drawing 20–22
triangles, interior and exterior angles 172–173, 180
triangles, isosceles 79
triangles, length of sides 274–285
triangles, similar 370–371, 374
triangles, sum of angles 78, 79
triangles, vertex 21
trigonometry 279–285
troughs and peaks 387

unequal probabilities 346
upper bound 317

vectors 200, 201, 206–207, 209
vertex 21
vertically opposite angles 78
volume, units of 143, 146

writing up findings 261, 263

Index